Songbird Behavior and Conservation in the Anthropocene

Editor

Darren S. Proppe
Research Director
Wild Basin Creative Research Center
St. Edward's University, Austin, Texas, USA

CRC Press
Taylor & Francis Group
Boca Raton   London   New York

CRC Press is an imprint of the
Taylor & Francis Group, an **informa** business

A SCIENCE PUBLISHERS BOOK

Cover Credit: Cover artwork and drawings on the front page of each chapter have been prepared by Jenna Atma. Jenna is a biologist and artist from southwest Michigan. Jenna can be reached at tigereye.jla@gmail.com.

First edition published 2022
by CRC Press
6000 Broken Sound Parkway NW, Suite 300, Boca Raton, FL 33487-2742

and by CRC Press
2 Park Square, Milton Park, Abingdon, Oxon, OX14 4RN

© 2022 Taylor & Francis Group, LLC

CRC Press is an imprint of Taylor & Francis Group, LLC

Library of Congress Cataloging-in-Publication Data

Names: Proppe, Darren S., 1975- editor.
Title: Songbird Behavior and Conservation in the
    Anthropocene/editor, Darren S. Proppe, Research Director,
    Wild Basin Creative Research Center, St. Edward's University, Austin,Texas, USA.
Description: First edition. | Boca Raton : CRC Press, 2021. | Includes
    bibliographical references and index. | Summary: "Learned and fixed behaviors underlie many of
    the patterns we observe in songbirds. But the environmental context in which these patterns occur
    is changing quickly, often to the detriment of the individual and species. The goal of this book is to
    weave concepts of behavior more tightly into our conservation strategies. Each chapter describes the
    current understanding of behavior in relation to a particular songbird life history trait. The authors
    then evaluate challenges that songbirds face in the Anthropocene, and explore the role of behavior in
    addressing these challenges. The future is uncertain for songbirds, but broadening our management
    toolkit will increase the potential for success"-- Provided by publisher.
Identifiers: LCCN 2021018816 | ISBN 9780367279288 (hardcover) | ISBN
    9781032058382 (paperback) | ISBN 9780429299568 (ebook)
Subjects: LCSH: Songbirds--Behavior. | Songbirds--Conservation.
Classification: LCC QL696.P2 S594 2021 | DDC 598.8--dc23
LC record available at https://lccn.loc.gov/2021018816

ISBN: 978-0-367-27928-8 (hbk)
ISBN: 978-1-032-05838-2 (pbk)
ISBN: 978-0-429-29956-8 (ebk)

DOI: 10.1201/9780429299568

Typeset in Times New Roman
by Shubham Creation

*This book is dedicated to
my wife and children who gave me the space and time to explore,
write and edit; to the many authors who put in countless hours to
write and edit each chapter; and to the songbirds that we aim to protect.*

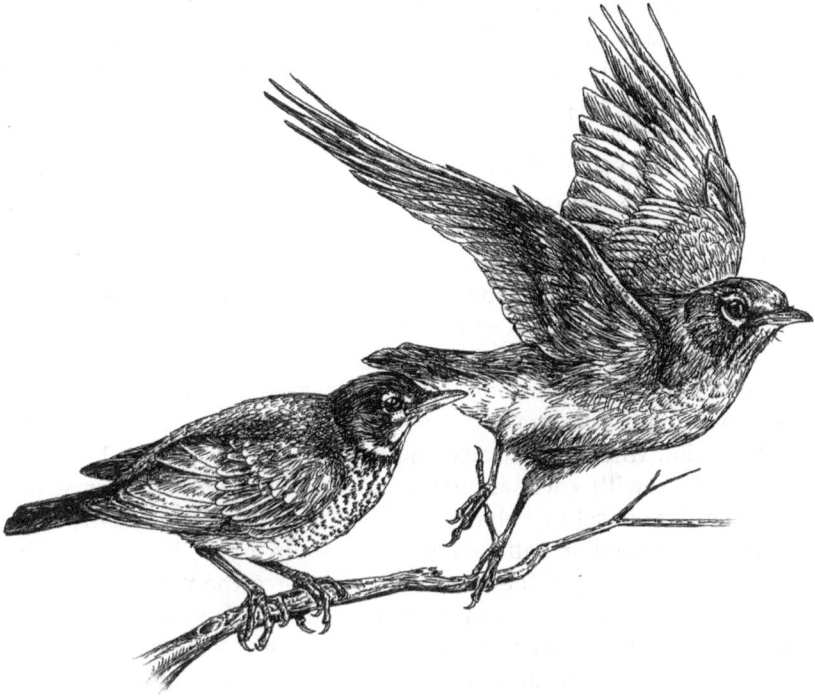

Preface: How to Utilize this Book

Darren S. Proppe[1]

This volume is dedicated to explaining the role that behavior plays in the function and ecology of free-living songbirds with a particular focus on its importance in conservation and management. Understanding songbird behavior is no small task. Indeed, many have dedicated their careers to studying particular aspects of songbird behavior. In addition, there is a long history of study on the behavior of animals—often located across disparate fields and academic journals. Although the integration of behavior into conservation and management is a relatively new endeavor, many of the underlying processes and systems have been explored in detail. In fact, developing a text that merely explored bird behavior as a scientific study might have been redundant. Donald and Lilian Stokes wrote several volumes on bird behavior in the 1980's [1], and in 2001, David Allen Sibley wrote an excellent and accessible text called *The Sibley Guide to Bird Life and Behavior* [2]. Other texts have extensively addressed the song system [3, 4] and elements of avian migration [5–7]. Still others have examined the impacts of urbanization on birds [8, 9]. In 2014, Dr. John Marzluff published a very informative and easy read for a lay audience called *Welcome to Subirdia* [10]. In part, I cite these texts here to direct

[1]Research Director, Wild Basin Creative Research Center, St. Edward's University

you to some of the excellent resources that are available. But I also wish to answer a question: Why another book? Our book is distinct from these texts because it explicitly addresses behavior in light of the conservation challenges we face in the human era—the Anthropocene. My goal in editing this book was to provide a 'one-stop shop' for those interested in integrating behavior into management or applied research programs. That is not to say that you will find every bit of known information contained within these pages—far from it. Rather, we aim to provide the needed groundwork to understand basic concepts and to recognize where research has led us at the time of publication.

Chapters are broken into particular behavioral topics (e.g., habitat selection, foraging, personality, etc.). In reality, however, animal behavior is not a discrete set of responses that relate only to one topic – or ecological system. Rather, there is much overlap. What an animal eats, for example, is related to energetics and foraging strategies, but also associated with predation and social information. In the same vein, anthropogenic changes will likely alter many behavioral systems. Human-produced noise might impact communication directly, but also impact foraging and sexual selection. As the editor, I have worked with each author team to keep their chapter within the bounds of their particular subset of behavioral systems. But do not be surprised when particular behaviors, or particular anthropogenic stressors are addressed within several chapters. In fact, this repetition of themes should reinforce the notion that behavioral responses operate along a continuum and within a multi-sensory input and response system. To suggest, for example, that predation and habitat selection are not connected, and that personality would not impact both of these systems would be misleading. Thus, chapters are designed to be separate, but overlapping. One positive aspect of interconnected themes is the potential that a few management techniques might go a long way towards preserving many behavioral systems.

To provide a foundational understanding of behavior, each chapter begins with a primer on the stated behavioral topic. Every chapter is written by experts in their particular niche of songbird behavior. Each author team is well published, actively engaged in scientific research, and well-regarded by their peers. However, each author was also instructed to write in a style that is accessible to fellow academic researchers, on-the-ground managers, and engaged community scientists alike (although we know that many of you will wear all of these hats). To get the creative juices flowing, each author has peppered their text with examples where particular species or systems display the behaviors being described. The intentional inclusion of examples was designed to display the phenomenal capabilities demonstrated by songbirds, but also to accelerate the application of information to your system of interest. These examples should also be particularly valuable for managing or conserving species with little published information. For example, Otter et al. (Chapter 5) describe that some lekking species might be more reliant on characteristics of ambient lighting than most birds [11]. Might the system or species that you manage be similar? What elements of your system might elevate, or not elevate, the role of light? Of course, most the time our examples will not directly address your particular species. But this is one of the beautiful elements about studying behavior–there is so much to learn! So, I challenge you to let the text push you to think more deeply, and more creatively, about your organisms and systems.

Once each chapter reviews behavioral fundamentals and the latest research, the authors move on to describe some of the primary threats stemming from the human alteration of native habitats and ecosystems. Perhaps this is the crux of the text. If we are going to successfully maintain diverse songbird populations and communities, we must recognize the threats to their established behavioral systems–which may go way beyond structural or vegetative changes to habitat. Again, specific examples from the scientific literature abound, allowing the reader to delve much more deeply into the topics they deem most pertinent or interesting. In each case, threats are followed by the presentation of potential solutions. On this front, some authors are fortunate to be able to report on protocols that have already been implemented, or experimental solutions that offer promise. Others must recognize that the solutions are yet forthcoming. But hopefully that is why you are reading this book – to play an active role in developing solutions that can sustain behavioral systems in songbirds, and thus their existence, in a world that is changing with incredible speed. Understanding the required habitats utilized by birds is critical, but failing to integrate the role of behavior into management is likely to produce subpar results. It is my hope that reading this book will lead you to agree, and that it will spur the development of holistic and novel techniques for managing songbirds.

LITERATURE CITED

[1] Stokes, D. and Stokes, L. 1983. Stokes Guide to Bird Behavior, Volume 1 (Reprint edition). Little, Brown and Company.

[2] Sibley, D.A. 2001. The Sibley Guide to Bird Life & Behavior. Knopf, New York.

[3] Marler, P.R. and H. Slabbekoorn. 2004. Nature's Music: The Science of Birdsong. Elsevier.

[4] Naguib, M., V. Janik, N. Clayton and K. Zuberbuhler. 2009. Vocal Communication in Birds and Mammals. Adv. Study Behav. 40(1). Academic Press.

[5] Wood, E.M. and J.L. Kellermann. 2015. Phenological Synchrony and Bird Migration: Changing Climate and Seasonal Resources in North America. CRC Press.

[6] Berthold, P. 2001. Bird Migration: A General Survey. (H.-G. Bauer and V. Westhead, Trans), 2nd Ed. Oxford University Press, Oxford, New York.

[7] Newton, I. 2007. The Migration Ecology of Birds. Academic Press, Amsterdam.

[8] Gil, D. and H. Brumm. 2014. Avian Urban Ecology. Oxford University Press, Oxford.

[9] Lepczyk, C.A. and P.S. Warren. 2012. Urban Bird Ecology and Conservation. University of California Press.

[10] Marzluff, J.M. 2014. Welcome to Subirdia: Sharing Our Neighborhoods with Wrens, Robins, Woodpeckers, and Other Wildlife. Yale University Press.

[11] Endler, J.A. and M. Thery. 1996. Interacting effects of lek placement, display behavior, ambient light, and color patterns in three neotropical forest-dwelling birds. Am. Nat. 148(3): 421–452. https://doi.org/10.1086/285934

Contents

Chapter 1

Static Organisms in a Changing System?

Darren S. Proppe[1]

Introduction

Domestic chickens are known for simple, predictable, and often stereotyped behavior. Eat, sleep, eat, sleep, eat, sleep... repeat. Although chickens are not songbirds, I open with this species here as an example of an organism with which many people are familiar; one that vividly portrays the types of behaviors that lead many to believe the actions of organisms are by-in-large static. But for those who have owned and raised chickens, or for that matter any domestic animal, the story is more complicated. Some chickens choose to feed primarily from a food cup while others forage predominantly

---

[1]Research Director, Wild Basin Creative Research Center, St. Edward's University

on the ground. Some individuals follow their owners incessantly, while others flee from human presence. Even behavior within an individual is not static. Not long ago, an unfortunate dog attack left one of my chickens severely injured. Although the chicken made a full recovery, her behavior was forever altered. The sound of a dog bark sent this free-ranging chicken in a full sprint back to the safety of the enclosed coop. Before her incident with the dog, she responded to an audible dog bark by merely raising, and sometimes cocking her head. Other chickens in the same flock continued to exhibit only this minimal response to the acoustic sounds of a potential canine predator. They had not experienced a fear-inducing interaction with a dog, and did not socially learn fear behavior from the injured chicken. Thus, they had no learned fear response as a result.

While an experienced chicken owner will likely recognize the presence of among-individual and within-individual variation in behavior, an educated owner also knows that some behaviors *are* largely static. No one expects that domestic chickens will suddenly become migratory. Nor will they begin building nests in trees, or nests at all for that matter. More narrowly, if an owner desires regular egg production, they must provide the right environmental conditions—which includes a relatively confined dry space with soft bedding. In fact, the importance of understanding and accommodating behavior is rarely lost on one who own pets or livestock. Dog owners speak regularly about good or bad behavior, and often elaborate about their pet's personality. Ranchers know which heifer or horse to be wary of, and farmers know which goat or rooster should not be left with children. Despite the clear acknowledgement that behavior plays a role in the care of captive animals, behavior has not received the same attention when it comes to the management of non-captive organisms.

I open this introductory chapter by providing the definition of behavior, and briefly describing the history of behavior as a scientific field of study. Next, I explore learning and cue-response systems as the mechanisms that underlie the behavioral responses we observe. This section is followed by an exploration of stability and flexibility in behavioral systems, with focus on relevant songbird responses to environmental cues and signals. I then place these systems within the context of anthropogenic change – addressing the need to bring an understanding of behavior to conservation and management plans.

The Foundations of Behavior

Behavior is broadly defined as, (1) the way in which one acts or conducts oneself, especially toward others, (2) the way in which an animal (non-human) or person acts in response to a particular situation or stimulus, or (3) the way in which a natural phenomenon or a machine works or functions [1]. For our purposes, we will focus on animal behavior, meaning that the second definition is most relevant. The beauty, and sometimes the curse, of animal behavior as a scientific study is that it covers a broad range of fields and expertise. For example, much of our initial understanding of learning and behavior was developed by prominent psychologists with an interest in explaining human behavior. More recently, entire branches of the psychological field have become dedicated to understanding the development of behavior and the proximal mechanisms that underlie non-human animal behavior. Lab and field studies often attempt to explain behavior under natural conditions – a particular scenario defined as *ethology*. Topics such as learning and cognition continue to be explored extensively among psychologists [2], producing many advances that are relevant to biologists focused on behavior and management. Animal behavior has become a focus of study for many ecologists as well, with foundations of the field established in the early to mid-1900's by scientists such as Nikolaas Tinbergen, Konrad Lorenz, and Karl von Frisch [3]. Perhaps one of Tinbergen's most significant contributions to the field was his description of the four questions (or arenas) for the study of animal behavior [4]. *Causation* (or mechanism) and *development* (or ontogeny) of behavior describes many of the topics investigated using captive animals. Ecologists that study behavior tend to focus on Tinbergen's second two questions: *function* (or adaptation) and *evolution* (or phylogeny) of behavior. These latter questions may be thought of as exploring why particular behaviors occur, which is often explained by the correlation between their performance costs and the benefits of their expression. This study of *ultimate* (as opposed to *proximate*) explanations of behavior is often defined as *behavioral ecology* [5].

While Tinbergen's designations have been extremely beneficial for studying and understanding behavior, they have also had the unintended consequence of creating somewhat siloed fields of study. This unfortunate situation is exacerbated by the difficulty of tracking the vast literature found across many field-specific journals, and the sometimes differing study techniques used by biologists, ecologists, and psychologist [6]. To fully understand behavior, and to manage animals accordingly, will require an integrated understanding of all four of Tinbergen's questions. The focus on collaborative management and ecosystem approaches in recent years has led to an exciting and unprecedented integration of information about the development and function of behavior [7–9]. This integration is timely and critical as we now work to sustain populations and communities of songbirds in landscapes that are changing rapidly. No doubt one of the keys to integrating behavior into management for songbirds will be understanding how, and how quickly, behavior can change to accommodate shifting environmental conditions.

Learning and the Development of Behavioral Response Systems

The development of particular behavioral systems (or responses) is shaped by the same processes of natural and sexual selection that leads to physical traits better suited for particular environments. If behaviors are thought of as responses to particular situations or environments, variation in these responses must exist for natural selection to shape the system. Within this range of responses, there must also be variation in how the resulting outcome impacts fitness (e.g., some responses provide better fitness gains than others). While heritability, a keystone of natural selection, also plays a role in the development and maintenance of many behaviors we observe, learning can also maintain some elements of behavioral systems. The basic cue-response system that leads to standardized behavioral responses, and to changes in behavioral responses, can be represented as a triangle with *cue*, *response*, and *outcome* at the three corners (Box 1). Although each term is singular here, in reality it may take multiple cues to elicit a response, or responses. Here the *cue* represents the environmental stimulus that could trigger a particular response. To understand this in human terms, simply imagine that someone is cooking bacon for breakfast (if you do not like bacon, imagine your favorite food here). As most readers are probably aware, bacon that is being cooked has a strong, distinctive smell. This smell could be classified as potential cue, which might trigger a *response* from nearby humans. The response to bacon is likely to differ among individuals, but for those that like bacon it is likely to trigger a desire to find and consume bacon. If this response does lead to consumption, the *outcome* in the bacon-lover will be positive—a pleasurable dining experience. Repeated experience with this particular cue-response system would establish a 'find-and-eat' response to the smell of bacon. If bacon was the only, or perhaps the most beneficial food source available, the fitness of individuals with the 'find-and-eat' response would increase, and the behavior would be selected for. If genetics were to underly this response, selection would increase the prevalence of this response in the population by favoring the responsible genotype. Even without genetic underpinnings, experience and teaching (a phenomenon that is less prevalent, but not absent, in non-human animals) might increase the bacon-philic response. While this example is somewhat silly and underestimates the complexity of behavioral development, it does provide the basic level of understanding needed to grasp the development of behavior (for a deeper understanding see [10, 11]). Our bacon example also allows for a few extensions that can explain how the use of cues and responses can change over time.

If bacon creates an automatic 'find-and-eat' response in many of us, that response is said to be an unconditioned stimulus (i.e., no training is needed to evoke this response). But through a procedure called *classical conditioning*, we could actively train individuals to exhibit the same 'find-and-eat' response to the sight of a pancake—if the sight of a pancake and smell of bacon are paired across several trials, and always result in the availability of bacon. Eventually, the 'find-and-eat' response would be learned and evoked by the sight of a pancake alone (conditioned stimulus; i.e., training was needed to evoke this response). The classic example of this process is Pavlov's dogs, who were trained to salivate for food in response to the sound of a metronome [12]. In songbirds, a similar process likely underlies the selection for particular habitats or tree species. While selected trees or habitats likely supply better food sources than non-selected locations, it is unlikely that birds directly assess insect populations prior to selection. More likely, experience (or selected genetic preferences) has facilitated an association between the preferred habitat characteristics and food availability [13, 14].

A particular response can also be extinguished by changing the outcome in the cue-response system [15]. For example, some alternative-fuel vehicles use recycled food grease as fuel [16]. The resulting exhaust can have a bacon-like smell. In this case, the 'find-and-eat' response to the smell of bacon described above could result in the negative outcome of being exposed to vehicle exhaust or the danger of approaching a moving vehicle. If a particular response is no longer rewarded with a positive outcome, or results in a negative outcome, the response might be extinguished. If bacon smell becomes more commonly associated with vehicles than a preferred food source, the 'find-and-eat' response may become extinct. Similarly, most songbirds exhibit a fear of humans, but fear behavior is diminishing in populations that inhabit human-dominated systems (e.g., cities [17, 18]). In this case, repeated neutral or positive (e.g., food provided) interactions with humans in urban systems are likely extinguishing fearful behavior. While the altered cue-response system here might appear benign, the impact of changing the outcome in a cue-response system can come with negative fitness impacts. For example, songbirds are killed regularly by wind-powered turbines during migration [19]. This is potentially to due to their attraction to light [20] which also serves as an important cue for migration, or through other established migratory behaviors such as low-altitude flight in large groups [21]. Given enough time and experience, songbirds may learn to associate particular characteristics of turbines with the danger that they represent (i.e., extinguish attraction). But for a time, there is a *mismatch* between the response and the expected outcome [22, 23]. During this mismatch period, there is a high likelihood that more songbirds will be injured or killed by turbine blades. Simply put, behavioral changes may be too slow to avoid the negative impacts of environmental change (i.e., the outcome).

Stability and Plasticity in Songbird Behavior

As with most organisms, songbird behavioral systems exhibit a range of plasticity. Some genetically coded behaviors have become *fixed*, which can indicate that strong selection has nearly, if not completely eliminated genetic variability in the loci responsible for producing the particular behavior, and thus eliminated variability in the associated behavior. Other behaviors are crystalized after a short learning period. Some more plastic behaviors are shaped in individuals and populations through experience, and some responses are subject to high levels of within-individual plasticity based on the particular set of environmental conditions. In this section I provide an explanation of, and examples for songbird behaviors that are a typically characterized as innate or largely stable, learned, shaped by experience, and plastic. However, much remains to be learned about each system, and these categories should not be considered to be set in stone or mutually exclusive. In addition, plasticity itself may be under selective pressures that can increase or decrease variability over time [24]. I conclude the section by explaining why multiple behavioral responses to the same stimuli often exist within songbird populations and species.

Innate or Largely Stable Behavior

Immediately after hatch, hungry Cactus wren (*Campylorhynchus brunneicapillus*) nestlings lift their heads upward with their mouth agape at any detection of movement or sound – which most likely represents an adult approaching the nest to feed [25]. Although critical for their survival, this 'gaping' behavior does not stem from experience or tutelage. Gaping, which is a trait shared across songbird species, is innate [26]; likely coded within the genes. This is not to say that there is not variability in gaping behavior, but rather that natural selection will strongly favor individuals that exhibit this behavior over those that do not. Gaping and begging correlate with hunger level [27], and parents likely perceive these behaviors to be an honest indicator of a nestlings need for nourishment. In keeping with this hypothesis, experimental work with tree swallows (*Tachycineta bicolor*) and great tits (*Parus major*) indicates that parents increase feeding rates as gaping and begging rates increase [28–30]. A nestling that does not gape, or gapes less often, is less likely to be fed. An unfed nestling is unlikely to survive and carry this trait into the next generation. Although gaping and begging are an honest indicator of hunger that is critical for survival, variation in begging rates among broods may also be indicator of health. For example, great tit nestlings that hatched from eggs laid by a female given supplemental carotenoids, an important but limited antioxidant, begged more intensely than control nestlings [31]. Because carotenoids are associated with nestling health, more intense begging is still an honest indicator of fitness, and increased feeding of these carotenoid-rich chicks is probably a good investment on the part of a parent. Again, selection favors stronger gaping and begging behavior.

Migration, the seasonal movement between spatially distinct habitats, is a common songbird life history trait. The role of genetics and experience in navigation between breeding and wintering grounds has received much attention in the scientific literature, and is addressed more broadly in Chapter 7. However, I want to visit this topic briefly here because elements of migration are thought to be innate [32]. For example, *zugunruhe* refers the restlessness observed in both experienced and naïve birds around the time when migration should occur [33]. Further the direction and distance to the wintering grounds appears to be genetically encoded in first-time migrants. Good confirmation of this came from crossbreeding studies where hybrids from two populations with unique migratory routes and destinations exhibited an intermediate migratory path [34]. Displacement experiments have also found that naive birds are not able to consistently correct the angle or distance of their migratory path to reach the correct wintering ground [35]. In adults, however, experience with a navigational map allows for such corrections to be made [36]. Despite the innate components of migration, plasticity is more common than might be expected. Recent work suggests that juvenile migrants do have some ability to compensate for strong winds and minor displacements [37]. The presence of *zugunruhe* in a resident species of stonechat (*Saxicola torquatus*; [38]) suggests that this species may have been migratory in the past, or could become migratory in the future [39]. Indeed, migratory populations of blackcaps (*Sylvia atricapilla*) in Europe are staying closer to home as conditions on the breeding grounds become more favorable year round [40] despite the presence of genetic differences between migratory and already resident populations in this species [41]. In contrast to gaping and begging, these studies suggest that it is plausible that migratory behavior could appear or disappear in response to changing conditions – a good reminder that innate underpinnings do not preclude rapid change.

Traditional thinking has been that elements of the nestbuilding process are also innate [42]. But, a role of experience has long been recognized, and experimental evidence to support claims of innate behavior were often lacking [43]. Evidence from hand-rearing experiments, field observations, and laboratory manipulations indicate that associative learning, social learning, and imprinting are important contributors to the nest building process [44]. While nest-building is probably not a hard-wired response [45], and thus not innate, the propensity to build nests is nearly ubiquitous across songbird species. This suggests that selection strongly favors nest-building as a strategy for increasing survival in offspring. This is not to say that alternative strategies are not plausible. A few passerines, including several cowbirds species (genus *Molothrus*), have abandoned nest building; instead laying parasitic eggs in the nests of other species [46]. Several non-passerine species, including a large percentage of shorebirds (order Charadriiformes) lay camouflaged eggs directly on the ground. This non-nest building strategy has been successful in many non-passerine bird species [47]. But in passerines, observing a shift from a nest-building to non-building strategy in response to management or environmental change is unlikely to occur. Nests provide many ecological benefits, including; protection from predation, providing a barrier for containment of nestlings, and to some extent buffering the variability in the environment (e.g., wind, rain, temperature) [48]. Nest characteristics, such as thickness, material, and even placement may be modified more quickly based on learning and experience. For example, orange-crowned warblers (*Vermivora celata*) nesting on Santa Catalina Island, California, USA, were more likely to place nests off the ground than their mainland counterparts, and offspring survival increased with nest height [49]. But when an aerial predator (blue jay, *Cyanocitta cristata*) found on the mainland was artificially presented to the island population during nest building, nests were moved closer to the ground and to more concealed locations.

Other songbird behaviors that exhibit innate components include; predator detection [50] and recognition [51], and some components of vocal production [52] and recognition [53]. In general, innate behaviors should exhibit relatively low levels of plasticity – especially in cases where alternative strategies are not available (e.g., begging). However, this does not preclude change. In the strictest situation, where a genetically encoded behavior has become fixed in a species, change will require genetic mutations that provide variability, and thus, the fodder on which selection can operate. But in most behavioral systems, genetic variation not so tightly constrained. Further, as exhibited through migration, innate behavior is often refined by experience.

Learned Behavior—Systematic

Here I address learned behavior as a systematic process that functions to maintain some level of stability within a range of potential behavioral responses. While experience also contributes to learned behavior, I will more explicitly address this situation in the next section. In songbirds, learning is often classified as open-ended (lifelong learning) or closed-ended (learning occurs during a set period) [54]. This system is often used to describe the development of song. The species-specific songs characteristic of most songbirds are learned only during development – being closed ended. Young birds listen to tutors (silent phase), practice (subsong), and establish adult song (crystallization) [55]. Once songs are crystallized, learning on the macro-scale ceases, although minor modifications to temporal and pitch characteristics can be made in many species [56, 57]. Without tutoring, some elements of song may marginally mimic adult song, which indicates that there may be some genetic components of song development, but these loose mimics are unlikely to function normally [58]. Some species, including those in the suborder Tyranni (i.e., suboscines) are an exception, because they tend to produce adult-like songs without tutoring [59, 60]. One the other hand, some species, such as mockingbirds and other members of the family Mimidae are open-ended learners [61, 62]. Song learning continues throughout adulthood. Still, species-specific characteristics, such as the number of times phrases are repeated, are maintained. In most songbird species, dramatic changes to song types can be made once per generation during the sensitive phase. The limited range of nearby tutors, which is not limited to the paternal parent [63, 64], serves to converge each generation of learned songs toward the species mean. Stabilizing selection is further supported by sexual selection, where females are more likely to pair with males that propagate songs that resemble the species mean [65]. Sexual selection may also serve to maintain more stable visual courtship displays, since females tend to prefer performances that exemplify the idealistic species-specific moves (e.g., highest leap, quickest flight) [66]. In manakins (genus *Manacus*) and other species with extreme courtship displays sexual selection may even drive the development of the physical traits required to perform otherwise atypical body movements including production of sound with other structures on their body like modified feathers [67].

While tutoring and sexual selection might place an outer limit on the rate of change in vocal or visual displays that will be observed within a particular individual or generation, it does not eliminate modification. A multi-decadal study of song types in the white-throated sparrow (*Zonotrichia albicollis*) recently revealed that a novel doublet-ending version of the species-specific song arose in western Canada and replaced the typical triplet-ending version across the much of the continent [68]. Geolocator tracking indicates that birds from spatially separate breeding regions mix on their wintering grounds, where tutoring is likely facilitating the broadscale vocal change. Experimental work in an isolated population of savannah sparrows (*Passerculus sandwichensis*) reveals the process by which a novel song can be integrated [69]. Researchers tutored five developing cohorts with a novel song exemplar played via loudspeakers. During six years of tutelage with the novel song type, 30 individuals incorporated the novel song type into their repertoires, and several second-generation birds learned the novel song from the previous generation. In this case, however, regulation from sexual selection may have been lessened because developing females were also exposed to the novel song type.

One learned behavior that may be less open to modification is imprinting. Young zebra finches selectively prefer their host nest via olfactory imprinting [70], which may also guide future nest building endeavors. Olfactory imprinting during development may also play a role in kin recognition [71, 72]. After fledging, most songbirds scout their breeding habitat before departure, imprinting on the habitat characteristics to which they will return [73, 74]. Although not confirmed in songbirds, pelagic birds [75], sea turtles [76, 77] and salmon [78] also imprint to the earth's magnetic field during development, allowing for long-distance navigation back to their natal habitats as adults for breeding. Imprinting on olfactory or magnetic cues may also facilitate return trips to the breeding grounds in songbirds. Because imprinted behaviors are solidified during development, they may be relatively non-flexible in adults. In these cases, management during development may be required if an alteration to behavior is desired. One famous example is the imprinting of naïve whooping cranes (*Grus americana*) on an ultralight aircraft as surrogate adult guide, and then using this craft to guide birds along the new migratory route. Since 2001, imprinted adult cranes continue to use this trained migratory path, and are now making their own adaptations to the route [79].

Behavior Shaped by Experience

In contrast with systematic learning, behavior that is shaped by experience is not always shared across individuals or populations. In humans, our experiences as a youth, our interactions with different cultures, and our exposure to particular traumatic or particularly rewarding events very much impact how we act as adults. Conversely, having a political figure or professional athlete in one's family increases the likelihood of this career path in the next generation – or more broadly, parental values directly and indirectly influence the aspirations of their offspring [80]. Pavlov's dogs, and extensions of classical conditioning make it clear that this process is not limited to the human realm.

Animals, including songbirds, regularly respond to environmental stimuli and these experiences shape future behavior. One particularly poignant example comes from American crows (*Corvus brachyrhynchos*), which are members of the Corvid family, a group of birds known for their cognitive abilities. Researchers from the University of Washington wore masks while trapping and banding crows [81]. When these same masks were worn on campus in future years, crows responded with harsh calls and mobbing, but neutral masks were not treated with equal disdain. Like humans, crows also transferred their disdain for particular masks to their young [82]. In a novel laboratory experiment, food-caching scrub jays (*Aphelocoma coerulescens*) that had pilfered the caches of other jays moved their own cache sites when other jays observed them during their initial cache [83]. However, jays without previous experience pilfering another's cache did not alter their own cache sites. While experience-dependent learning is particularly impressive in Corvids, it is certainly not limited to this family. Many songbird species become less leery of people when they experience humans regularly [84]. Urban black-capped chickadees (*Poecile atricapillus*) show less neophobia towards novel objects [85] than their rural counterparts, but are less likely to return to a feeder when a house cat model is present [86]. Both differences are likely due to a differing set of past experiences; more benign interactions with novel objects and more adverse interactions with cats in urban-dwelling chickadees.

McGrath and colleagues provided a particularly compelling example of shaping behavior via experience in a field experiment with wild superb fairy-wrens (*Malurus cyaneus*) [87]. Ten individuals were exposed to novel alarm calls accompanied with a gliding predator model. Although none of the birds fled in response to the novel alarm prior to experiment pairing with a predator model, nine of ten individuals fled to the novel alarm after training. These experimental results clearly display the role of experience in shaping critical behavioral responses in songbirds, but they also reveal a potentially important conservation tool - managers may be able to intentionally modify cue-response systems in songbirds via artificial experiences that alter future behavioral responses.

Within-individual Behavioral Plasticity

In many cases, individuals possess the ability to respond and behave differently based on variation in a particular set of external cues. This is broadly known as within-individual plasticity. Foraging demonstrates this concept well in that individuals will select prey species based on the combination of prey availability, energetic gain, and time required to capture and consume prey [88]. This idea, formalized as optimal foraging theory [89], proposes that variation in the aforementioned criteria will lead to a constantly shifting diet composition. For example, great tits offered profitable and unprofitable prey in a lab setting were not selective when both prey densities were low [90]. But when the encounter rate of profitable prey was increased, tits ceased to ingest unprofitable prey. Continued research indicates that optimal foraging is also shaped by competition from heterospecific species [91, 92], and by predation pressure (e.g., landscape of fear [93, 94]). In the urban setting, perceived risk – and therefore foraging behavior – may also be impacted by connectivity, where isolated patchy habitats are less likely to be used by foragers [95].

Plastic behavior may be thought of as possessing a suite of if-then tactics [96]. For the foraging example used above, this might translate to 'if profitable prey is abundant, avoid unprofitable prey'. Alternatively, 'if profitable prey is not abundant, consume any available prey'. Tactics can be distinct responses, or represent a range of responses along a continuum. A number of songbird species are cooperative breeders, such that related or unrelated, non-parental individuals will assist in raising offspring. In this case, a young, reproductively capable songbird must select categorically between having a nest of its own or helping at the parental nest. Which tactic is chosen is dependent on the costs and benefits of each option. In the well-studied Seychelles warbler cooperative system [97], the decision to help at a nest is based upon the availability of new territories, the quality of available territories, and the likelihood of finding a mate [98, 99]. The decision to breed or help will be revisited each year, with helping to raise genetically similar offspring selected when the likelihood of producing one's own offspring is low. Habitat selection is another case where discrete tactics are employed each year in migratory species: settle or do not settle [100]. A number of interacting variables likely play a role in the selection of breeding and wintering habitats [101]; including patch size, vegetation, and the presence of conspecific and heterospecific species [102, 103]. In this case, vocal conspecific cues have been used to train several species to utilize newly created [104], or alternative [105, 106], habitats. Here conservation objectives might include manipulating behavior by carefully altering flexible cue-response systems.

In contrast to these discrete sets of tactics, a number of studies have demonstrated that songbirds can adjust the frequency of their adult songs along a continuum in response to the level of human produced noise, which tends to be low-frequency. While higher frequency songs that reduce masking by low-frequency anthropogenic noise could arise from selection for individuals that always sing higher-frequency songs, a number of studies indicate that individuals can shift their frequency in relation to current ambient conditions [107, 108]. In this case, individual songbirds may be balancing the need to be heard over elevated ambient noise, with the need to produce vocal cues of quality or dominance. Accordingly, we found that black-capped chickadees increased their frequency throughout the morning as traffic levels (and noise) increased [57]. In addition, chickadees sang at generally lower frequencies on weekends when traffic levels were lower. Here the tactics would likely be 'if noise level is high, sing at high frequencies to be heard', but 'if noise level is low, sign at low frequencies to convey information about quality'. While behavioral plasticity might be a welcome trait in a rapidly changing world, it is important to remember that the range of tactics available is not boundless. Physical and physiological limits will limit capabilities. For example, song production is limited by the vocal tract and beak shape, while foraging will be limited by the ability to handle, digest, and procure sufficient nutrients from available food sources. Further, sexual and natural selection will tend to eliminate extreme diversions from typical, species-specific behavior.

Different Behavioral Strategies Among-individuals

If individual songbirds alter their behavior in ways to maximize gain (benefits-costs), the existence of multiple tactics is not surprising. However, we also observe several behavioral patterns that are consistent within individuals, but differ between individuals. For example, wintering ovenbirds (*Seiurus aurocapilla*) in the West Indies were classified as sedentary [109] (e.g., holding stable territories) or floaters (e.g., holding small or no territories with regular foraging excursions). Only a small portion of the population exhibited a floating strategy, and artificially altering food availability did not produce changes in foraging strategies. So, the question might arise as to which strategy is best, and why both continue to persist. The answer here is that the fitness value of each strategy was dependent upon the environment. In years when food resources were high, sedentary, territory holders had higher body mass – an indicator of health. But, when food was scarce, floaters – who were more likely to search for and exploit novel food resources – had higher body mass. Thus, a fluctuating environment will maintain both behavioral strategies within the population.

Even in a stable environment, multiple behavioral strategies can coexist. In the above example, floaters were always less common than sedentary birds. If the percentage of floaters in the population increases, there will be fewer territory holders. Under this scenario, defending a high-quality territory might become easier since fewer birds are doing so – and selection will favor territory holders. Conversely, if most birds hold territories, defense of new territories becomes costly, and intrusion by a few individuals into territories may go unnoticed. Under this scenario, individuals with a floating strategy become favored. This is known as frequency-dependent selection, where the most stable situation will favor the existence of multiple behavioral strategies within the population [110]. At times, the relative frequency of multiple strategies can even be predicted. For example, floating should be consistently less common than a sedentary strategy. If a majority, or even a strong minority of individuals utilize a floating strategy, territory holders will likely strengthen their defenses to avoid a substantial loss of resources. This equilibrium of strategies is known as an Evolutionary Stable Strategy (ESS [111]). As with species and genetics, maintaining a diversity of behavioral strategies is often important for maintaining ecosystem function. For example, an environment that eliminates the floating strategy in wintering ovenbirds, will increase risk of population, or even species extinction during years with low foraging resources.

Box 1 Plasticity in behavior

- Behavior is the way in which an animal acts in response to a particular situation or stimulus
- Behavior is shaped by selection through a *cue*, *response*, and *outcome*

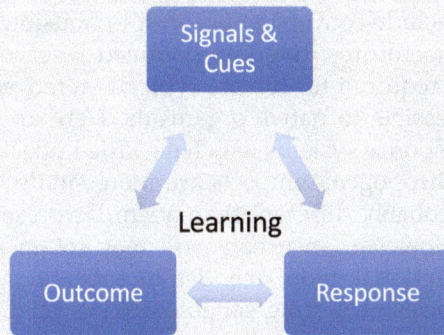

- Modified environments can misalign cue-response systems, potentially resulting in mismatches where behavior becomes maladaptive (e.g., evolutionary mismatch)
- Behavioral systems can also shift to accommodate altered systems to some extent
- Flexibility in a behavioral system depends on the strength of selection and the presence of variation or plasticity in responses

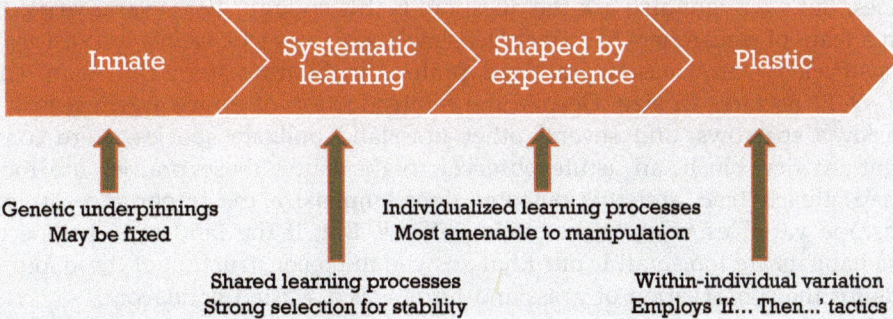

- An evolutionarily stable strategy (ESS) might include multiple behavioral strategies, which are maintained by:
  - Variable habitats
  - Frequency dependent selection

Managing Behavior in the Midst of Anthropogenic Change

Why Integrating Behavior is Important

The structure and function of earth's ecosystems are changing rapidly – perhaps more rapidly than at any time in history. The mere fact that ecosystems are changing is neither abnormal nor problematic. How many times have we been astonished to find ancient find marine fossils in high deserts [112], or signs of lush ancient forests in frozen ice cores [113]. The issue is not change, but rather, rate of change. In the examples above we are observing the remnants of dramatic changes that occurred over Millenia. Today, we are witnessing equally dramatic changes that occur over only a few years, or sometimes just days. While natural events, such as hurricanes or volcanoes can dramatically reshape landscapes over a short duration, the majority of the changes we are witnessing are anthropogenic – or human produced. Andy Sih and colleagues have labelled this pervasive phenomenon as Human Induced Rapid Environmental Change, abbreviated as HIREC [23, 114]. In its most extreme form, HIREC results in complete habitat conversion – which is typically equivalent to habitat destruction for native organisms. Examples include conversion of forest to housing developments and parking lots, grasslands to monocultures, rivers to dammed reservoirs, etc. In these cases, the fundamental resources required for life are typically removed, or altered to a point that they are no longer accessible to native organisms. Here an understanding of behavior and flexibility in cue-response systems may have little impact on a management strategy. But in many cases, anthropogenic impacts are more subtle. The food web in an urban park, for example, is probably different than a similar preserve located within a larger undeveloped habitat. However, an urban park may retain a number of trophic levels and a different, but functional, food web. Yet, as urbanization increases, these habitats often contain fewer species of nesting songbirds [115]. Why?

Classical concepts from landscape ecology would suggest that managers consider larger-scale spatial variables, such as patch size, landscape configuration, and connectivity with other habitat patches [116–118]. There is no doubt that fragmenting habitat, restricting movement, reducing and isolating resources, etc. will directly impact fitness and population persistence. I am confident that every author in this text would advocate for assessing resource and landscape needs as priority objectives for managing songbird species. But we might also ask the question: Is this enough? For several years, I worked with a team of researchers to attract Henslow's sparrows (*Ammodramus henslowii*) to four reconstructed warm-season grassland prairies in Southwestern Michigan. Each field was ~9–12 hectares in size. Despite the reintroduction of native warm-season grasses, Henslow's sparrows, and several other grassland obligate species, were consistently absent. At first blush, an astute observer might state, "these prairies are too small." Perhaps this is true, and this outcome does emphasize the importance of evaluating landscape variables prior to restoration efforts. But, if the landscape perspective that these habitats are too small is our final answer, the reconstruction of these four ~10 acre fields for the conservation of grassland birds was a wasted endeavor.

Might an understanding of behavior yet open these fields to use by obligate grassland birds? At <0.5 hectares [119], a Henlow's nesting territory could fit easily within these fields with ample room for foraging. Our fields were flush with a diverse insect community, ample water was available in nearby habitats, and grasses provided the cover thought necessary for nest placement and concealment. If the fundamental resource provisions are in place, perhaps an overlooked cue-response system is not functional – or has prevented the selection of these habitats. Alternatively, interspecific interaction might differ. For example, the greater composition of edge habitat might result in higher populations of avian or terrestrial nest predators. In this particular case, the landowner and managing organization wanted to investigate further. One hypothesis was that conspecific cues were missing. It is not uncommon for birds to use vocal cues that indicate conspecific presence when selecting breeding territories [120, 121]. But, two years of song playback resulted in only one nesting Henslow's sparrow. Not zero, but not the resounding success that has been seen using this technique elsewhere [122].

Dr. Rob Keys, a colleague at a university in Grand Rapids, Michigan, hypothesized that these northern populations of grassland obligate birds were preferentially selecting old fields comprised of non-native, cool-seasons grasses (as opposed to our fields planted with warm-season grasses). While this would contrast with research from other parts of the range [123], a few anecdotal, but notable, observations supported this hypothesis. First, Henslow's sparrows had been reported more regularly in our field before their conversion to warm-season grasses. Second, a broad survey of other potential grassland sites in the region indicated that locations dominated by cool-season grasses were for more likely to house populations of grassland obligates; including Henslow's sparrows, Bobolinks (*Dolichonyx oryzivorus*), Eastern meadowlarks (*Sturnella magna*), savannah sparrows (*Passerculus sandwichensis*), and grasshopper sparrows (*Ammodramus savannarum*). In 2019, experimental mowing and replanting resulted in territorial establishment by eight individuals from two obligate species (Henslow's and savannah sparrows; unpublished data). An analogous study in the historically forested piedmont region of North Carolina, USA, found that obligate grassland birds showed no preference for warm or cool-season grasses [124]. It is plausible that grassland obligate songbirds now breeding in regions that historically contained minimal native grassland habitat are cueing in on the cool-season grasses found more commonly in abandoned farm fields. If so, 'restoring' cool-season grasslands in these regions that accommodate this potentially altered cue-response system might be of value. Alternatively, facultative grassland bird species, such as the field (*Spizella pusilla*) and song sparrows (*Melospiza melodia*) that arrive earlier in the spring could be preferentially selecting warm-season grasses and excluding obligates from these territories – a case of heterospecific competition. Research on this particular topic is ongoing and it may be too early to alter management regimes, but it illustrates well the point that understanding behavior might lead to tangible shifts in management strategies.

Potential Anthropogenic Changes to Cue-response Systems

Some cue-response systems in songbirds are clearly more flexible than others. But all are to some extent engrained in individual birds and populations. As such, we must expect that any time human activities alter the environment, there is likely to be an impact on the established behavioral responses, and therefore, on fitness. But what anthropogenic impacts should be monitored beyond the obvious impacts of habitat destruction and modification? To answer this question conclusively requires an intimate knowledge of the species and behaviors present in the ecosystem of interest – indicating that location-specific field studies and natural history scholars will remain paramount as we move through the Anthropocene. However, there are a few common, but often overlooked, disturbance regimes that I will address here. Specifically, I will briefly address sensory ecology, range shifts and species invasions, and temperature change. Each topic is explored in more depth in the chapters that follow.

Sensory ecology deals with how birds perceive and respond to cues through vision, olfaction, acoustics, taste, touch, and magnetoreception. These topics are explored in depth in the text, *The Sensory Ecology of Birds* [125]. Despite their critical role in eliciting particular behaviors, many of these changes are not quickly perceived by human observers. For example, because most of us have become accustomed to light and sound pollution near our homes and places of work, we are less aware of the potential impact of the noise from airplanes as they regularly pass over otherwise remote habitats [126] or the effect of the pervasive, ever-present glow of the city skyline in exurban landscapes [127]. If we fail to measure, mitigate, and manage the sensory environment, we may create visually pleasing habitats that are devoid of the species we aim to protect. Dominoni et al. [128] nicely describe how sensory ecology can contribute to conservation biology. In their perspective paper, they suggest that sensory pollutants impact cue-response systems in three ways. First, *masking* reduces an organism's ability to detect or discriminate cues. Second, *distraction* interferes with an organism's ability to process cues because other overlapping cues have captured the individual's attention. Third, *misleading* cues result from conflation of sensory pollutants with natural cues, leading to an inappropriate response. The path towards resolution differs for each of these sensory pollution mechanisms. For example, spatial and/or temporal separation between competing cues may be required when distraction is the primary mechanism. As the literature on sensory ecology grows and summative reviews become available [7, 129], I recommend that managers incorporate the latest developments whenever possible.

As the physical parameters that limit songbird establishment shift, we should expect that changes in species ranges and niches will respond accordingly – which will alter which set of species are interacting with each other in any given area. The impacts of climate change on regional and global temperature is one clear example that I will address in the next paragraph. But even small-scale changes to habitats can alter the distribution and interaction levels of heterospecific species. For example, oil extraction in Northern Michigan has perforated forest interiors with small anthropogenic openings. Between these openings, which are referred to as 'pads', the native forests remain largely contiguous. In many cases interior songbird species continue to nest near the edges of these pads. But research by some of my former students indicated that the insertion of small openings and edge habitats has increased the presence of edge-associated avian nest predators near pad sites, which resulted in significantly higher predation rates in an artificial nest study [130]. Competition from novel, non-predatory heterospecific species may also impact native species as ranges shift. The succession of old field habitat to shrub ecosystems in portions of the northern United States has facilitated the recent establishment of blue-winged warblers (*Vermivora pinus*) in locations previously settled by golden-winged warblers (*V. chrysoptera*). Although the mechanisms are not entirely clear, novel overlap in the range of these two species predictably results in the reduction or extirpation of the golden-winged warbler genetic phenotype [131]. Taken together, these studies serve as a reminder that understanding the interactions between predators and competitors is required when maintaining or reintroducing particular species, and that we must take into account the shifting ranges of the these heterospecific players in light of anthropogenic changes.

Climate changes may be the most concerning driver of external change to songbird cue-response systems in the near future. Temperature, and associated moisture levels, drive changes in vegetative ranges [132] and phenology [133, 134], insect emergence [135], and may directly alter the physiological parameters that facilitate songbird survival. Thus, changes in temperature and moisture are likely to drive alterations in songbird species ranges and heterospecific interactions, and may introduce foraging mismatches during critical life stages such as migration and nestling provisioning. Although attempting to prevent climate change may be largely futile at the local management level, planning accordingly is critical. For example, the conservation-reliant Kirtland's warbler (*Setophaga kirtlandii*) breeds primarily in early-successional jack pine stands (*Pinus banksiana*), and most of their habitat is managed through intentional planting programs in Michigan, USA [136]. But some climate models indicate that future jack pine stands will exist largely to the north in the Canadian provinces [137]. Management must either recognize the Kirtland's warblers reliance on the shifting jack pine range and act accordingly [138], or consider facilitating use of other habitats [105].

The Rate of Change Problem

Anthropogenic processes are quickly changing our landscapes. A moderately zoomed out aerial map of almost any location will reveal signs of human development. But ecosystems are not static, and neither is songbird behavior. In the preceding sections, I have outlined several instances where changes in behavior have been observed naturally in response to anthropogenic alteration [49], or as a result of human manipulation [87]. The real question though, is whether songbird behavior can change fast enough to accommodate rapid anthropogenic change. In many cases the answer may be no. Migration research, for example, has indicated that shifting arrival and departure dates do not necessarily mitigate rising temperatures on the breeding grounds [139]. Correspondingly, we found that the arrival temperature at a fall stopover site in Western Michigan increased over the decades despite significantly later arrival times in four commonly captured short-distance migrants [142]. To estimate the impact of anthropogenic change *a*-priori, I recommend assessing the rate of change in, (1) the anthropogenic variable of interest and, (2) in critical behavioral systems. A dynamic and flexible cue-response system might be better pre-adapted to accommodate change than largely innate or heavily selected behaviors. But it is good to remember that all cue-response systems have rate of change limitations and absolute limits to the range of potential changes on the short-term timescale.

Behavioral Conservation and Management

Where anthropogenic processes threaten to disrupt cue-response systems, mangers have two primary options: decrease the rate of anthropogenic change *or* increase the rate of behavioral change in animals by manipulating the learning process. Where possible, decreasing the rate of anthropogenic change is generally the preferred method. This may rightly sound intuitive, but it does contrast with the stark assessment that either human development or functional ecosystems must cease completely. If academics, managers, politicians, and communities can join forces to slow the rate of land transformation and sensory pollution we may be able to buy time for cue-response systems to redevelop and adjust to novel cues and shifting outcomes. Practical guidelines might include developing neighborhoods and urban landscapes slowly, preserving native vegetation whenever possible, reducing edge or patchy ecotones by maintaining the integrity of the tree canopy, maintaining areas of grass-shrub pollinator supporting patches, minimizing and dimming lighting, abating noise, etc. A parallel principle might also be to contain anthropogenic impacts within limited spatial areas, allowing organisms to accommodate these changes within a larger intact landscape. As a young adult who loved to hike and backpack in the United States, I was taught to leave the trail to urinate. The goal (in addition to privacy) was to spread the human impact across the landscape. As a graduate student in Canada, however, the advice of many in the outdoor community was to urinate directly on the trail – with the goal being to limit human impact to a small portion of the landscape. Urban centers are an example of this concept on a much larger scale. High impact in a spatially compact area. Taking this development principle to heart might also slow the overall rate of change across our native landscapes.

Slow change might also be a pertinent principle for restoration efforts. Is converting large swaths of old field into restored prairie in one instance disruptive to communities that have already established there? Might a few acres at a time better increase diversity? Does a quick removal of all non-native species from a forest decimate vegetative cover to the point that songbirds are not able, or not inclined, to nest in these habitats? Perhaps an incremental invasive removal program will allow birds to continue nesting until natives can take hold, even if productivity is somewhat lower.

As ecologists we often have to recognize that there are many situations where we simply do not have the power to moderate the rate of development. Further, restoration of native habitats or the removal of sensory pollutants may be an impossibility in many circumstances. In these cases, it may be worth asking whether we can increase the rate of behavioral change in songbirds themselves. Specifically, can we use learning paradigms to alter cue-response systems in songbirds to reduced or eliminate maladaptive responses? Manipulating behavior, just like manipulating genetics, may be unappealing to some. Some may consider this to be meddling in biological systems, which can have the potential for unexpected outcomes. But often the alternative is to be a casual observer of slowly (or rapidly) declining populations as evolutionary mismatches fail to resolve themselves quickly enough to facilitate recovery. For example, the recovery of small populations in many species may be inhibited by the Allee effect – which points to a reliance on social interaction for successful breeding [140]. Can managers artificially create the appearance of social groups in the field by manipulating the sensory environment (e.g., visual models and vocal playback)? Could reintroduction programs use similar cues in captive settings paired with live models from similar species? Certainly, the successful development of a new migratory route in a captive-raised group of whooping cranes was a heavily manipulated management regime. Are there other areas, other ways, that we can manipulate behavior to enhance or rescue songbird populations?

Several years ago, I worked with students to see if we could draw songbirds into forested areas near low-use road in Northern Michigan. Previous data suggested that vehicular noise reduced bird abundance and diversity up to a kilometer from roadsides [141]. But my own observations were that noise levels were quite low beyond a few hundred meters from the roadway. I hypothesized that in these areas where noise levels were low but not absent, entrenched cue-response systems prohibited settlement because preferred habitats were historically quiet [8]. To over-ride aversion to low-level noise, we attempted to place another cue on the landscape known to enhance site settlement – conspecific song. Although our experimental progress has been slow, we were able to draw several species into roadside habitats using song playback [106]. I close with this example to demonstrate an area where I see promise for using behavioral manipulation to advance conservation goals, but also to demonstrate the complexity and need for caution when manipulating behavior. The first point is that the trained or manipulated behavior (settlement in low-noise areas in this case) must confer a fitness benefit. Or at the very least, it must not result in a fitness decrease (e.g., a population sink). Perhaps there is another impact of low-level noise we are not accounting for (e.g., distraction, increased predators). In our case, if playback draws birds from high-quality areas into an area that is unsuitable, we will likely decrease productivity, and thus population sustainability. Conversely, if non-territory holders (i.e., floaters) settle and breed successfully, populations may increase. We must examine productivity and survivorship before using such a technique on a broader scale. Second, if birds are retrained to use a novel cue-response system, how will it be maintained? The positive outcome in response to the behavioral choice must be reinforced. Our hypothesis was that natal imprinting might facilitate site return by some individuals, and provide conspecific song for addition generations – reinforcing establishment without artificial song playback. But this hypothesis remains untested.

I invite you to consider how you might decrease the rate of anthropogenic changes *and* how you might increase the rate of behavioral change in the systems and species where you work. The challenges we face in the Anthropocene are profound. Those whose job and passion is to see songbirds thrive for generations to come will need to be creative. We must be ready to employ an array of conservation tools – and managing habitats and species in relation to behavior must be among these tools.

Box 2 Anthropogenic and Behavioral Change

- Anthropogenic changes will impact birds directly by removing required resources
- Anthropogenic changes will also impact bird behavior by altering cue-response systems
- Ignoring behavioral impacts might lead to subpar management. Two examples, include:
  - Ignoring sensory ecology, which drives behavior and ultimately physical functions
  - Ignoring changes to species ranges, which drives novel heterospecific interactions
- Anthropogenic change moves on a rapid scale
- Changes in songbird cue-response systems may not change rapidly enough to mitigate mismatches before irreversible negative impacts on populations and species are incurred

Box 2 (Contd.)   Anthropogenic and Behavioral Change

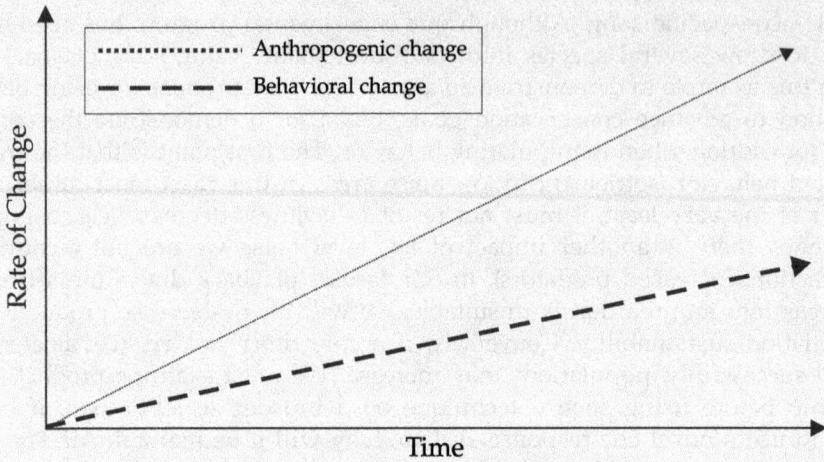

- Where preventing change is not an option, slowing the rate of change might provide benefits by allowing behavioral cue-response systems time to adjust or shift
- Behavioral management falls into two primary categories

Managing Habitat
Managing habitats to retain critical cues and signals that facilitate successful behavioral systems in songbirds

Managing Behavior
Manipulating cue-response systems to intentionally alter songbird responses to novel cues and signals

LITERATURE CITED

[1]   Behaviour. 2020. Oxford Online Dictionary. Retrieved from https://www.lexico.com/definition/behaviour

[2]   Domjan, M.P. 2014. The Principles of Learning and Behavior, 7th Ed. Stamford, CT: Cengage Learning.

[3]   Rubenstein, D.R. and J. Alcock. 2018. Animal Behavior, 11th Ed. Sunderland, Massachusetts: Sinauer Associates, imprint of Oxford University Press.

[4]   Tinbergen, N. 1963. On aims and methods of Ethology. Zeitschrift für Tierpsychologie. 20(4): 410–433. https://doi.org/10.1111/j.1439-0310.1963.tb01161.x

[5]   Davies, N.B., J.R. Krebs and S.A. West. 2012. An Introduction to Behavioural Ecology, 4th Ed. Oxford: Wiley-Blackwell.

[6] Hutchinson, J.M.C. and G. Gigerenzer. 2005. Simple heuristics and rules of thumb: Where psychologists and behavioural biologists might meet. Behav. Process. 69(2): 97–124. https://doi.org/10.1016/j.beproc.2005.02.019

[7] Blumstein, D.T. and O. Berger-Tal. 2015. Understanding sensory mechanisms to develop effective conservation and management tools. Curr. Opin. Behav. Sci. 6: 13–18. https://doi.org/10.1016/j.cobeha.2015.06.008

[8] Proppe, D.S., N. McMillan, J.V. Congdon and C.B. Sturdy. 2017. Mitigating road impacts on animals through learning principles. Anim. Cogn. 20(1): 19–31. https://doi.org/10.1007/s10071-016-0989-y

[9] Robinson, J.K. and W.R. Woodward. 1989. The convergence of behavioral biology and operant psychology: toward an interlevel and interfield science. Behav. Analyst. 12(2): 131–141. https://do i.org/10.1007/BF03392490

[10] Pearce, J.M. 2008. Animal Learning and Cognition: An Introduction, 3rd Ed. Hove, New York: Psychology Press.

[11] Shettleworth, S.J. 2009. Cognition, Evolution, and Behavior, 2nd Ed. Oxford, New York: Oxford University Press.

[12] Yerkes, R.M. and S. Morgulis. 1909. The method of Pawlow in animal psychology. Psychol. Bull. 6(8): 257–273. https://doi.org/10.1037/h0070886

[13] McGrath, L.J., C. van Riper and J.J. Fontaine. 2009. Flower power: tree flowering phenology as a settlement cue for migrating birds. J. Anim. Ecol. 78(1): 22–30.

[14] Wolfe, J.D., M.D. Johnson and C.J. Ralph. 2014. Do birds select habitat or food resources? Nearctic-Neotropic migrants in northeastern Costa Rica. PLoS One. 9(1): e86221. https://doi.org/10.1371/journal.pone.0086221

[15] Bouton, M.E. 2004. Context and behavioral processes in extinction. Learning & Memory. 11(5): 485–494. https://doi.org/10.1101/lm.78804

[16] Kulkarni, M.G. and A.K. Dalai. 2006. Waste cooking oil an economical source for biodiesel: a review. Ind. Eng. Chem. Res. 45(9): 2901–2913. https://doi.org/10.1021/ie0510526

[17] Battle, K., S.L. Foltz and I.T. Moore. 2016. Predictors of flight behavior in rural and urban songbirds. Wilson J. Ornithol. 128(3): 510–519. https://doi.org/10.1676/1559-4491-128.3.510

[18] Møller, A.P. 2008. Flight distance of urban birds, predation, and selection for urban life. Behav. Ecol. Sociobiol. 63(1): 63–75. https://doi.org/10.1007/s00265-008-0636-y

[19] Dai, K., A. Bergot, C. Liang, W.-N. Xiang and Z. Huang. 2015. Environmental issues associated with wind energy – A review. Renew. Energ. 75, 911–921. https://doi.org/10.1016/j.renene.2014.10.074

[20] Hüppop, O., J. Dierschke, K.-M. Exo, E. Fredrich and R. Hill. 2006. Bird migration studies and potential collision risk with offshore wind turbines. Ibis. 148(s1): 90–109. https://doi.org/10.1111/j.1474-919X.2006.00536.x

[21] Kikuchi, R. 2008. Adverse impacts of wind power generation on collision behaviour of birds and anti-predator behaviour of squirrels. J. Nat. Conserv. 16(1): 44–55. https://doi.org/10.1016/j.jnc.2007.11.001

[22] Sih, A. 2013. Understanding variation in behavioural responses to human-induced rapid environmental change: a conceptual overview. Anim. Behav. 85(5): 1077–1088. https://doi.org/10.1016/j.anbehav.2013.02.017

[23] Sih, A., P.C. Trimmer and S.M. Ehlman. 2016. A conceptual framework for understanding behavioral responses to HIREC. Curr. Opin. Behav. Sci. 12: 109–114. https://doi.org/10.1016/j.cobeha.2016.09.014

[24] Nussey, D.H., E. Postma, P. Gienapp and M.E. Visser. 2005. Selection on heritable phenotypic plasticity in a wild bird population. Science. 310(5746): 304–306. https://doi.org/10.1126/science.1117004

[25] Ricklefs, R.E. 1966. Behavior of young cactus wrens and curve-billed thrashers. Wilson Bull. 78(1): 47–56.

[26] Tinbergen, N. 1939. On the analysis of social organization among vertebrates, with special reference to birds. Am. Midl. Nat. 21(1): 210–234. https://doi.org/10.2307/2420381

[27] Heist, C.A. and G. Ritchison. 2016. Effects of variation in nestling hunger levels and begging on the provisioning behavior of male and female eastern phoebes (*Sayornis phoebe*). Wilson J. Ornithol. 128(1): 132–143. https://doi.org/10.1676/wils-128-01-132-143.1

[28] Bengtsson, H. and O. Rydén. 1983. Parental feeding rate in relation to begging behavior in asynchronously hatched broods of the great tit Parus major. Behav. Ecol. Sociobiol. 12(3): 243–251. https://doi.org/10.1007/BF00290777

[29] Clark, A.B. and W.-H. Lee. 1998. Red-winged blackbird females fail to increase feeding in response to begging call playbacks. Anim. Behav. 56(3): 563–570. https://doi.org/10.1006/anbe.1998.0831

[30] Leonard, M.L. and A.G. Horn. 2001. Begging calls and parental feeding decisions in tree swallows (*Tachycineta bicolor*). Behav. Ecol. Sociobiol. 49(2): 170–175. https://doi.org/10.1007/s002650000290

[31] Helfenstein, F., A. Berthouly, M. Tanner, F. Karadas and H. Richner. 2008. Nestling begging intensity and parental effort in relation to prelaying carotenoid availability. Behav. Ecol. 19(1): 108–115. https://doi.org/10.1093/beheco/arm103

[32] Wiltschko, R. and W. Wiltschko. 2015. Avian navigation: a combination of innate and learned mechanisms. pp. 229–310. *In*: M. Naguib, H.J. Brockmann, J.C. Mitani, L.W. Simmons, L. Barrett, S. Healy, et al. (eds). Advances in the Study of Behavior, Vol. 47. Academic Press. https://doi.org/10.1016/bs.asb.2014.12.002

[33] Farner, D.S. 1950. The annual stimulus for migration. Condor. 52(3): 104–122. https://doi.org/10.2307/1364895

[34] Delmore, K.E. and D.E. Irwin. 2014. Hybrid songbirds employ intermediate routes in a migratory divide. Ecol. Lett. 17(10): 1211–1218. https://doi.org/10.1111/ele.12326

[35] Perdeck, A.C. 1958. Two types of orientation in migrating starlings, *Sturnus yulgaris* L., and *Chaffinches, Fringilla coelebs* L., as revealed by displacement experiments. Ardea. 46. https://doi.org/10.5253/arde.v1i2.p1

[36] Thorup, K., I.-A. Bisson, M.S. Bowlin, R.A. Holland, J.C. Wingfield, M. Ramenofsky, et al. 2007. Evidence for a navigational map stretching across the continental U.S. in a migratory songbird. P. Natl. Acad. Sci. USA. 104(46): 18115–18119. https://doi.org/10.1073/pnas.0704734104

[37] Thorup, K., T.E. Ortvad, J. Rabøl, R.A. Holland, A.P. Tøttrup and M. Wikelski. 2011. Juvenile songbirds compensate for displacement to oceanic islands during autumn migration. PLoS One. 6(3): e17903. https://doi.org/10.1371/journal.pone.0017903

[38] Helm, B. and E. Gwinner. 2006. Migratory restlessness in an equatorial nonmigratory bird. PLoS Biol. 4(4): e110. https://doi.org/10.1371/journal.pbio.0040110

[39] Gross, L. 2006. Remnants of the past or ready to move? resident birds display migratory restlessness. PLoS Biol. 4(4): e130. https://doi.org/10.1371/journal.pbio.0040130

[40] Pulido, F. and P. Berthold. 2010. Current selection for lower migratory activity will drive the evolution of residency in a migratory bird population. P. Natl. Acad. Sci. USA. 107(16): 7341–7346. https://doi.org/10.1073/pnas.0910361107

[41] Delmore, K., J.C. Illera, J. Pérez-Tris, G. Segelbacher, J.S. Lugo Ramos, G. Durieux, et al. 2020. The evolutionary history and genomics of European blackcap migration. eLife. 9: e54462. https://doi.org/10.7554/eLife.54462

[42] Sargent, T.D. 1965. The role of experience in the nest building of the zebra finch. Auk. 82(1): 48–61. https://doi.org/10.2307/4082794

[43] Hansell, M. 2000. Bird Nests and Construction Behaviour. Cambridge University Press.

[44] Breen, A.J., L.M. Guillette and S.D. Healy. 2016. What can nest-building birds teach us? Comparative Cognition & Behavior Reviews. 11: 83-102. http://comparative-cognition-and-behavior-reviews.org/2016/vol11_breen_guillette_healy/

[45] Guillette, L.M. and S.D. Healy. 2015. Nest building, the forgotten behaviour. Curr. Opin. Behav. Sci. 6: 90–96. https://doi.org/10.1016/j.cobeha.2015.10.009

[46] Ortega, C.P. 1998. Cowbirds and Other Brood Parasites. University of Arizona Press.

[47] Méndez, V., J.A. Alves, J.A. Gill and T.G. Gunnarsson. 2018. Patterns and processes in shorebird survival rates: a global review. Ibis. 160(4): 723–741. https://doi.org/10.1111/ibi.12586

[48] Collias, N.E. and E.C. Collias. 1984. Nest-site selection and the physical environment. pp. 86–100. *In*: Nest Building and Bird Behavior. Princeton University Press. https://doi.org/10.2307/j.ctt7zvc5n.11

[49] Peluc, S.I., T.S. Sillett, J.T. Rotenberry and C.K. Ghalambor. 2008. Adaptive phenotypic plasticity in an island songbird exposed to a novel predation risk. Behav. Ecol. 19(4): 830–835. https://doi.org/10.1093/beheco/arn033

[50] Amo, L., M.E. Visser, and K. van Oers. 2011. Smelling out predators is innate in birds. Ardea. 99(2): 177–184. https://doi.org/10.5253/078.099.0207

[51] Veen, T., D.S. Richardson, K. Blaakmeer and J. Komdeur. 2000. Experimental evidence for innate predator recognition in the Seychelles warbler. Proceedings of the Royal Society of London. Series B: Biological Sciences. 267(1459): 2253–2258. https://doi.org/10.1098/rspb.2000.1276

[52] Mooney, R. 2020. The neurobiology of innate and learned vocalizations in rodents and songbirds. Curr. Opin. Neurobiol. 64: 24–31. https://doi.org/10.1016/j.conb.2020.01.004

[53] Marler, P. 1990. Innate learning preferences: signals for communication. Dev. Psychobiol. 23(7): 557–568. https://doi.org/10.1002/dev.420230703

[54] Beecher, M.D. and E.A. Brenowitz, E.A. 2005. Functional aspects of song learning in songbirds. Trends Ecol. Evol. 20(3): 143–149. https://doi.org/10.1016/j.tree.2005.01.004

[55] Marler, P.R. 1990. Song learning: the interface between behaviour and neuroethology. Philosophical Transactions of the Royal Society of London. Series B: Biological Sciences. 329(1253): 109–114. https://doi.org/10.1098/rstb.1990.0155

[56] Ota, N. and M. Soma. 2014. Age-dependent song changes in a closed-ended vocal learner: elevation of song performance after song crystallization. J. Avian Biol. 45(6): 566–573. https://doi.org/10.1111/jav.00383

[57] Proppe, D.S., C.B. Sturdy and C.C.S. Clair. 2011. Flexibility in animal signals facilitates adaptation to rapidly changing environments. PLoS One. 6(9): e25413. https://doi.org/10.1371/journal.pone.0025413

[58] Nottebohm, F. 1970. Ontogeny of bird song. Science. 167(3920): 950–956.

[59] Kroodsma, D.E. 1984. Songs of the alder flycatcher (*Empidonax alnorum*) and willow flycatcher (*Empidonax traillii*) are Innate. Auk. 101(1): 13–24. https://doi.org/10.1093/auk/101.1.13

[60] Touchton, J.M., N. Seddon and J.A. Tobias. 2014. Captive rearing experiments confirm song development without learning in a tracheophone suboscine bird. PLoS One. 9(4): e95746. https://doi.org/10.1371/journal.pone.0095746

[61] Eens, M., R. Pinxten and R.F. Verheyen. 1992. Song learning in captive European starlings, *Sturnus vulgaris*. Anim. Behav. 44(6): 1131–1143. https://doi.org/10.1016/S0003-3472(05)80325-2

[62] Gammon, D.E. 2020. Are northern mockingbirds classic open-ended song learners? Ethology. 126(11): 1038–1047. https://doi.org/10.1111/eth.13080

[63] Beecher, M.D. 2017. Birdsong learning as a social process. Anim. Behav. 124: 233–246. https://doi.org/10.1016/j.anbehav.2016.09.001

[64] Brenowitz, E.A. and M.D. Beecher. 2005. Song learning in birds: diversity and plasticity, opportunities and challenges. Trends Neurosci. 28(3): 127–132. https://doi.org/10.1016/j.tins.2005.01.004

[65] Searcy, W.A. and M. Andersson. 1986. Sexual selection and the evolution of song. Annu. Rev. Ecol. Evol. Syst. 17(1): 507–533. https://doi.org/10.1146/annurev.es.17.110186.002451

[66] Candolin, U. 2003. The use of multiple cues in mate choice. Biological Reviews. 78(4): 575–595. https://doi.org/10.1017/S1464793103006158

[67] Fusani, L., J. Barske, L.D. Day, M.J. Fuxjager and B.A. Schlinger. 2014. Physiological control of elaborate male courtship: female choice for neuromuscular systems. Neurosci. Biobehav. Rev. 46: 534–546. https://doi.org/10.1016/j.neubiorev.2014.07.017

[68] Otter, K.A., A. Mckenna, S.E. LaZerte and S.M. Ramsay. 2020. Continent-wide shifts in song dialects of white-throated sparrows. Curr. Biol. 30(16): 3231–3235 https://doi.org/10.1016/j.cub.2020.05.084

[69] Mennill, D.J., S.M. Doucet, A.E.M. Newman, H. Williams, I.G. Moran, I.P. Thomas, et al. 2018. Wild birds learn songs from experimental vocal tutors. Curr. Biol. 28(20): 3273-3278. e4. https://doi.org/10.1016/j.cub.2018.08.011

[70] Caspers, B.A., J.I. Hoffman, P. Kohlmeier, O. Krüger and E.T. Krause. 2013. Olfactory imprinting as a mechanism for nest odour recognition in zebra finches. Anim. Behav. 86(1): 85–90. https://doi.org/10.1016/j.anbehav.2013.04.015

[71] Bonadonna, F. and A. Sanz-Aguilar, 2012. Kin recognition and inbreeding avoidance in wild birds: the first evidence for individual kin-related odour recognition. Anim. Behav. 84(3): 509–513. https://doi.org/10.1016/j.anbehav.2012.06.014

[72] Krause, E.T., O. Krüger, P. Kohlmeier and B.A. Caspers. 2012. Olfactory kin recognition in a songbird. Biol. Lett. 8(3): 327–329. https://doi.org/10.1098/rsbl.2011.1093

[73] Camacho, C., D. Canal and J. Potti. 2016. Natal habitat imprinting counteracts the diversifying effects of phenotype-dependent dispersal in a spatially structured population. BMC Evol. Biol. 16(1): 158. https://doi.org/10.1186/s12862-016-0724-y

[74] Morton, M.L., M.W. Wakamatsu, M.E. Pereyra and G.A. Morton. 1991. Postfledging dispersal, habitat imprinting and philopatry in a montane, migratory sparrow. Ornis Scandinavica (Scandinavian Journal of Ornithology). 22(2): 98–106. https://doi.org/10.2307/3676540

[75] Wynn, J., O. Padget, H. Mouritsen, C. Perrins and T. Guilford. 2020. Natal imprinting to the Earth's magnetic field in a pelagic seabird. Curr. Biol. 30(14): 2869–2873.e2. https://doi.org/10.1016/j.cub.2020.05.039

[76] Brothers, J.R. and K.J. Lohmann. 2015. Evidence for geomagnetic imprinting and magnetic navigation in the natal homing of sea turtles. Curr. Biol. 25(3): 392–396. https://doi.org/10.1016/j.cub.2014.12.035

[77] Lohmann, K.J., N.F. Putman and C.M.F. Lohmann. 2008. Geomagnetic imprinting: a unifying hypothesis of long-distance natal homing in salmon and sea turtles. P. Natl. Acad. Sci. USA. 105(49): 19096–19101. https://doi.org/10.1073/pnas.0801859105

[78] Putman, N.F., E.S. Jenkins, C.G.J. Michielsens and D.L.G. Noakes. 2014. Geomagnetic imprinting predicts spatio-temporal variation in homing migration of pink and sockeye salmon. Journal of The Royal Society Interface. 11(99): 20140542. https://doi.org/10.1098/rsif.2014.0542

[79] Teitelbaum, C.S., S.J. Converse, W.F. Fagan, K. Böhning-Gaese, R.B. O'Hara, A.E. Lacy, et al. 2016. Experience drives innovation of new migration patterns of whooping cranes in response to global change. Nat. Commun. 7(1): 12793. https://doi.org/10.1038/ncomms12793

[80] Jodl, K.M., A. Michael, O. Malanchuk, J.S. Eccles and A. Sameroff. 2001. Parents' roles in shaping early adolescents' occupational aspirations. Child Development. 72(4): 1247–1266. https://doi.org/10.1111/1467-8624.00345

[81] Marzluff, J.M., J. Walls, H.N. Cornell, J.C. Withey and D.P. Craig. 2010. Lasting recognition of threatening people by wild American crows. Anim. Behav. 79(3): 699–707. https://doi.org/10.1016/j.anbehav.2009.12.022

[82] Cornell, H.N., J.M. Marzluff and S. Pecoraro. 2012. Social learning spreads knowledge about dangerous humans among American crows. Proceedings of the Royal Society B: Biological Sciences. 279(1728): 499–508. https://doi.org/10.1098/rspb.2011.0957

[83] Emery, N.J. and N.S. Clayton. 2001. Effects of experience and social context on prospective caching strategies by scrub jays. Nature. 414(6862): 443–446. https://doi.org/10.1038/35106560

[84] Samia, D.S.M., D.T. Blumstein, M. Díaz, T. Grim, J.D. Ibáñez-Álamo, J. Jokimäki, et al. 2017. Rural-urban differences in escape behavior of european birds across a latitudinal gradient. Front. Ecol. Evol. 5: 66. https://doi.org/10.3389/fevo.2017.00066

[85] Jarjour, C., J.C. Evans, M. Routh and J. Morand-Ferron. 2020. Does city life reduce neophobia? a study on wild black-capped chickadees. Behav. Ecol. 31(1): 123–131. https://doi.org/10.1093/beheco/arz167

[86] Van Donselaar, J.L., J.L. Atma, Z.A. Kruyf, H.N. LaCroix and D.S. Proppe. 2018. Urbanization alters fear behavior in black-capped chickadees. Urban Ecosyst. 21(6): 1043–1051. https://doi.org/10.1007/s11252-018-0783-5

[87] Magrath, R.D., T.M. Haff, J.R. McLachlan and B. Igic. 2015. Wild birds learn to eavesdrop on heterospecific alarm calls. Curr. Biol. 25(15): 2047–2050. https://doi.org/10.1016/j.cub.2015.06.028

[88] Emlen, J.M. 1966. The role of time and energy in food preference. Am. Nat. 100(916): 611–617.

[89] MacArthur, R.H. and E.R. Pianka. 1966. On optimal use of a patchy environment. Am. Nat. 100(916): 603–609.

[90] Krebs, J.R., J.T. Erichsen, M.I. Webber and E.L. Charnov. 1977. Optimal prey selection in the great tit (*Parus major*). Anim. Behav. 25, 30–38. https://doi.org/10.1016/0003-3472(77)90064-1

[91] Milinski, M. 1982. Optimal foraging: The influence of intraspecific competition on diet selection. Behav. Ecol. Sociobiol. 11(2): 109–115. https://doi.org/10.1007/BF00300099

[92] Shochat, E., S.B. Lerman, M. Katti and D.B. Lewis. 2004. Linking optimal foraging behavior to bird community structure in an urban-desert landscape: field experiments with artificial food patches. Am. Nat. 164(2): 232–243. https://doi.org/10.1086/422222

[93] Laundré, J.W., L. Hernández and K.B. Altendorf. 2001. Wolves, elk, and bison: reestablishing the "landscape of fear" in Yellowstone National Park, U.S.A. Can. J. Zool. 79(8): 1401–1409. https://doi.org/10.1139/z01-094

[94] Lima, S.L. and L.M. Dill. 1990. Behavioral decisions made under the risk of predation: a review and prospectus. Can. J. Zool. 68(4): 619–640. https://doi.org/10.1139/z90-092

[95] Visscher, D.R., A. Unger, H. Grobbelaar and P.D. DeWitt. 2018. Bird foraging is influenced by both risk and connectivity in urban parks. J. Urban Ecol. 4(1): juy020. https://doi.org/10.1093/jue/juy020

[96] Gross, M.R. 1996. Alternative reproductive strategies and tactics: diversity within sexes. Trends Ecol. Evol. 11(2): 92–98. https://doi.org/10.1016/0169-5347(96)81050-0

[97] Komdeur, J. 1994. The effect of kinship on helping in the cooperative breeding Seychelles warbler (*Acrocephalus sechellensis*). Proceedings of the Royal Society of London. Series B: Biological Sciences. 256(1345): 47–52. https://doi.org/10.1098/rspb.1994.0047

[98] Komdeur, J. 1992. Importance of habitat saturation and territory quality for evolution of cooperative breeding in the Seychelles warbler. Nature. 358(6386): 493–495. https://doi.org/10.1038/358493a0

[99] Komdeur, J. 1994. Experimental evidence for helping and hindering by previous offspring in the cooperative-breeding Seychelles warbler *Acrocephalus sechellensis*. Behav. Ecol. Sociobiol. 34(3): 175–186. https://doi.org/10.1007/BF00167742

[100] Jones, J. 2001. Habitat selection studies in avian ecology: a critical review. Auk. 118(2): 557–562. https://doi.org/10.1093/auk/118.2.557

[101] Klopfer, P.H. and J.P. Hailman. 1965. Habitat Selection in Birds. pp. 279–303. *In*: D.S. Lehrman, R.A. Hinde and E. Shaw (eds). Advances in the Study of Behavior, Vol. 1. Academic Press. https://doi.org/10.1016/S0065-3454(08)60060-1

[102] Szymkowiak, J., R.L. Thomson and L. Kuczyński. 2017. Interspecific social information use in habitat selection decisions among migrant songbirds. Behav. Ecol. 28(3): 767–775. https://doi.org/10.1093/beheco/arx029

[103] Ward, M.P. and S. Schlossberg. 2004. Conspecific attraction and the conservation of territorial songbirds. Conserv. Biol. 18(2): 519–525. https://doi.org/10.1111/j.1523 -1739.2004.00494.x

[104] Ahlering, M.A. and J. Faaborg. 2006. Avian habitat management meets conspecific attraction: if you build it, will they come? Auk. 123(2): 301–312. https://doi.org/10.1093/auk/123.2.301

[105] Anich, N.M. and M.P. Ward. 2017. Using audio playback to expand the geographic breeding range of an endangered species. Divers. Distrib. 23(12): 1499–1508. https://doi.org/10.1111/ddi.12635

[106] Schepers, M.J. and D.S. Proppe. 2017. Song playback increases songbird density near low to moderate use roads. Behavioral Ecology. 28(1): 123–130. https://doi.org/10.1093/beheco/arw139

[107] Bermúdez-Cuamatzin, E., A.A. Ríos-Chelén, D. Gil and C.M. Garcia. 2011. Experimental evidence for real-time song frequency shift in response to urban noise in a passerine bird. Biology Letters. 7(1): 36–38. https://doi.org/10.1098/rsbl.2010.0437

[108] Potvin, D.A. and R,A. Mulder. 2013. Immediate, independent adjustment of call pitch and amplitude in response to varying background noise by silvereyes (*Zosterops lateralis*). Behav. Ecol. 24(6): 1363–1368. https://doi.org/10.1093/beheco/art075

[109] Brown, D.R. and T.W. Sherry. 2008. Alternative strategies of space use and response to resource change in a wintering migrant songbird. Behav. Ecol. 19(6): 1314–1325. https://doi.org/10.1093/beheco/arn073

[110] Ayala, F.J. and C.A. Campbell. 1974. Frequency-dependent selection. Annual Review of Ecology and Systematics. 5(1): 115–138. https://doi.org/10.1146/annurev.es.05.110174.000555

[111] Maynard-Smith, J. and G.R. Price. 1973. The logic of animal conflict. Nature. 246(5427): 15–18. https://doi.org/10.1038/246015a0

[112] O'Leary, M.A., M.L. Bouaré, K.M. Claeson, K. Heilbronn, R.V. Hill, J.A. McCartney, et al. 2019. Stratigraphy and paleobiology of the Upper Cretaceous-Lower Paleogene sediments from the Trans-Saharan Seaway in Mali. Bulletin of the American Museum of Natural History. 436: 1–183. http://digitallibrary.amnh.org/handle/2246/6950.

[113] Willerslev, E., E. Cappellini, W. Boomsma, R. Nielsen, M.B. Hebsgaard, T.B. Brand and M.J. Collins. 2007. Ancient biomolecules from deep ice cores reveal a forested southern greenland. Science. 317(5834): 111–114. https://doi.org/10.1126/science.1141758

[114] Sih, A., M.C.D. Ferrari and D.J. Harris. 2011. Evolution and behavioural responses to human-induced rapid environmental change. Evol. Appl. 4(2): 367–387. https://doi.org/10.1111/j.1752-4571.2010.00166.x

[115] Melles, S., S. Glenn and K. Martin. 2003. Urban bird diversity and landscape complexity: species–environment associations along a multiscale habitat gradient. Conserv. Ecol. 7(1): 5. https://www.jstor.org/stable/26271915

[116] Burrough, P.A. 1981. Fractal dimensions of landscapes and other environmental data. Nature. 294(5838): 240–242. https://doi.org/10.1038/294240a0

[117] Crooks, K.R. and M. Sanjayan. (Eds). 2006. Connectivity Conservation. Cambridge: Cambridge University Press. https://doi.org/10.1017/CBO9780511754821

[118] Pickett, S.T.A. and J.N. Thompson. 1978. Patch dynamics and the design of nature reserves. Biol. Conserv. 13(1): 27–37. https://doi.org/10.1016/0006-3207(78)90016-2

[119] Robins, J.D. 1971. A study of Henslow's sparrow in michigan. Wilson Bull. 83(1): 39–48.

[120] Reed, J.M. and A.P. Dobson. 1993. Behavioural constraints and conservation biology: Conspecific attraction and recruitment. Trends Ecol. Evol. 8(7): 253–256. https://doi.org/10.1016/0169-5347(93)90201-Y

[121] Stamps, J.A. 1988. Conspecific attraction and aggregation in territorial species. Am. Nat. 131(3): 329–347. https://doi.org/10.1086/284793

[122] Vogel, J.A., R.R. Koford and D.L. Otis. 2011. Assessing the role of conspecific attraction in habitat restoration for Henslow's sparrows in Iowa. Prairie Naturalist. 43(1–2): 23–28.

[123] Giuliano, W.M. and S.E. Daves. 2002. Avian response to warm-season grass use in pasture and hayfield management. Biol. Conserv. 106(1): 1–9. https://doi.org/10.1016/S0006-3207 (01)00126-4

[124] Moorman, C.E., R.L. Klimstra, C.A. Harper, J.F. Marcus and C.E. Sorenson. 2017. Breeding songbird use of native warm-season and non-native cool-season grass forage fields. Wildl. Soc. Bull. 41(1): 42–48. https://doi.org/10.1002/wsb.726

[125] Martin, G.R. 2017. The Sensory Ecology of Birds. Oxford University Press.

[126] Lynch, E., D. Joyce and K. Fristrup. 2011. An assessment of noise audibility and sound levels in U.S. National Parks. Landscape Ecol. 26(9): 1297–1309. https://doi.org/10.1007/s10980-011-9643-x

[127] Adams, C.A., A. Blumenthal, E. Fernández-Juricic, E. Bayne and C.C. St. Clair. 2019. Effect of anthropogenic light on bird movement, habitat selection, and distribution: a systematic map protocol. Environ. Evidence. 8(1): 1–16. https://doi.org/10.1186/s13750-019-0155-5

[128] Dominoni, D.M., W. Halfwerk, E. Baird, R.T. Buxton, E. Fernández-Juricic, K.M. Fristrup, et al. 2020. Why conservation biology can benefit from sensory ecology. Nat. Ecol. Evol. 4(4): 502–511. https://doi.org/10.1038/s41559-020-1135-4

[129] Holland, R.A., K. Thorup, A. Gagliardo, I.A. Bisson, E. Knecht, D. Mizrahi and M. Wikelski. 2009. Testing the role of sensory systems in the migratory heading of a songbird. J. Exp. Biol. 212(24): 4065–4071. https://doi.org/10.1242/jeb.034504

[130] Valentine, E.C., C.A. Apol and D.S. Proppe. 2019. Predation on artificial avian nests is higher in forests bordering small anthropogenic openings. Ibis. 161(3): 662–673. https://doi.org/10.1111/ibi.12662

[131] Gill, F.B. 2004. Blue-Winged Warblers (*Vermivora pinus*) Versus Golden-Winged Warblers (*V. chrysoptera*). Auk. 121(4): 1014–1018. https://doi.org/10.1093/auk/121.4.1014

[132] Fei, S., J.M. Desprez, K.M. Potter, I. Jo, J.A. Knott and C.M. Oswalt. 2017. Divergence of species responses to climate change. Sci. Adv. 3(5): e1603055. https://doi.org/10.1126/sciadv.1603055

[133] Badeck, F.-W., A. Bondeau, K. Böttcher, D. Doktor, W. Lucht, J. Schaber, et al. 2004. Responses of spring phenology to climate change. New Phytol. 162(2): 295–309. https://doi.org/10.1111/j.1469-8137.2004.01059.x

[134] Richardson, A.D., T.F. Keenan, M. Migliavacca, Y. Ryu, O. Sonnentag and M. Toomey. 2013. Climate change, phenology, and phenological co+ntrol of vegetation feedbacks to the climate system. Agric. For. Meteorol. 169: 156–173. https://doi.org/10.1016/j.agrformet.2012.09.012

[135] Forrest, J.R. 2016. Complex responses of insect phenology to climate change. Curr. Opin. Insect Sci. 17: 49–54. https://doi.org/10.1016/j.cois.2016.07.002

[136] Probst, J.R. and J. Weinrich. 1993. Relating Kirtland's warbler population to changing landscape composition and structure. Landscape Ecol. 8(4): 257–271. https://doi.org/10.1007/BF00125132

[137] Walker, K.V., M.B. Davis and S. Sugita. 2002. Climate change and shifts in potential tree species range limits in the great lakes region. J. Great Lakes Res. 28(4): 555–567. https://doi.org/10.1016/S0380-1330(02)70605-9

[138] Donner, D.M., D.J. Brown, C.A. Ribic, M. Nelson and T. Greco. 2018. Managing forest habitat for conservation-reliant species in a changing climate: the case of the endangered Kirtland's Warbler. For. Ecol. Manage. 430: 265–279. https://doi.org/10.1016/j.foreco.2018.08.026

[139] Saino, N., R. Ambrosini, D. Rubolini, J. von Hardenberg, A. Provenzale, K. Hüppop, et al. 2011. Climate warming, ecological mismatch at arrival and population decline in migratory birds. Proceedings of the Royal Society B: Biological Sciences. 278(1707): 835–842. https://doi.org/10.1098/rspb.2010.1778

[140] Stephens, P.A. and W.J. Sutherland. 1999. Consequences of the Allee effect for behaviour, ecology and conservation. Trends Ecol. Evol. 14(10): 401–405. https://doi.org/10.1016/S0169-5347(99)01684-5

[141] Reijnen, R. and R. Foppen. 2006. Impact of road traffic on breeding bird populations. pp. 255–274. *In*: J. Davenport and J.L. Davenport (eds). The Ecology of Transportation: Managing Mobility for the Environment. Dordrecht: Springer, Netherlands. https://doi.org/10.1007/1-4020-4504-2_12

[142] VanTol, S.D., C.R. Koehn, R. Keith, B. Keith, and D.S. Proppe. (2021). Avian migrants encounter higher temperatures but continue to add mass at an inland stopover site in the Great Lakes region. J. Avian Biol. 52(4). https://doi.org/10.1111/jav.02626

Chapter 2

Habitat Selection in Human-Dominated Landscapes

Desiree L. Narango[1]

Drivers of Habitat Selection

How does a bird choose where to live? The ways that birds select habitat is a question that has been asked for nearly a century of ornithological history. Habitat selection, i.e., the behavioral choices an animal makes on what type of habitat to use [1], has broad implications for individual fitness. As such, habitat selection has evolved so that selection ensures birds can find, identify, and take advantage of habitats with the resources necessary to support populations. As habitat selection is a far-reaching topic with countless studies across disciplines and taxa, this chapter aims to be a broad overview of the topic's theoretical underpinnings and hopefully serve as an introduction to the topic to guide future studies. This chapter's central aim is to link how our

[1]Department of Biology, University of Massachusetts, Amherst, dnarango@gmail.com

understanding of the habitat selection theory can guide management decisions geared toward bird conservation, emphasizing the management, preservation, and restoration of avian habitat in novel, human-dominated landscapes. Because this chapter focuses on conservation opportunities for birds in shared human-dominated landscapes, most of the evidence presented here will be from terrestrial temperate, boreal, and tropical ecosystems. However, there is a vast body of knowledge on avian habitat selection in marine and freshwater ecosystems that is not covered here. The reader should explore the following discussions for comprehensive reviews of habitat selection that include additional information and ecosystems [1–3].

Habitat selection is the behavioral choices made by animals on what types of habitat (or physical environment) to use and is closely related to environmental features such as structure, vegetation, food, and nest site availability. A species' habitat selection is an adaptive response over evolutionary and contemporary time to match individuals with environments that maximize resources and lifetime fitness (i.e., adaptive selection). The type of habitats a species has adapted to select aligns closely with a species' life history, including their development, behavior, nutritional needs, physiological limitations, and life cycle.

Central to identifying habitat selection, or preference, is the ability to demonstrate evidence of choice. In other words, habitats are chosen disproportionately relative to alternative unselected areas more often than can be expected by chance. Selection is measured by quantifying bird densities, occupancy (i.e., presence), or abundance within habitats using passive survey techniques and relative comparisons among sites. When some habitats of interest are unoccupied, comparing use versus availability can reveal insight into habitat characteristics or landscape features preferred. Pairing with behavioral data such as activity budgets or movement data from behavioral observation or tracking techniques can complement survey data with details about selection occurring at finer scales within a patch or a territory. In addition to individuals' presence and behavior, consistency of occupancy, or the variance in occupancy, may also yield insights into habitat selection over longer time scales.

Selection occurs through a combination of genetic and learned behavioral components. Species can be genetically predisposed to prefer certain habitat types because of evolutionary-scale selection pressures. In an experiment with hand-raised Parids, Blue Tits (*Cyanistes caeruleus*), and Coal Tits (*Periparus ater*) maintained innate selection for tree species from their natal habitat [4]. Habitat selection may also occur as learned behaviors based on internal or external information. For example, habitat features exposed to juveniles may be selected during the later stages of an individual's life [5]. Prior experience within a habitat, such as whether a nest was successful, or foraging opportunities, may also influence selected habitat at subsequent decision-making [6, 7]. Selection may be a plastic (i.e., flexible) behavior such that choices of which habitat to use may change throughout an individual's life and in different contexts. Learned selection, plastic selection, and genetic variation in selection may be advantageous for species that occupy dynamic, unstable, or ephemeral habitats as well as long-lived species [8].

Habitat selection is a process by which information about habitat quality must be weighed with information about the costs of settling in a habitat. High-quality individuals may not spend time using low-quality habitat, and lower-quality individuals may not take opportunities to occupy the best habitat because of perceived low benefit to cost ratios [9, 10]. When most habitats are occupied, evaluating settlement phenology (i.e., which habitats are settled first), proportions of time spent within a habitat (i.e., activity budgets), as well as identifying the characteristics of individuals in various occupied habitats (e.g., body size) can also reveal information on which habitats may be preferred and are the highest quality [3]. For example, habitats occupied by dominant individuals may be the most preferred by the species, and suboptimal individuals may opportunistically move into preferred locations when dominant individuals are removed [11].

There are numerous ways that individuals are driven to use optimal and suboptimal habitat due to environmental and social drivers. Thus, measuring true habitat selection requires a comprehensive approach that quantifies both choice and performance [1]. If habitats are strongly preferred, and fitness is high, then habitat is adaptively selected. A central theory in habitat selection is ideal free distribution; birds select habitat features that maximize fitness as a function of habitat quality and conspecific density [2]. As habitat becomes more saturated, habitat quality, and thus realized fitness, also declines, leading individuals to select lower quality habitat with less competition. An alternative but complementary theory is ideal despotic distribution (also known as the ideal dominance distribution): dominant individuals who are competitively superior occupy higher quality habitats and exclude subordinate individuals to lower quality habitat regardless of density [12]. In this model, individual densities are higher in lower quality habitat, but other intrinsic factors attributed to individuals (e.g., age, body size, plumage quality) indicate that the less-dense habitats are superior.

Selection Cues

To accurately assess and select habitat, birds use direct and indirect environmental cues to provide information. **Structural cues**, such as vegetation characteristics, can provide information on a habitat's suitability for particular life history events or resource availability [2]. For example, grassland birds use structural cues like vegetation density, vegetation height, and ground cover to select habitat suitable for breeding territories, and different bird species vary in preferred features [13]. Vegetation structure and complexity can serve as a proxy for resources available in the habitat, such as prey or nesting sites, competitors, environmental conditions, or habitat suitability and quality. Proximate visual cues such as leaf or flowering phenology may also indicate arthropod prey resources and are strongly selected for by foraging migratory insectivores [14]. In many cases, vegetation attributes selected for at habitat settlement may signal resources later in the season. For instance, female Yellow-headed blackbirds (*Xanthocephalus xanthocephalus*) used marsh stem densities to select territories with high cover that were productive in dragonflies which are preferred prey for nestlings [15]. Structural cues may be especially crucial for migratory species that settle on territories earlier than the flux of resources necessary for rearing nestlings like the seasonal peak of Lepidopteran larvae [16].

Species also recognize and select habitats because of direct **resource cues** such as food, cover, or water. For example, specialized frugivorous species may use fruit abundance or foliage densities as signals for ideal foraging substrates [17, 18], whereas availability of nesting locations may be the primary cue for settlement in cavity-nesting species [19]. In some cases, birds may not assess resources directly or accurately and instead rely on indirect evidence of food availability. For instance, plant damage (i.e., herbivory) provides visual cues of locations with insect prey through mechanical damage or leaf light reflectance [20, 21]. Birds also select individual trees with particular chemical cues because trees emit volatile organic compounds when damaged by herbivorous insects [22]. Resource cues that predict successful foraging in particular locations may provide prior information that reinforces selection for certain habitat types and foraging strata where the probability of food capture is high [23, 24].

Social cues from con- and hetero-specifics convey public information about a site that assists in habitat selection [25]. The presence of other individuals can signify whether a habitat is suitable for settlement [26]. Public information can also signal aspects of habitat quality, such as whether a territory was successful. For example, Black-throated Blue Warblers (*Setophaga caerulescens*) used vocalizations during the post-fledging period to select territories the following year [27]. Northern cardinals (*Cardinalis cardinalis*) use social information about neighbors' nesting success to select nesting locations [7]. Social cues may be most relevant for naïve birds settling into a territory than experienced birds, which have individual prior information available to select and retain territories [25]. Conspecific attraction may result in clumped distributions that may or may not indicate habitat quality, particularly if new individuals are attracted to social cues because of mate availability over habitat attributes *per se*.

Social cues may also be a mechanism for habitat selection because signals or behaviors indicate individual quality or resource availability, such as females selecting mates using plumage quality or song. In American Redstarts (*Setophaga ruticilla*), song playbacks strongly attracted conspecifics to settle new territories; on average, four more individuals than non-playback locations [28]. Interestingly, heterospecific songs are also an informative social cue. In Collared Flycatchers (*Ficedula albicollis*), experimental playback of high-quality Great tit (*Parus major*) song induced settlement by older, dominant female flycatchers, suggesting that information about individual-level quality translates across species [29]. On the other hand, social cues may discourage habitat use by additional individuals because of density-dependent drivers due to limits in a habitat's carrying capacity. As higher-quality habitats become saturated and social cues increase, subordinate individuals may occupy lower-quality habitats to avoid competition for resources.

Finally, birds may select or avoid habitat based on **risk cues** from predators and parasites. The importance of predators as a mechanism for driving adaptive selection has been demonstrated experimentally using predator exclusion and playbacks. Fontaine and Martin [30] found that locations with reduced predator risk from predator exclusion had more birds recruited into the population, were settled more densely, and experienced higher nesting success. Moreover, these individuals sang more often, suggesting risk cues could result in a feedback loop of increased or decreased settlement via additive effects from social cues. Birds and predators may similarly respond to gradients of habitat, and individuals may select or avoid habitat based on reducing these risks. For example, forest interior species may be particularly susceptible to increases in mesopredators or nest parasites in fragmented habitats within developed landscapes, and as such, avoid nesting in locations or experience pronounced population declines [31].

The Role of Scale in Habitat Selection

Birds select habitat selection features along a gradient of spatial and temporal scales, from local scales characterized by decisions within short time periods to regional scales that occur over evolutionary time [32]. Scales of habitat selection can have different costs and benefits to fitness dependent on the scale of decision as well [32, 33]. Identifying the scale where birds are most limited is essential for understanding how best to conserve habitat needs for a species. Still, the most limiting habitat scale may vary between individuals, populations, and species [32]. For example, in Cerulean Warblers (*Setophaga cerulea*), a declining songbird of the Eastern US Appalachian Mountains, all populations select deciduous hardwood forests dominated by *Quercus* sp., however, populations in low-forested regions were most limited by the availability of closed-canopy forest, whereas populations in high-forested areas strongly selected forests with high disturbance [34]. Although scales of habitat selection occur along a continuum and descriptions of scale vary depending on the species and mode of inference [32], here, I describe two broad categories of selection related to specific management objectives.

- **Macrohabitat selection** is the broader habitat categories selected by an individual, such as a territory, home range, or species distribution [2, 3, 32]. Descriptions of macrohabitat may include coarse habitat categories or ecosystems, vegetation communities, biomes, climates, abiotic conditions, environments or gradients. Macrohabitat selection can occur across multiple scales: from the broad ecological features of a habitat type selected by individuals (e.g., forest or wetland), as well as where habitat is located within a landscape, its relation to other habitat patches, and distribution across land use or cover gradients. Macrohabitat selection can also be described at larger scales, such as how a species is distributed regionally or globally. At these coarse scales of selection, a bird is expected to select habitat features necessary for fitness but may not maximize performance.

- **Microhabitat selection** is the selection of finer-scale features directly related to the day-to-day activities within an individual's home range, such as providing food, nesting locations, or shelter [2, 3, 32]. Microhabitat features are related to resource availability at a local scale and may include features such as where to forage within a territory, microclimates for maximizing energy efficiency, or nest placement for cover. The availability of selected microhabitat features can be directly related to species occupancy, density, and demographic responses like survival and reproduction; thus, there is strong selective pressure for microhabitat selection that maximizes fitness [35].

  For foraging birds, microhabitat selection costs are related to the optimal foraging theory, i.e., foraging strategies and decisions related to the optimal acquisition of resources and subsequent fitness [2]. Even if a habitat type seems appropriate at broad scales, it cannot be used if it does not contain adequate resources to reproduce and survive at local scales, such as foraging substrates, cover, and nesting sites.

Both macro- and microhabitat selection are also impacted by interactions with other individuals, such as competitors and predators. When species share preferences for habitats at macroscales, they may partition habitat at microscales such as foraging strata, or diet. A classic example of partitioning was demonstrated in Macarthur's study of warblers of the Northern coniferous Forest [36] in which species forage in different locations within trees (i.e., competitive exclusion); for example, Cape May Warblers (*Setophaga tigrina*) forage primarily on the outside of the highest canopy, while Yellow-rumped Warbler (*Setophaga coronata*) occupying the lower interior strata. Likewise, if species overlap in microhabitat selection by sharing preferences for foraging or nesting resources, they may partition at macrohabitat scales, by occupying different ecosystems, climates, elevations, or geographies.

In this way, birds occupy different **ecological niches** (i.e., the position of a species within resource or environmental gradients) across ecosystems by using different habitats, exploiting different resources within habitats, or employing different behavioral strategies that maximize fitness given intense selective pressure on individuals for optimal micro- and macro-habitat selection. The range, or variation, in conditions that a species is observed using is considered its **niche breadth**. To manage imperiled or culturally important species, understanding the ecological niche through a lens of habitat selection is essential for conserving habitat features that maximize species presence and fitness at local and landscape scales. For example, at landscape scales, bird species may require broad characteristics such as hardwood forest or canopy cover; however, at local scales, birds preferentially choose to forage in some plant species that maximize arthropod prey or fruit quality [23, 37, 38]. Predation can also drive habitat selection, as birds selectively place nests in plants or locations that maximize cover and protection to young or forage in structurally complex areas with high cover for safety. Nest placement and foraging decisions are always a tradeoff of maximizing resources while reducing predator exposure and time and energy costs [18].

Variation in Habitat Selection

Habitat selection is often not static and can vary based on **extrinsic** temporal and spatial factors. For example, time of day, season, year, life stage, weather patterns, and habitat age, can all temporally affect the availability of resources, competitors, and risk, as well as individual needs, that drive habitat selection [1, 2, 15, 39, 40]. For example, habitat selection may vary by season for migratory species because individuals select different habitat across the annual cycle. Bicknell's thrush (*Catharus bicknelli*), a boreal breeding songbird, uses exclusively high-elevation spruce-fir forest during the breeding season and migrates to Hispaniola to winter in variable elevations of dry broadleaf forests [41]. For migratory species, breeding season habitat may be selected to maximize food resources or nesting substrates for young, while during migration, individuals may flexibly select resources and prioritize habitat that provides fast refueling or immediate cover as protection from predation [18]. During the non-breeding season, birds may be less tied to particular habitat types and instead seek out habitats, foraging locations, and resources that reduce competition from resident competitors. However, the selection of high-quality habitat is essential across all seasons because carry-over effects can impact subsequent stages of the annual cycle through effects on survival and reproduction [42].

Even within the breeding season, birds may select different habitats to maximize fitness at different life stages. For example, juvenile interior forest songbirds travel to regenerating forest cuts during the fledging period, presumably to take advantage of increased cover or food resources during vulnerable stages such as molt [43]. Fledging birds can also exhibit daily habitat selection changes as they age because birds become more independent, and use larger, riskier areas for foraging [44]. Differences in habitat selection between nesting, fledging, and independence suggest that for many species, a mosaic of habitat types is necessary for conservation even when a species commonly breeds within one habitat type. Without habitats ideal for fledgling survival, population growth could be impacted by higher mortality or lower conditions during one of the most sensitive life stages.

Migration is another period of the annual cycle where mortality is high, and habitat selection may differ from other stages of the annual cycle. During migration, many birds travel thousands of kilometers to breeding and non-breeding areas and need to stopover in unfamiliar locations to rest and refuel along the way. Because of the time-sensitivity of migration, selective pressure must drive birds to make decisions from reliable cues over relatively short time frames. At first arrival, habitat selection may be more relaxed because decisions must be based on limited information on habitat availability, quality, and configuration. However, as birds acquire additional information about the location, such as food, predators, and competitor density, habitat selection may become more refined to target locations that maximize refueling rates [45]. During migration, multiscale selection may be particularly apparent. In Ovenbirds (*Seiurus aurocapilla*), migrating birds select hardwood cover at landscape scales, but at finer patch-level scales, arthropod and fruit abundance were primary indicators of habitat use [46]. Because migration directly impacts arrival on the breeding grounds via changes in individual condition, survival, or phenology, stopover habitat selection can have an inordinate influence on population persistence.

Habitat selection can also vary spatially within different regional and local contexts [1, 15, 47]. Habitat selection can depend on where a population occurs within a species distribution due to regional differences in resources or risks, abiotic or biotic environments, or species-level differences in niche specialization or conservatism. For example, birds on the leading or trailing edge of a distribution may be more flexible in habitat preferences or take advantage of new habitat types due to abundant prey resources or the absence of direct competition [48]. As species move northward due to human-driven climate change, individuals or species on the leading edge may be more flexible in habitat selection than individuals in the distribution's interior. At broad scales, habitat selection can be related to the arrangement of habitats that may vary spatially. For example, there is a robust body of literature on birds' negative responses to forest fragmentation at landscape scales despite the availability of resources at local scales within small fragments [49]. In all cases, habitat selection can vary spatially due to the hierarchical nature of selection across scales that vary from the landscape to the foraging location [47].

Habitat selection can also vary among **intrinsic characteristics** of individuals within and between species. For example, species-level selection differences may be driven by differences in diet, physiological condition, or movement, while individual-level differences may be due to status, life stage, or reproductive status. In some cases, habitat selection differences may be due to inherent behavioral differences between individuals within a species, such as when both habitat (or resource) specialists and generalists exist within a population. For example, in the United Kingdom, a population of European Nightjars (*Caprimulgus europaeus*) was composed of individuals that strongly selected few habitats for foraging and individuals that broadly used a broader range of managed and unmanaged habitat types [50]. Thus, in this case, a mosaic of habitats is necessary to support a population of individuals exhibiting behavioral diversity in foraging habitat selection. When social dominance drives differences, habitat selection may vary by individuals' body size, age, or experience. For instance, juvenile birds may be subordinate to older, more experienced individuals [12, 51]. Social dominance may also drive interspecific settlement choices. Migratory species can often be subordinate to resident species that have more experience with a location or have settled territories for more extended periods of time [52].

At times of the year, when birds of either sex are not dependent on one another (i.e., nonbreeding), habitat selection may also vary by sex. In migratory warblers wintering in Mexico, habitat differences are linked to sexual segregation and social dominance, whereby dominant males exclude females and subordinate individuals from high-quality primary habitat into suboptimal disturbed habitat [51]. For other species, intersexual differences in habitat selection may be inherent and not driven by explicit segregation or intersexual conflict [53]. During the nonbreeding season, sexes can segregate among many numerous habitat dimensions, including geography, ecosystem, plant community, diet, or foraging locations [54]. Understanding how habitat selection varies by sex at both macro- and micro-scales is critical for effective conservation strategies, especially if management or sampling efforts are biased toward features that disproportionately support males and exclude females [55].

Box 1   Habitat selection summary

Behavior
- Habitat selection describes the behavioral choices among habitats that vary in conditions to match individuals with environments that maximize reproduction and survival.
- Birds use structural, resource, social, and risk cues to select appropriate habitat.
- Selection occurs along gradients of scale, from microhabitats (e.g., nest site) to macrohabitat (e.g., ecosystem type).
- Habitat selection within and among species can vary extrinsically over space and time and intrinsically between individuals that differ in characteristics.

Habitat Consequences of Land Use Change
- Habitat loss can eliminate preferred habitat or selective cues resulting in birds avoiding land uses.
- Habitat degradation can reduce habitat quality and resource availability resulting in birds avoiding land uses unless higher-quality habitat is unavailable.
- Habitat alteration can create novel habitats that may have resources to sustain populations or may be unrecognizable as appropriate habitat.
- Ecological traps can occur if birds prefer low-quality habitats that reduce performance over highquality habitats because cues are unreliable, resulting in population declines.

Land Use Changes and Habitat Selection

As landscapes become dominated by human land uses, numerous mechanisms drive changes in habitat availability, quality and contribute to some of the most significant negative impacts on biodiversity and ecosystem functioning [56]. Understanding how habitat selection and use varies across species, scales, space, and time in these systems will help inform how different land-use change mechanisms can be mitigated to sustain habitat availability and quality to conserve bird populations in the face of rapid global change without compromising human needs.

Habitat loss and degradation are the primary drivers of biodiversity declines for birds and other wildlife worldwide [57], resulting in loss of ecological interactions and services [58]. In highly developed ecosystems, habitat availability may be deficient, particularly for our most habitat or resource-specialized species such as insectivores [59, 60]. Habitat may also remain too fragmented or isolated to attract dispersing individuals to occupy these locations, or if species are present, may represent an 'extinction lag' whereby species persist but do not maintain fitness to contribute to population stability [61]. When habitat is available, quality may be diminished because of direct land degradation, edge effects from the surrounding matrix, pollution, altered abiotic conditions or biotic relationships, lack of resources or trophic relationships, or loss of colonization from landscape-scale fragmentation. Degraded habitats may remain unoccupied because they lack food resources, nesting locations, or the availability of mates to sustain populations, the risks of predation or parasitism are too high, or these locations function as sinks to the regional metapopulation with high rates of turnover [62].

If a habitat is selected in altered landscapes, it may be suboptimal and does not maintain a fitness advantage; a situation called **maladaptive selection**. Anthropogenic changes to habitat can induce maladaptive selection by altering the resources or risks associated with preferred habitats. For example, along edges in fragmented landscapes, increases in generalist predators can increase predation risk despite similar food resources [47]. Alternatively, maladaptive selection could also occur if suitable habitat is avoided due to unreliable cues. For example, the so-called 'predation paradox' in urban systems describes how bird populations maintain similar or reduced predation rates despite increases in predators [63]. Thus, if the cues to avoid a habitat are unreliable, suitable habitats may remain unoccupied despite advantageous conditions.

An extreme form of maladaptive selection is called **ecological (or evolutionary) traps** [64]. In these cases, habitats are highly preferred, and the cues for selection are present and recognized but give unreliable signals of habitat quality or fitness outcomes. Although ecological traps are often challenging to detect, there are some examples of this phenomenon in anthropogenic landscapes. In central Ohio, Northern Cardinals (*Cardinalis cardinalis*) select territories dominated by invasive Bush Honeysuckle (*Lonicera* sp.) despite these locations having higher nesting failure, reduced fledgling productivity, and inadequate resources for plumage quality [65, 66]. Another example is in Red-Backed Shrikes (*Lanius collurio*) breeding in agricultural landscapes, where birds select low-quality agricultural and cut areas despite breeding productivity remaining highest in forested habitats [67].

Habitats can also become traps by abiotic influences on habitat quality due to other anthropogenic drivers such as climate change that influence the phenology or abundance of resources. In Europe, farmland birds that advance their laying date due to increased spring temperatures are more susceptible to nest failures by machinery during farm sowing [68]. In other cases, human transformation of habitat that reduces natural resources and increases novel resources such as supplementary food, exotic plants or nesting boxes may lead birds to use habitats that are suboptimal for breeding. For example, in UK cities, bird feeders can increase densities of foraging and nesting of Blue Tits (*Cyanistes caeruleus*) to areas with low insect prey [69], as well as attract predators that increase adult and nestling mortality [70].

It has been suggested that the occurrence of ecological traps, or at least those with population-level consequences, are relatively rare [64], in part because intense selection pressure exists against habitat preferences where traps persist. In either case, traps may also be rare because they are understudied, and more work is needed to understand demographic responses in anthropogenic habitats in addition to occupancy and abundance. In many situations, even at high densities, the presence of a species may not necessarily indicate a habitat is suitable or provides the resources for sustainable populations [71]. Lack of information on fitness attributes like reproductive performance leads to habitat suitability guidelines or recommendations that may not contribute to population maintenance [72]. Care should be taken to evaluate direct management interventions to ensure that management actions improve selection and performance and do not contribute to ecological traps.

With burgeoning human populations encroaching on wildlife habitat, our conservation success relies on creating and restoring adjacent or embedded habitat within landscapes that simultaneously support people. Land development like urbanization, agriculture, forestry, and energy can degrade habitat availability, and yet, also contain examples of green spaces that can serve as potential habitat to support populations. However, success will rely on management that maximizes selection and performance is informed by behavioral data.

Urban and Urbanizing Landscapes

Urban areas make up ~10% of the globe but support >50% of people and are among the fastest-growing land use globally [73, 74]. Urban areas are often characterized by the loss and degradation of natural habitat from anthropogenic development [74], which results in declines in bird condition, survival, and reproduction, as well as bird communities filtered to species that can successfully colonize urban areas [71, 75]. However, urban areas are composed of a heterogeneous mosaic of green spaces, which also have the potential to support bird habitats, including parks, cemeteries, playgrounds, residential yards, botanical gardens, commercial land, utility right of ways, vacant lots, green roofs, roadsides, and verges [76]. For example, in South America, Burrowing Owls (*Athene cunicularia*) have higher reproduction in urban areas and strongly select these habitats because of the low predator abundance suggesting urban grasslands can serve as a conservation refuge for this species [77]. However, studies that evaluate whether selection of these green spaces also contribute to increases in fitness are relatively rare and we lack a comprehensive understanding of the extent to which specific management decisions such as native plantings, outdoor pets, or mowing frequency impact habitat quality and selection [76]. Land management can disrupt habitat relationships in urban areas from the reduction in the availability of natural food resources such as decreases in arthropod prey by introducing nonnative plants, proliferate pesticide use, low plant densities, and fragmentation [79, 80]. However, urban areas are also supplemented by resource subsidies of anthropogenic foods such as feeders and refuse, which shift food web dynamics and the composition of bird species that occupy these areas [78]. Differences in the abiotic properties such as warmer temperatures and low air and water conditions may make conditions poor for some species, but other species may remain unaffected. High degrees of habitat fragmentation in urban areas can also disrupt birds' ability to disperse within urbanized landscapes [80], resulting in reduced occupancy of otherwise suitable locations. Relative to nearby natural areas, urban areas may be depauperate in bird species, functionally homogenous, and characterized by high densities of generalists and nonnative bird species [75, 81]. Nevertheless, because of the high heterogeneity in available green space, the potential to increase connectivity to broader landscapes, and the high benefit of management to support wildlife without subsequent costs to people, urban areas have been increasing focal areas of habitat restoration for mobile wildlife like birds [82, 83].

Agriculture and Farmland

Agriculture is one of the most dominant land uses, occupying almost 40% of the total worldwide land area, and is one of the primary drivers of deforestation of natural habitat and ecosystem change [84]. Agriculture is also implicated in bird population declines, especially specialized species like insectivores, due to habitat loss, direct mortality from harvesting and pesticides, and reductions in food supply [85, 86]. Despite agriculture's vast dominance, several agricultural management examples can support wildlife while maintaining adequate crop yields [87]. In California, USA, for example, conservation organizations like The Nature Conservancy have provided payment to farmers for altering water management that creates temporary wetlands for waterbirds during migration, creating a 'dynamic conservation' system that also sustains continued agricultural land use [88]. In agriculture, both high-yield farming with natural area conservation, as well as low-intensity farming practices, are potential conservation strategies and often, combinations of both approaches yield the best outcomes for conserving the widest variety of bird species [89]. Habitat in agricultural landscapes can simultaneously support bird populations by retaining specific features such as natural area buffers, uncultivated crop margins, no-till harvesting, reducing pesticides, and altering sowing and cutting regimes that benefit farmland birds [90, 91]. For example, populations of Bobolink (*Dolichonyx oryzivorus*) and Savannah Sparrows (*Passerculus sandwichensis*), both declining grassland species, regularly use managed hayfields during the breeding season, however, population declines can be curbed by small changes to the timing of mowing [87]. Surrounding farms with natural habitat can result in 'spillover' of birds and other wildlife that provide essential ecosystem services that increase crop yield, like pest control from insectivorous birds or seed dispersal [92, 93].

Forestry

Almost 20% of the world's 3.5 billion ha of forest are actively managed for services to people and, as such, are human-dominated ecosystems [94]. In addition to providing goods and services like fuel and wood, cleaning the air, and storing tremendous carbon quantities, forests are also critical habitat for tremendous biodiversity worldwide [94]. Logging, harvesting, and silviculture decisions occur along a continuum of intensities which directly impact local characteristics and habitat selection: from harvesting that ranges from clear-cuts to selective thinning, and post-harvest management that varies from no management to clearing and burning [49, 94]. Birds that rely on burned forests may be particularly affected by disruptions to natural fire regimes into novel and highly managed programs. Forest harvest decisions can impact the colonization of bird species, the succession of habitat, the availability and synchrony of prey, and connectivity of habitats due to spatial and temporal variation in harvest mosaics and forest fragmentation effects on fitness [49, 95–97].

One well studied forest-agricultural system is agroforestry practices, whereby crops are grown underneath a canopy of shade, for example, in shade-grown coffee. Numerous studies have identified that resident and migratory birds select shade coffee over sun-dominated coffee agriculture [98]. For example, migratory warblers (family: Parulidae), can be found in up to 14x higher densities in shade coffee than primary forest and were in improved body condition [99]. For this species, and evidence from other Neotropical migrants, shade coffee provides habitat comparable to natural areas. In some cases, resident birds (i.e., birds that breed in the Neotropics) may select and perform better in forested land uses compared to coffee; however, fitness varies strongly among species and farm management styles, suggesting more information is necessary to understand the extent that coffee management can contribute towards supplementary habitat for resident tropical birds [100]. Because shade coffee also varies in habitat quality within and among farms, occupancy and abundance of species can be strongly driven by farm-scale features that can be directly manipulated by farmers, such as canopy cover, floristics, and vegetative complexity [101–103]. Understanding these finer-scale features, and to what extent different species and individuals select them, can inform restoration for 'bird-friendly' management practices that can be implemented at larger scales and be successful for other agro-forestry crops (i.e., chocolate, tea, vanilla) or livestock-forestry systems (i.e., silvopasture) [104].

Renewable Energy

An understudied land use with the potential to positively and negatively impact bird populations is renewable energy infrastructure. Renewable energy sources like solar and wind are growing in use with higher demand and disadvantages of relying on fossil fuels for energy, resulting in increased land use, land cover changes, and fragmentation to support infrastructure. Renewable energy is currently being developed within numerous different ecosystems and implemented at larger scales than previously experienced. However, current understanding of the effects of energy infrastructure on bird habitat is minimal and primarily focused on direct mortality of species from collisions with infrastructure [105, 106]. Initial and prolonged energy development may degrade habitat quality, reduce recognition, or dampen cues for settlement, particularly for habitat specialists. For example, infrastructure development can facilitate dispersal of invasive plant and animal species, disrupt trophic relationships and food availability, introduce sensory pollution from anthropogenic noise, increase predator abundance or contribute to population declines by maintaining habitat selection cues while reducing fitness [106]. In addition, studies of renewable energy land uses are limited to short-term studies of few systems, and there is a lack of understanding of how birds and other wildlife use renewable energy land uses throughout the year and across time [105].

Strategic development that considers the placement of renewable energy in the landscape relative to bird behavior and habitat needs, the type of infrastructure, the sensitivity of local species, and land management within and around the infrastructure may provide opportunities to mitigate adverse effects [106]. For example, solar arrays with reduced mowing and native plants may support, or at least reduce negative effects on, grassland and open habitat birds and pollinators. It is possible that vegetation within energy infrastructure may be more predictive of selection and performance for some species than the density of infrastructure itself [107]. However, at present, we lack comprehensive studies of which birds select or avoid energy infrastructure for habitat, the mechanisms behind selection, and whether the reproductive benefits outweigh the mortality costs of settling these land uses.

Leveraging Principles of Habitat Selection to Inform Land Management

Preserving, Creating, and Restoring Habitat in Human-dominated Landscapes—If you Build it, will they Come?

Both habitat preservation of natural areas (i.e., land sharing) and restoration approaches within human land uses (i.e., land sharing) are used in human-dominated systems. There is an ongoing debate surrounding which approach is most beneficial for effective conservation outcomes; however, in reality, both approaches have benefits from a habitat selection perspective and are complementary in synergistic ways because of the multiple scales at which birds select and avoid habitat. For instance, in many cases, both local-scale variables (e.g., native plant abundance, patch type) and landscape-scale attributes (e.g., building densities, canopy cover, land uses) attract and repel occupancy of native bird species [108, 109]. To improve bird habitat, informed management is required at a local, proximate scale and a broader patch- and landscape- (or municipality) scale. Understanding the features driving habitat selection for a given species or community at multiple scales will help inform the agency that can impact avian biodiversity in a given system and establish practical conservation priorities and strategies. For example, consider habitats within a growing city. Neighborhood and city expansion by planners and developers may have a more considerable impact on the presence, location, type, and arrangement of macrohabitat at a landscape scale. In contrast, householders, park managers, and landscapers may have a more substantial influence on specific microhabitat features that can be manipulated to attract individual birds [110]. Moreover, at smaller scales, individual manipulations that support bird habitat will be more apparent in aggregate (i.e., decisions by multiple cooperating parcel-owners) rather than at a single-parcel scale [111]. Quantitative comparisons of the specific features selected for by species, guilds of species, or regional metacommunities can provide direct recommendations that can be used to restore habitat availability and quality and the scale necessary for success [111]. For example, within the urban landscape, landscaping decisions that include regionally native plants also maximize the availability of foraging substrates [112], arthropod prey [113], nesting locations [114], sustain population growth [113], and increase species diversity for insectivorous birds within residential yards [115, 116], and preserved parkland [117].

The presence, density, and abundance of bird species are used to monitor different habitats to assess whether human-dominated areas support species relative to nearby natural areas. In the case of habitat- or resource-specialist species, birds may be unlikely to select human-dominated land uses as preferred habitat because they lack essential resources necessary for fitness or are fundamentally unrecognized as habitat. For species able to colonize human-dominated systems, relative comparisons of habitat selection can identify patch characteristics within human-dominated areas with the highest chance at success. When comparing natural area fragments within a developed landscape, remaining habitats may differ in size, shape, configuration, proximity to other habitats, and connectivity, all of which may influence habitat selection. The proximity of high-quality natural patches or greenspaces may also greatly enhance selection for nearby habitats, as does the quality of the surrounding matrix (i.e., the background ecosystem in a landscape) around a habitat [109]. Assessing birds' spatial arrangement and their distribution in the landscape can help prioritize which habitats might be most valuable for preservation versus restoration and create conservation targets for patches that would have a disproportionate effect on local populations.

Box 2 Examples of risk and habitat opportunities in urban and agricultural land uses

Urban and Suburban areas

- *Risks*: Fragmentation and low habitat connectivity, sensory pollution masking habitat cues, reduced resource availability, invasive species, increased predators and nest parasites, altered microclimates.
- *Microhabitat management example*: Householders restoring properties with native vegetation landscaping that increases prey availability for insectivores.
- *Macrohabitat management*: City planners preserving intact natural areas suitable for habitat specialists.

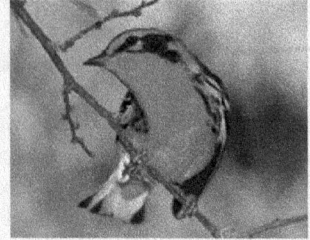

Figure 1 Insectivorous Blackbumian warblers (*Serophogo fusco*) use urban stopover habitats during migration and will forage in native trees to find food.

Agriculture

- *Risks*: Habitat loss, pesticide exposure, mortality from mechanical management or direct control, reduced resource availability.
- *Microhabitat management*: Farmers employ 'no-till' or modified mowing schedules to maintain cover for nesting and foraging habitat.
- *Macrohabitat management*: NGOs provide monetary incentives to farmers to maintain temporary wetlands during migration.

Figure 2 Bobolinks (*Ddichonyx oryziivorus*) select agricultural farms for breeding habitat and benefit from practices that increase cover and reduce mortality.

Determining the microhabitat features within human-dominated areas selected for or avoided by birds can also inform how individual landowners can promote increased resources on private and public land that otherwise suffer from a paucity of foraging, nesting, and roosting sites. For example, coniferous trees may be heavily selected by birds in the winter and may buffer resident species against harsh weather and cold temperatures in any developed habitat with low vegetation density [118]. Fruiting plants provide essential food resources for fall migrating birds and can also be a visually appealing addition to landscaping projects or natural buffers [38]. Retention of woodpiles, leaf litter, deadwood, and water bodies may also be necessary components of novel habitats to provide food and cover for birds and their arthropod prey. For programs that promote retention of wildlife habitat on private land, like the agricultural Conservation Reserve Program [119] or the National Wildlife Federation's certified residential yard program [120], evaluating which habitats are selected among locations that vary in specific mechanistic features can inform explicit guidelines for management that will drive program success.

Across human-dominated habitats, assessments of the habitat occupied by species are only one means of identifying selection. As previously stated, demonstrations of increased fitness are also required to understand habitat selection fully. Bird surveys yield cheap, easy-to-acquire information about birds' selected habitats, yet, the most easily detected individuals (e.g., unpaired males that sustain high singing rates throughout the season [121]) may be least likely to contribute young to sustainable populations. For example, Wood thrush (*Hylocichla mustelina*) wintering in Honduras occupied coffee farms at high densities but experienced lower survival, and higher movements [122], suggesting these farms as presently managed did not provide adequate habitat. Nevertheless, despite that species presence does not guarantee habitat quality, demographic studies in these systems are rare. New research in human-dominated systems must go beyond comparisons between land use types (i.e., forest vs. farm) and focus on mechanisms within land-use types that drive habitat selection and individual fitness to identify specific management goals that increase high-quality habitat availability.

Box 3 Examples of risk and habitat opportunities in forestry and energy land uses

Forestry
- *Risks*: Habitat fragmentation and edge effects, intensive cutting and post-harvest clearing. increased invasive species and predators, altered habitat heterogeneity.
- *Microhabitat management*: Forest managers retaining snags and woody debris for nesting habitat.
- *Macrohabitat management*: Logging companies selectively creating forest corridors to sustain landscape habitat connectivity.

Figure 3 Retaining snags supports nesting substrates for cavity-nesting woodpeckers like Red-bellied Woodpeckers (*Melonerpes cardinus*).

Renewable Energy
- *Risks*: Direct mortality from collisions, habitat fragmentation for infrastructure, sensory pollution, invasive species.
- *Microhabitat management*: Reducing mowing around solar farms to promote native plant meadows.
- *Macrohabitat management*: Strategic placement of renewable energy in locations that avoid sensitive species at high risk of mortality or habitats.

Figure 4 Reduced mowing around solar arrays can create meadows for birds and pollinators.
(photo credit: MarylandGovPics).

A particular challenge of habitat conservation in human-dominated landscapes is that some features may influence selection that cannot be directly manipulated, at least at local scales. Novel landscapes may have similar structural or vegetative characteristics but different abiotic attributes (e.g., temperature, humidity, precipitation, microclimates, light, and sound) that may affect habitat selection in predictable ways. For example, species may select habitat within particular light or noise gradients that facilitate successful foraging [123] or sound propagation [124]. Increasing sensory pollution from artificial light, traffic, or industrial noise can degrade habitat to the extent that birds select habitat but experience reproductive declines across species [125]. Another aspect is that land management that increases selection may put species at risk to other immediate novel threats that must be mitigated to be successful. For example, native plantings that attract birds to forage may also increase the risk of window collisions in high-density urban areas, mortality from feral cats around infrastructure, or expose birds to pesticides in agricultural landscapes. In some cases, restoring a system's biotic properties via vegetation manipulation or land acquisition may not yet fully capture the specific conditions needed for a specific species to thrive. Another consideration is that human-dominated landscapes may naturally select individuals with different phenotypes or adaptations than nearby natural systems [126]. In this way, selective behaviors or performance quantified in natural areas may not always translate seamlessly to human-dominated areas. If individuals have different behavioral flexibility, neophobia, or experience different costs associated with suboptimal habitats, they may be more or less likely to select and thrive in human-dominated habitats.

Finally, preserving or restoring habitat may not be enough if the necessary cues to attract birds are not present or actively discourage use. Restoration may enhance structural or resource cues that indicate appropriate habitat but lack the social cues and con- or hetero-specific attraction necessary to encourage settlement by territorial birds [127] (for more information, see Chapter 4). Isolated fragments and small populations, which are characteristic of typical human-dominated landscapes, may be especially prone to low recruitment from a failure to attract conspecifics. Increasing recruits by utilizing social information may also bolster genetic diversity when populations are too small to attract dispersing individuals adequately. Likewise, as in the "predation paradox" mentioned previously [128], increased risk cues from competitors, predators, and parasites in higher densities may discourage birds from occupying habitats even if there is no apparent change in risk. Increased densities of people and their activities may also pose an apparent risk that may cause birds to avoid otherwise appropriate habitats.

Informed Plant Selection for Foraging Habitat in Novel Landscapes

In a hierarchical process of habitat selection, vegetation is a dominant force, and birds prefer foraging locations that are the most profitable or predictable for finding food resources. Accordingly, birds across ecological communities disproportionately prefer plant species over others because plants vary widely in provided resources throughout the year [112, 129, 130]. For example, Birch (*Betula*) [23], Oaks (*Quercus*) [37, 112, 131], and Hickory (*Carya*) [132], were preferred foraging trees for insectivorous birds during breeding and migration seasons. A characteristic shared among these particular taxa is that these plants are disproportionately crucial for supporting caterpillar diversity [133], a preferred prey item for insectivorous birds [134].

By studying microhabitat selection in managed systems through local bird species' foraging behavior on selected trees, tree preference information can be used to prioritize specific tree plantings in cultivated ecosystems that maximize species interactions and bird habitat [129]. Planting native, insect-producing trees as tree canopies or natural area buffers can benefit local bird communities by providing necessary insect prey to reproduce and survive in novel landscapes that are not provided to the same extent by nonnative species [113]. While planting preferred trees can have local impacts on foraging availability, scaling up plantings to the broader neighborhood and municipality scales could also have more significant effects on bird community composition [111, 116]. In forests of the Eastern United States, for example, oak-dominated communities support higher diversity and abundance of birds than maple-dominated forests [135]. Although maple (*Acer*) is one of the most commonly planted native trees in developed areas, we lack comparisons of how plant community composition drives bird community composition within cultivated ecosystems, even among native species. Information on specific plant species that support foraging habitat for target species can help guide decision-making in current urban- and agro-forestry initiatives such as street tree planting, land restoration, and horticultural landscaping to easily include biodiversity benefits alongside other concerns such as shade, water runoff, or aesthetics [112, 131].

Informed tree selection can also be implemented in other tree-dominated landscapes such as agroforestry [103, 136] and timber harvesting [137]. In timber systems, selective harvesting to promote oak (*Quercus*) and elm (*Ulmus*) regeneration would retain trees highly preferred for foraging and nesting across bird species and provide mast for other wildlife. In tropical ecosystems, legumes such as *Inga, Enterolobium, and Erythrina* [102, 103, 136] are highly preferred foraging locations by both resident and migratory bird communities and are disproportionate suppliers of food resources. Farms dominated by legumes also tend to support a higher abundance and diversity of birds [136]. Because legumes also provide farmers ecological services to improve soil quality (via nitrogen-fixation), attract pollinators, and provide additional revenue from lumber [138]. Promoting trees that provide complementary services that benefit farmers and imperiled species will ease the acceptance of wildlife-friendly management tactics at broader scales.

Understanding Nest Site Selection for Breeding Birds

Management decisions at local scales can influence the availability of selected nesting locations to support breeding populations. Selective cutting and planting of trees and shrubs, mowing and clearing regimes, retention of snags and woody debris, and artificial substrates can all contribute to the availability of nesting habitats within novel landscapes.

Managing human-dominated landscapes for breeding bird populations may require a complex approach that considers all features that impact nesting selection and success. Increases in bird densities from higher availability of nesting sites, or stronger stimuli of selected cues, could result in density-dependent negative impacts that require other ecological features to be manipulated to maintain performance, such as native or invasive predators, parasites, or prey. For example, although selective timber harvesting mimics natural processes like fire, Olive-sided Flycatcher (*Contopus cooperi*) nesting densities were higher in selectively harvested forests than naturally burned forests; however, nest success was lower due to predator increases in managed forests [139]. To avoid selectively harvested forests or other human-dominated habitats from becoming population sinks, predators may have to be controlled or excluded from nests at least until populations become established.

Artificial nesting locations such as nest boxes, platforms, and holes can be implemented by municipalities and the public to enhance habitat attractiveness and encourage birds to nest in anthropogenic areas [140]. For example, the provision of nest boxes in California's agricultural areas for nesting Western bluebirds (*Sialia mexicana*) increased the bird density with additional benefits of heightened pest control [141]. Adding or retaining holes incorporated within buildings can conserve nesting and roosting habitat for swift (family: Apodidae) colonies in many urban areas. Platforms for Peregrine Falcons (*Falco peregrinus*) and other birds of prey on buildings and near farms have supported populations that take advantage of abundant prey resources in urban and agricultural areas. For some species, appropriate nesting locations may not be available in anthropogenic locations, or birds may use novel habitats for which nesting was not previously known from life history. Birds nesting on or within novel human-made structures such as buildings may experience higher nesting success because of protection from predators or nest in densities too high for optimal performance due to an inordinate number of competitors. More research into adaptive selective behaviors in novel ecosystems is necessary to fully understand the potential for artificial nesting resources to support bird populations [140] adequately or whether natural resources are preferred.

Information about nest site selection can also help discourage birds from nesting in locations that are detrimental to individuals or people. For example, the Monk Parakeet (*Myiopsitta monachus*) is a colonial nesting parrot that has colonized agricultural areas and cities across the globe and can have detrimental impacts on buildings and electric infrastructure [142]. Understanding locations this species prefers, and the proximate features selected (e.g., edible fruits, nesting trees), could help mitigate damage by this species at early stages before colonies are established and reduce their impacts on native species. For native species with sensitive populations, understanding the interactions between nest site selection and nest success can help management efforts deter nesting in preferred but suboptimal nest site locations that do not benefit the population in favor of more successful locations. For example, ground-nesting birds that favored early nest sites on unsown fields were more likely to experience nest failure and low seasonal productivity from destruction by farm machinery [68]. To protect these birds, farmers would need to provide alternative sites that are equally favored relative to farm locations, change the timing of sowing to discourage nesting, or harvest at later dates to minimize nest failures.

In some cases, features selected for nesting may be maladaptive in some contexts yet adaptive in others. For example, invasive bush honeysuckle (*Lonicera* sp.), which reduces habitat quality and fledgling survival during the breeding season [65], provides suitable cover for fledgling survival [143] which may also influence population growth. Thus, before invasive plant management, careful assessment of the mechanisms underlying selection should be considered. In the context of honeysuckle, the selection for dense, structural components of the invasive plant is unrelated to plant origin and should be replaced by native plants with similar attributes to increase demographic benefits without compromising habitat for fledglings.

Conclusion

Habitat selection is a complex behavioral process by which individuals assess their environment at multiple scales to make adaptive decisions about where to settle, forage, and breed. Nevertheless, individuals' behavior can give critical insights into the types of habitat and their specific attributes that can be preserved and restored in human-dominated landscapes to better support bird populations. At a microscale, behavioral choices of where to forage and where nests are placed can provide managers with specific features that can be manipulated to enhance habitat availability and attractiveness. At a macroscale, settlement decisions, habitat occupancy, and densities can guide conservation strategies aimed at land preservation and restoration at patch and landscape scales by providing data on environments necessary for a species or insight into land uses with high potential quality or connectivity. At both scales, conservation success can be evaluated and updated based on birds' responses to manipulations, which can serve as an ecological indicator for habitat quality more broadly. By informing the restoration of human-dominated landscapes with avian behavior, we have a better chance of successfully improving ecological function and creating habitats that work for birds and people.

LITERATURE CITED

[1]   Jones, J. 2001. Habitat selection studies in avian ecology: a critical review. Auk. 118 (2): 557–562.

[2]   Cody, M.L. 1985. Habitat Selection in Birds. Academic Press.

[3]   Fuller, R.J. 2012. Birds and Habitat: Relationships in Changing Landscapes. Cambridge University Press.

[4]   Partridge, L. 1974. Habitat selection in titmice. Nature. 247(5442): 573–574.

[5]   Davis, J.M. and J.A. Stamps. 2004. The effect of natal experience on habitat preferences. Trends. Ecol. Evol. 19(8): 411–416.

[6]   Piper, W.H. 2011. Making habitat selection more "familiar": a review. Behav. Ecol. Sociobiol. 65(7): 1329–1351.

[7]   Kearns, L.J. and A.D. Rodewald. 2013. Within-season use of public and private information on predation risk in nest-site selection. J. Ornithol. 154(1): 163–172.

[8]   Kokko, H. and W.J. Sutherland. 2001. Ecological traps in changing environments: ecological and evolutionary consequences of a behaviourally mediated Allee effect. Evol. Ecol. Res. 3(5): 603–610.

[9]   Gunnarsson, T.G., J.A. Gill, J. Newton, P.M. Potts and W.J. Sutherland. 2005. Seasonal matching of habitat quality and fitness in a migratory bird. Proc. Biol. Sci. 272(1578): 2319–2323.

[10]   Forsman, J.T., M.B. Hjernquist, J. Taipale and L. Gustafsson. 2008. Competitor density cues for habitat quality facilitating habitat selection and investment decisions. Behav. Ecol. 19(3): 539–545.

[11]   Marra, P.P., T.W. Sherry and R.T. Holmes. 1993. Territorial exclusion by a long-distance migrant warbler in Jamaica: a removal experiment with American Redstarts (*Setophaga ruticilla*). Auk. 110(3): 565–572.

[12]   Petit, L.J. and D.R. Petit. 1996. Factors governing habitat selection by Prothonotary warblers: field tests of the Fretwell-Lucas models. Ecol. Monogr. 66(3): 367–387.

[13]   Fisher, R.J. and S.K. Davis. 2010. From Wiens to Robel: a review of grassland-bird habitat selection. J. Wildl. Manage. 74(2): 265–273.

[14]   McGrath, L.J., C. Van Riper III and J.J. Fontaine. 2009. Flower power: tree flowering phenology as a settlement cue for migrating birds. J. Anim. Ecol. 78(1): 22–30.

[15]   Orians, G.H. and J.F. Wittenberger. 1991. Spatial and temporal scales in habitat selection. Am. Nat. 137: S29–S49.

[16]   Cornell, K.L. and T.M. Donovan. 2010. Scale-dependent mechanisms of habitat selection for a migratory passerine: an experimental approach. Auk. 127(4): 899–908.

[17]   Sapir, N., Z. Abramsky, E. Shochat and I. Izhaki. 2004. Scale-dependent habitat selection in migratory frugivorous passerines. Naturwissenschaften. 91(11): 544–547.

[18] McCabe, J.D. and B.J. Olsen. 2015. Tradeoffs between predation risk and fruit resources shape habitat use of landbirds during autumn migration. Auk. 132(4): 903–913.

[19] Cockle, K.L., K. Martin and M.C. Drever. 2010. Supply of tree-holes limits nest density of cavitynesting birds in primary and logged subtropical Atlantic forest. Biol. Conserv. 143(11): 2851–2857.

[20] Koski, T.M., C. Lindstedt, T. Klemola, J. Troscianko, E. Mantyla, E. Tyystjarvi, et al. 2017. Insect herbivory may cause changes in the visual properties of leaves and affect the camouflage of herbivores to avian predators. Behav. Ecol. Sociobiol. 71(6): 1–12.

[21] Mantyla, E., S. Kipper and M. Hilker. 2020. Insectivorous birds can see and smell systemically herbivore-induced pines. Ecol. Evol. 10(17): 9358–9370.

[22] Hiltpold, I. and W.G. Shriver. 2018. Birds bug on indirect plant defenses to locate insect prey. J. Chem. Ecol. 44(6): 576–579.

[23] Holmes, R.T. and S.K. Robinson. 1981. Tree species preferences of foraging insectivorous birds in a northern hardwoods forest. Oecologia. 48(1): 31–35.

[24] Holmes, R.T. and J.C. Schultz. 1988. Food availability for forest birds: effects of prey distribution and abundance on bird foraging. Can. J. Zool. 66(3): 720–728.

[25] Muller, K.L., J.A. Stamps, V.V. Krishnan and N.H. Willits. 1997. The effects of conspecific attraction and habitat quality on habitat selection in territorial birds (*Troglodytes aedon*). Am. Nat. 150(5): 650–661.

[26] Ahlering, M.A. and J. Faaborg. 2006. Avian habitat management meets conspecific attraction: if you build it, will they come? Auk. 123(2): 301–312.

[27] Betts, M.G., A.S. Hadley, N. Rodenhouse and J.J. Nocera. 2008. Social information trumps vegetation structure in breeding-site selection by a migrant songbird. Proc. Biol. Sci. 275(1648): 2257–2263.

[28] Hahn, B.A. and E.D. Silverman. 2006. Social cues facilitate habitat selection: American redstarts establish breeding territories in response to song. Biol. Lett. 2(3): 337–340.

[29] Morinay, J., J.T. Forsman and B. Doligez. 2020. Heterospecific song quality as social information for settlement decisions: an experimental approach in a wild bird. Anim. Behav. 161: 103–113.

[30] Fontaine, J.J. and T.E. Martin. 2006. Parent birds assess nest predation risk and adjust their reproductive strategies. Ecol. Lett. 9(4): 428–434.

[31] Ladin, Z.S., V. D'Amico, J.M. Baetens, R.R. Roth and W.G. Shriver. 2016. Long-term dynamics in local host–parasite interactions linked to regional population trends. Ecosphere. 7(8): 01420.

[32] Mayor, S.J., D.C. Schneider, J.A. Schaefer and S.P. Mahoney. 2009. Habitat selection at multiple scales. Ecosci. 16(2): 238–247.

[33] Morris, D.W. 1987. Ecological scale and habitat use. Ecology. 68(2): 362–369.

[34] Boves, T.J., D.A. Buehler, J. Sheehan, P.B. Wood, A.D. Rodewald, J.L. Larkin, et al. 2013. Spatial variation in breeding habitat selection by Cerulean Warblers (*Setophaga cerulea*) throughout the Appalachian Mountains. Auk. 130(1): 46–59.

[35] Martin, T.E. 1998. Are microhabitat preferences of coexisting species under selection and adaptive? Ecology. 79(2): 656–670.

[36] MacArthur, R.H. 1958. Population ecology of some warblers of northeastern coniferous forests. Ecology. 39(4): 599–619.

[37] Wood, E.M., A.M. Pidgeon, F. Liu and D.J. Mladenoff. 2012. Birds see the trees inside the forest: the potential impacts of changes in forest composition on songbirds during spring migration. For. Ecol. Manage. 280: 176–186.

[38] Gallinat, A.S., R.B. Primack and T.L. Lloyd-Evans. 2020. Can invasive species replace native species as a resource for birds under climate change? A case study on bird-fruit interactions. Biol. Conserv. 241: 108268.

[39] Borgmann, K.L., C.J. Conway and M.L. Morrison. 2013. Breeding phenology of birds: mechanisms underlying seasonal declines in the risk of nest predation. PLoS One. 8(6): 65909.

[40] Nelson, S.B.M., J.J. Coon and J.R. Miller. 2020. Do habitat preferences improve fitness? Contextspecific adaptive habitat selection by a grassland songbird. Oecologia. 193: 15–26.

[41] Rimmer, C.C. and K.P. McFarland. 2001. Known breeding and wintering sites of a Bicknell's Thrush. Wilson J. Ornithol. 113(2): 234–236.

[42] Marra, P.P., E.B. Cohen, S.R. Loss, J.E. Rutter and C.M. Tonra. 2015. A call for full annual cycle research in animal ecology. Biol. Lett. 11(8): 20150552.

[43] Vitz, A.C. and A.D. Rodewald. 2007. Vegetative and fruit resources as determinants of habitat use by mature-forest birds during the postbreeding period. Auk. 124(2): 494–507.

[44] Raybuck, D.W., J.L. Larkin, S.H. Stoleson and T.J. Boves. 2020. Radio-tracking reveals insight into survival and dynamic habitat selection of fledgling Cerulean Warblers. Condor. 122(1): 1–15.

[45] Moore, F.R. and D.A. Aborn. 2000. Mechanisms of en route habitat selection: How do migrants make habitat decisions during stopover? Studies in Avian Biology 20: 34–42.

[46] Buler, J.J., F.R. Moore and S. Woltmann. 2007. A multi-scale examination of stopover habitat use by birds. Ecology. 88(7): 1789–1802.

[47] Chalfoun, A.D. and K.A. Schmidt, K.A. 2012. Adaptive breeding-habitat selection: Is it for the birds? Auk. 129(4): 589–599.

[48] Rolstad, J., B. Loken and E. Rolstad. 2000. Habitat selection as a hierarchical spatial process: the green woodpecker at the northern edge of its distribution range. Oecologia. 124(1): 116–129.

[49] Thompson, F.R., T.M. Donovan, R.M. DeGraff, J. Faaborg and S.K. Robinson. 2002. A multi-scale perspective of the effects of forest fragmentation on birds in eastern forests. pp. 8–19 *In*: George, T.L. and S.D. Dobkin. [eds]. Effects of Habitat Fragmentation on Birds in Western Landscapes: Contrasts with Paradigms from the Eastern United States. Studies in Avian Biology.

[50] Mitchell, L.J., T. Kohler, P.C. White and K.E. Arnold. 2020. High interindividual variability in habitat selection and functional habitat relationships in European nightjars over a period of habitat change. Ecol. Evol. 10: 5932–5945.

[51] Marra, P.P. and R.T. Holmes, R.T. 2001. Consequences of dominance-mediated habitat segregation in American Redstarts during the nonbreeding season. Auk. 118(1): 92–104.

[52] Powell, L.L., E.M. Ames, J.R. Wright, J. Matthiopoulos and P.P. Marra. 2020. Interspecific competition between resident and wintering birds: experimental evidence and consequences of coexistence. Ecology. e03208.

[53] Wunderle Jr, J.M. 1992. Sexual habitat segregation in wintering Black-throated Blue Warblers in Puerto Rico. pp 299–307. *In*: Hagan J.M. and D.W. Johnston [eds]. Ecology and Conservation of Neotropical Migrant Landbirds. Smithsonian Institution Press, Washington, DC.

[54] Catry, P., R.A. Phillips, J.P. Croxall, K. Ruckstuhl and P. Neuhaus. 2006. Sexual segregation in birds: patterns, processes and implications for conservation. pp. 351–378. *In*: Ruckstuhl, K.E. and P. Neuhau [eds]. Sexual Segregation in Vertebrates: Ecology of the Two Sexes. Cambridge University Press, Cambridge.

[55] Bennett, R.E., A.D. Rodewald and K.V. Rosenberg. 2019. Overlooked sexual segregation of habitats exposes female migratory landbirds to threats. Biol. Conserv. 240: 108266.

[56] Vitousek, P.M., H.A. Mooney, J. Lubchenco and J.M. Melillo. 1997. Human domination of earth's ecosystems. Science. 277: 494–499.

[57] Butchart, S.H., M. Walpole, B. Collen, A. Van Strien, J.P. Scharlemann, R.E. Almond, et al. 2010. Global biodiversity: indicators of recent declines. Science. 328(5982): 1164–1168.

[58] Dobson, A., D. Lodge, J. Alder, G.S. Cumming, J. Keymer, J. McGlade, et al. 2006. Habitat loss, trophic collapse and the decline of ecosystem services. Ecology. 87(8): 1915–1924.

[59]   Devictor, V., R. Julliard, D. Couvet, A. Lee and F. Jiguet. 2007. Functional homogenization effect of urbanization on bird communities. Conserv. Biol. 21(3): 741–751.

[60]   Wretenberg, J., A. Lindstrom, S. Svensson, T. Thierfelder and T. Part. 2006. Population trends of farmland birds in Sweden and England: similar trends but different patterns of agricultural intensification. J. Appl. Ecol. 43(6): 1110–1120.

[61]   Warren, P.S., S.B. Lerman, R. Andrade, K.L. Larson and H.L. Bateman. 2019. The more things change: species losses detected in Phoenix despite stability in bird–socioeconomic relationships. Ecosphere. 10(3): 02624.

[62]   Padilla, B.J. and A.D. Rodewald. 2015. Avian metapopulation dynamics in a fragmented urbanizing landscape. Urban Ecosyst. 18(1): 239–250.

[63]   Fischer, J.D., S.H. Cleeton, T.P. Lyons and J.R. Miller. 2012. Urbanization and the predation paradox: the role of trophic dynamics in structuring vertebrate communities. Biosci. 62(9): 809–818.

[64]   Robertson, B.A. and R.L. Hutto. 2006. A framework for understanding ecological traps and an evaluation of existing evidence. Ecology. 87(5): 1075–1085.

[65]   Rodewald, A.D., D.P. Shustack and L.E. Hitchcock. 2010. Exotic shrubs as ephemeral ecological traps for nesting birds. Biol. Invasions. 12(1): 33–39.

[66]   Rodewald, A.D., D.P. Shustack and T.M. Jones. 2011. Dynamic selective environments and evolutionary traps in human-dominated landscapes. Ecology. 92(9): 1781–1788.

[67]   Hollander, F.A., N. Titeux, M.J. Holveck and H. Van Dyck. 2017. Timing of breeding in an ecologically trapped bird. Am. Nat. 189(5): 515–525.

[68]   Santangeli, A., A. Lehikoinen, A. Bock, P. Peltonen-Sainio, L. Jauhiainen, M. Girardello, et al. 2018. Stronger response of farmland birds than farmers to climate change leads to the emergence of an ecological trap. Biol. Conserv. 217: 166–172.

[69]   Pollock, C.J., P. Capilla-Lasheras, R.A. McGill, B. Helm and D.M. Dominoni. 2017. Integrated behavioural and stable isotope data reveal altered diet linked to low breeding success in urbandwelling blue tits (*Cyanistes caeruleus*). Sci. Rep. 7(1): 1–14.

[70]   Hanmer, H.J., R.L. Thomas and M.D. Fellowes. 2017. Provision of supplementary food for wild birds may increase the risk of local nest predation. Ibis. 159(1): 158–167.

[71]   Chamberlain, D.E., A.R. Cannon, M.P. Toms, D.I. Leech, B.J. Hatchwell and K.J. Gaston. 2009. Avian productivity in urban landscapes: a review and meta-analysis. Ibis. 151(1): 1–18.

[72]   Titeux, N., O. Aizpurua, F.A. Hollander, F. Sarda-Palomera, V. Hermoso, J.Y. Paquet, et al. 2020. Ecological traps and species distribution models: a challenge for prioritizing areas of conservation importance. Ecography. 43(3): 365–375.

[73]   [UN] United Nations Population Fund. 2007. State of World Population 2007: Unleashing the Potential of Urban Growth. United Nations Population Fund, New York, NY.

[74]   Grimm, N.B., D. Foster, P. Groffman, J.M. Grove, C.S. Hopkinson, K.J. Nadelhoffer, et al. 2008. The changing landscape: ecosystem responses to urbanization and pollution across climatic and societal gradients. Front. Ecol. Environ. 6(5): 264–272.

[75]   Lerman, S.B, D.L. Narango, R. Andrande, P.W. Warren, A. Grade and K. Straley. 2020. Wildlife in the city: human drivers and human consequences. pp. 37–66. *In*: Barbosa P. [ed.]. Urban Ecology: Its Nature and Challenges. CAB International.

[76]   Lepczyk, C.A., M.F. Aronson, K.L. Evans, M.A. Goddard, S.B. Lerman and J.S. MacIvor. 2017. Biodiversity in the city: fundamental questions for understanding the ecology of urban green spaces for biodiversity conservation. Biosci. 67(9): 799–807.

[77]   Rebolo-Ifran, N., J.L. Tella and M. Carrete, M., 2017. Urban conservation hotspots: predation release allows the grassland-specialist burrowing owl to perform better in the city. Sci. Rep. 7(1): 1–9.

[78]   Faeth, S.H., P.S. Warren, E. Shochat and W.A. Marussich. 2005. Trophic dynamics in urban communities. Biosci. 55(5): 399–407.

[79] Tallamy, D.W., D.L. Narango and A. Mitchell. 2020. Do non-native plants contribute to insect population declines? Ecol. Entomol. 12973.

[80] Evans, B.S., A.M. Kilpatrick, A.H. Hurlbert and P.P. Marra. 2017. Dispersal in the urban matrix: assessing the influence of landscape permeability on the settlement patterns of breeding songbirds. Front. Ecol. Evol. 5: 63.

[81] Chace, J.F. and J.J. Walsh. 2006. Urban effects on native avifauna: a review. Landsc. Urban Plan. 74(1): 46–69.

[82] Marzluff, J. and A.D. Rodewald. 2008. Conserving biodiversity in urbanizing areas: nontraditional views from a bird's perspective. Cities Environ. (CATE) 1(2): 6.

[83] Soanes, K., M. Sievers, Y.E. Chee, N.S. Williams, M. Bhardwaj, A.J. Marshall, et al. 2019. Correcting common misconceptions to inspire conservation action in urban environments. Conserv. Biol. 33(2): 300–306.

[84] [FAO] Food and Agriculture Organization of the United Nations. www.fao.org. Accessed November 29, 2020

[85] Donald, P.F., R.E. Green and M.F. Heath. 2001. Agricultural intensification and the collapse of Europe's farmland bird populations. Proc. Biol. Sci. 268(1462): 25–29.

[86] Stanton, R.L., C.A. Morrissey and R.G. Clark. 2018. Analysis of trends and agricultural drivers of farmland bird declines in North America: a review. Agric. Ecosyst. Environ. 254: 244–254.

[87] Kremen, C. and A.M. Merenlender. 2018. Landscapes that work for biodiversity and people. Science. 362(6412).

[88] Reynolds, M.D., B.L. Sullivan, E. Hallstein, S. Matsumoto, S. Kelling, M. Merrifield, et al. 2017. Dynamic conservation for migratory species. Sci Adv 3(8): 1700707.

[89] Finch, T., S. Gillings, R.E. Green, D. Massimino, W.J. Peach and A. Balmford. 2019. Bird conservation and the land sharing-sparing continuum in farmland-dominated landscapes of lowland England. Conserv. Biol. 33(5): 1045–1055.

[90] Hole, D.G., A.J. Perkins, J.D. Wilson, I.H. Alexander, P.V. Grice and A.D. Evans. 2005. Does organic farming benefit biodiversity? Biol. Conserv. 122(1): 113–130.

[91] Perlut, N.G., A.M. Strong, T.M. Donovan and N.J. Buckley. 2008. Regional population viability of grassland songbirds: Effects of agricultural management. Biol. Conserv. 141(12): 3139–3151.

[92] Blitzer, E.J., C.F. Dormann, A. Holzschuh, A.M. Klein, T.A. Rand and T. Tscharntke. 2012. Spillover of functionally important organisms between managed and natural habitats. Agric. Ecosyst. Environ. 146(1): 34–43.

[93] Karp, D.S., C.D. Mendenhall, R.F. Sandi, N. Chaumont, P.R. Ehrlich, E.A. Hadly, et al. 2013. Forest bolsters bird abundance, pest control and coffee yield. Ecol. Lett. 16(11): 1339–1347.

[94] Noble, I.R. and R. Dirzo. 1997. Forests as human-dominated ecosystems. Science. 277(5325): 522–525.

[95] Schieck, J. and S.J. Song. 2006. Changes in bird communities throughout succession following fire and harvest in boreal forests of western North America: literature review and meta-analyses. Can. J. For. Res. 36(5): 1299–1318.

[96] Perry, R.W., J.M. Jenkins, R.E. Thill and F.R. Thompson III. 2018. Long-term effects of different forest regeneration methods on mature forest birds. For. Ecol. Manage. 408: 183–194.

[97] Kellner, K.F., P.J. Ruhl, J.B. Dunning Jr, J.K. Riegel and R.K. Swihart. 2016. Multi-scale responses of breeding birds to experimental forest management in Indiana, USA. For. Ecol. Manage. 382: 64–75.

[98] Perfecto, I., R.A. Rice, R. Greenberg and M.E. Van der Voort. 1996. Shade coffee: a disappearing refuge for biodiversity: shade coffee plantations can contain as much biodiversity as forest habitats. Biosci. 46(8): 598–608.

[99] Bakermans, M.H., A.C. Vitz, A.D. Rodewald and C.G. Rengifo. 2009. Migratory songbird use ofshade coffee in the Venezuelan Andes with implications for conservation of Cerulean Warbler. Biol. Conserv. 142(11): 2476–2483.

[100] Sanchez-Clavijo, L.M., N.J. Bayly and P.F. Quintana-Ascencio. 2020. Habitat selection in transformed landscapes and the role of forest remnants and shade coffee in the conservation of resident birds. J. Anim. Ecol. 89(2): 553–564.

[101] Johnson, M.D. 2000. Effects of Shade-Tree Species and Crop Structure on the Winter Arthropod and Bird Communities in a Jamaican Shade Coffee Plantation. Biotropica. 32(1): 133–145.

[102] Bakermans, M.H., A.D. Rodewald, A.C. Vitz and C. Rengifo, 2012. Migratory bird use of shade coffee: the role of structural and floristic features. Agrofor. Syst. 85(1): 85–94.

[103] Narango, D.L., D.W. Tallamy, K.J. Snyder and R.A. Rice. 2019. Canopy tree preference by insectivorous birds in shade-coffee farms: Implications for migratory bird conservation. Biotropica. 51(3): 387–398.

[104] McDermott, M.E. and A.D. Rodewald. 2014. Conservation value of silvopastures to Neotropical migrants in Andean forest flocks. Biol. Conserv. 175: 140–147.

[105] Moorman, C.E., S.M. Grodsky and S. Rupp. 2019. Renewable Energy and Wildlife Conservation. JHU Press.

[106] Smith, J.A. and J.F. Dwyer. 2016. Avian interactions with renewable energy infrastructure: an update. Condor. 118(2): 411–423.

[107] Mahoney, A. and A.D. Chalfoun. 2016. Reproductive success of Horned Lark and McCown's Longspur in relation to wind energy infrastructure. Condor. 118(2): 360–375.

[108] Pennington, D.N. and R.B. Blair. 2011. Habitat selection of breeding riparian birds in an urban environment: untangling the relative importance of biophysical elements and spatial scale. Divers. Distrib. 17(3): 506–518.

[109] Dunford, W. and K. Freemark. 2005. Matrix matters: effects of surrounding land uses on forest birds near Ottawa, Canada. Landsc. Ecol. 20(5): 497–511.

[110] Hostetler, M. 2001. The importance of multi-scale analyses in avian habitat selection studies in urban environments. pp. 139–154. *In*: Marzluff, J.M., R. Bowman and R. Donnelly [eds]. Avian Ecology and Conservation in an Urbanizing World. Springer, Boston, MA.

[111] Goddard, M.A., A.J. Dougill and T.G. Benton. 2010. Scaling up from gardens: biodiversity conservation in urban environments. Trends Ecol. Evol. 25(2): 90–98.

[112] Narango, D.L., D.W. Tallamy and P.P. Marra. 2017. Native plants improve breeding and foraging habitat for an insectivorous bird. Biol. Conserv. 213: 42–50.

[113] Narango, D.L., D.W. Tallamy and P.P. Marra. 2018. Nonnative plants reduce population growth of an insectivorous bird. Proc. Natl. Acad. Sci. USA. 115(45): 11549–11554.

[114] Borgmann, K.L. and A.D. Rodewald. 2004. Nest predation in an urbanizing landscape: the role of exotic shrubs. Ecol. Appl. 14(6): 1757–1765.

[115] Burghardt, K.T., D.W. Tallamy and G.W. Shriver. 2009. Impact of native plants on bird and butterfly biodiversity in suburban landscapes. Conserv. Biol. 23(1): 219–224.

[116] Lerman, S.B. and P.S. Warren. 2011. The conservation value of residential yards: linking birds and people. Ecol. Appl. 21(4): 1327–1339.

[117] Donnelly, R. and J.M. Marzluff. 2004. Importance of reserve size and landscape context to urban bird conservation. Conserv. Biol. 18(3): 733–745.

[118] Mcclure, C.J., B.W. Rolek and G.E. Hill. 2012. Predicting occupancy of wintering migratory birds: is microhabitat information necessary? Condor. 114(3): 482–490.

[119] Herse, M.R., M.E. Estey, P.J. Moore, B.K. Sandercock and W.A. Boyle. 2017. Landscape context drives breeding habitat selection by an enigmatic grassland songbird. Landsc. Ecol. 32(12): 2351–2364.

[120] Widows, S.A. and D. Drake. 2014. Evaluating the National Wildlife Federation's Certified Wildlife Habitat™ program. Landsc. Urban Plan. 129: 32–43.

[121] Gibbs, J.P. and D.G. Wenny. 1993. Song output as a population estimator: effect of male pairing status. J. Field Ornithol. 64(3): 316–322.

[122] Bailey, B.A. and D.I. King. 2019. Habitat selection and habitat quality for wintering wood thrushes in a coffee growing region in Honduras. Glob. Ecol. Conserv. 20: 00728.

[123] Ausprey, I.J., F.L. Newell, S.K. Robinson. 2020. Adaptations to light predict the foraging niche and disassembly of avian communities in tropical countrysides. Ecology. 03213.

[124] Goodwin, S.E. and W.G. Shriver. 2011. Effects of traffic noise on occupancy patterns of forest birds. Conserv. Biol. 25(2): 406–411.

[125] Senzaki, M., J.R. Barber, J.N. Phillips, N.H. Carter, C.B. Cooper, M.A. Ditmer, et al. 2020. Sensory pollutants alter bird phenology and fitness across a continent. Nature. 587: 605–609

[126] Atwell, J.W., G.C. Cardoso, D.J. Whittaker, S. Campbell-Nelson, K.W. Robertson and E.D. Ketterson. 2012. Boldness behavior and stress physiology in a novel urban environment suggest rapid correlated evolutionary adaptation. Behav. Ecol. 23(5): 960–969.

[127] Ahlering, M.A. and J. Faaborg. 2006. Avian habitat management meets conspecific attraction: if you build it, will they come? Auk. 123(2): 301–312.

[128] Fischer, J.D., S.H. Cleeton, T.P. Lyons and J.R. Miller. 2012. Urbanization and the predation paradox: the role of trophic dynamics in structuring vertebrate communities. Biosci. 62(9): 809–818.

[129] Peters, V.E., T.A. Carlo, M.A. Mello, R.A. Rice, D.W. Tallamy, S.A. Caudill, et al. 2016. Using plant–animal interactions to inform tree selection in tree-based agroecosystems for enhanced biodiversity. Biosci. 66(12): 1046–1056.

[130] Johnson, M.D. and T.W. Sherry. 2001. Effects of food availability on the distribution of migratory warblers among habitats in Jamaica. J. Anim. Ecol. 70(4): 546–560.

[131] Wood, E.M. and S. Esaian. 2020. The importance of street trees to urban avifauna. Ecol. Appl. 30(7): 02149.

[132] Gabbe, A.P., S.K. Robinson and J.D. Brawn. 2002. Tree-species preferences of foraging insectivorous birds: implications for floodplain forest restoration. Conserv. Biol. 16(2): 462–470.

[133] Narango, D.L., D.W. Tallamy and K.J. Shropshire, K.J. 2020. Few keystone plant genera support the majority of Lepidoptera species. Nat. Commun. 11: 5751.

[134] Cooper, R.J. 1988. Dietary relationships among insectivorous birds of an eastern deciduous forest. Ph.D. Dissertation, West Virginia University, Morgantown, WV.

[135] Rodewald, A.D. and M.D. Abrams. 2002. Floristics and avian community structure: implications for regional changes in eastern forest composition. Forest Sci. 48(2): 267–272.

[136] Newell, F.L. and A.D. Rodewald. 2011. Role of topography, canopy structure and floristics in nest-site selection and nesting success of canopy songbirds. For. Ecol. Manage. 262(5): 739–749.

[137] Newell, F.L., T.A. Beachy, A.D. Rodewald, C.G. Rengifo, I.J. Ausprey and P.G. Rodewald. 2014. Foraging behavior of migrant warblers in mixed-species flocks in Venezuelan shade coffee: interspecific differences, tree species selection and effects of drought. J. Field Ornithol. 85(2): 134–151.

[138] Soto-Pinto, L., V. Villalvazo-Lopez, G. Jimenez-Ferrer, N. Ramirez-Marcial, G. Montoya and F.L. Sinclair. 2007. The role of local knowledge in determining shade composition of multistrata coffee systems in Chiapas, Mexico. Biodivers. Conserv. 16(2): 419–436.

[139] Robertson, B.A. and R.L. Hutto. 2007. Is selectively harvested forest an ecological trap for Olivesided Flycatchers?. Condor. 109(1): 109–121.

[140] Reynolds, S.J., J.D. Ibáñez-Álamo, P. Sumasgutner and M.C. Mainwaring. 2019. Urbanisation and nest building in birds: a review of threats and opportunities. J. Ornithol. 160(3): 841–860.

[141] Jedlicka, J.A., R. Greenberg and D.K. Letourneau. 2011. Avian conservation practices strengthen ecosystem services in California vineyards. PLoS One. 6(11): 27347.

[142] Burgio, K.R., M.A. Rubega and D. Sustaita. 2014. Nest-building behavior of Monk Parakeets and insights into potential mechanisms for reducing damage to utility poles. PeerJ. 2: 601.

[143] Ausprey, I.J. and A.D. Rodewald. 2011. Postfledging survivorship and habitat selection across a rural-to-urban landscape gradient. Auk. 128(2): 293–302.

Chapter 3

Migration in the Anthropocene

Kevin C. Fraser[1]

Introduction

Most of the world's 5,000 species of songbird are migratory, with some species travelling thousands of kilometres per year. These remarkable long-distance migrations require a complex suite of behaviours, which have been shaped by evolution over millennia. Rapid, human-driven environmental change may push the limits of phenotypic plasticity in some species or populations, and some evolved strategies may be incompatible with new conditions. We have much to learn about the degree to which songbirds are able to respond adaptively to environmental change, where this will result in winners and losers in future climate change scenarios, and how conservation and management can best address these challenges. This chapter summarizes how we currently understand the way in which migration timing, navigation, and stopover behaviour is formed and maintained in songbirds. Next, these behaviours are considered within the context of

[1]Department of Biological Sciences, University of Manitoba, Kevin.Fraser@umanitoba.ca

new challenges faced by migrants, where climate change, habitat loss, and light pollution may lead to mistimed migrations or disconnected routes. This includes exploration of the evidence for phenotypic plasticity of migration behaviour as well as where and how this may limit responses to rapid environmental change. Lastly, this chapter investigates the future role of conservation and management strategies, including the potential to manage for adaptive migration behaviour and to re-introduce lost migratory populations. Most research has focused on passerine migrant species breeding in the temperate zone, therefore most of this chapter will draw on examples from these systems. However, it is anticipated these examples have relevance to other migratory systems as well.

Migration Timing and Phenotypic Plasticity in the Context of Climate Change

Migration Timing Background

Extensive laboratory and field-based research has revealed that migration timing in songbirds is directed mostly by endogenous factors, but is fine-tuned by local environmental or physiological cues. The initiation, or 'starting gun,' for each journey is the result of an evolved, species- or population-specific circannual routine, that is most often entrained to changes in local daylength. Over millennia, this combination of genetic and environmental factors has provided a consistent and reliable cue of seasonal change. However, climate change creates new migratory conditions, and to understand the potential for flexible individual, population, or species-specific responses, we have to consider the underlying mechanisms that govern migration timing.

Endogenous, circannual routines cued by photoperiod (i.e., day length) may be particularly important for long-distance migrants (Box 1). These species often begin their journeys thousands of kilometres away from their destinations in the absence of other external, environmental cues (such as temperature) to indicate seasonal change at their distant breeding, migratory, or nonbreeding areas [1, 2]. Some endogenous routines can be very stable, even in the absence of external cues. Studies using Palearctic-Afrotropical migratory birds, demonstrated that individuals exposed in the lab to constant photoperiod can continue to use a circannual routine to replicate the appropriate timing of migration and other seasonal activities for 12 years or more [2, 3]. This may be advantageous to species, for example, that overwinter in areas near the equator, where daylength exhibits little seasonal change [4]. However, without the cue of seasonal change in photoperiod, timing begins to drift and even the longest-running circannual programs exhibited variation between 9–13 months [3]. Shorter-distance migrants tend to show much more drift in their timing in the absence of photoperiod cues [3]. Thus, we might anticipate a gradient across species in reliance upon circannual routines as well as variation in the degree of plasticity to external conditions. However, we still have much to learn regarding the flexibility of timing programs [5] and whether there are species- or migration system-specific patterns in these responses.

Selection on timing behaviours has resulted in a remarkable degree of fine-tuning of migration timing even within different populations of widely-distributed species. Laboratory-based studies of Garden Warblers (*Sylvia borin*), a species that breeds from Europe to western Siberia and overwinters across a range of latitudes in equatorial Africa, demonstrated that spring migration timing is cued to migration distance. Birds exposed to photoperiods simulating overwintering areas that were more distant from breeding sites showed earlier migratory restlessness (*zugunruhe*) in the lab than those exposed to photoperiods reflecting wintering sites that were closer to breeding sites [2]. This adaptation may serve to synchronize the breeding arrival timing of birds that have wintered further away with birds that had less distance to travel, thus ensuring that individuals with greater travel distances are not at a disadvantage in terms of mate and territory choice and are able to capitalize on peak insect abundance for raising nestlings [2].

There is also evidence that birds respond adaptively to photoperiodic variation experienced during migration. Such a mechanism is advantageous, as it would allow birds to adjust their speed and compensate for delays during migration. First-year, Long-tailed Tits (*Aegithalos c. caudatus*) were captured during their autumn journeys between Russia and southern Europe and brought into the lab [6]. After acclimatizing to the experimental set up, birds exposed to shorter daylengths that simulated a later date exhibited greater locomotor activity, suggesting they were compensating for being 'late' on migration with more migratory activity. Similarly, Stonechats (*Saxicola torquata*) varied the duration of spring migratory activity in the lab in relation to photoperiod [7]. If birds can adjust their speed along the way in response to local photoperiod, this could be an important mechanism to assist birds in adjusting to new vagaries of weather and conditions they may now find with climate change. This apparent en-route flexibility, to speed up or slow down relative to their position, requires further exploration in other species and systems. However, if 'speeding-up' across a full migration is required, some recent evidence suggests this may not be possible as birds may already be flying at or near their maximum speed limit [8, 9]. This may be a major limitation with climate change, where trends are toward earlier, warmer springs, requiring that birds arrive and nest earlier to maximize fitness [10].

While the influence of light duration (photoperiod) has received much research attention, there is some evidence that the intensity of ambient light may also play a role in seasonal timing. In an equatorial resident population of African Stonechats, gonadal development and moult became synchronized with an experimental manipulation of light intensity, suggesting a dual role for photoperiod and intensity on the seasonal timing of breeding [11]. Many migrants overwinter at tropical latitudes and the start of their spring journeys coincides in some cases with the beginning of the dry season, where there is greater light intensity. However, the impact of light intensity on migratory cueing has not been examined. This is an important area for future research, as anthropogenic change to both daytime and nighttime light intensity (and duration) is growing.

Other environmental cues such as seasonal change in weather or habitat conditions can be important, fine-scale modulators of migration timing. Compared to photoperiod cues, these play a much smaller role in migration timing in many species, perhaps because year-to-year variation makes these cues less consistent, and therefore it would not be adaptive to rely heavily upon them. For example, overwintering rainfall patterns and moisture gradients impacted spring migration departure timing of American Redstarts (*Setophaga ruticilla*) [12]. Birds experiencing drier, overwinter habitat conditions delayed departure by 3–5 days as compared to birds from wetter sites. Moisture gradients may also drive timing patterns at broader spatial scales. Wood Thrushes (*Hylocinchla mustelina*) overwintering at the southern edge of the range where conditions were wetter departed later and travelled to more distant breeding areas than populations overwintering further north in drier sites [13]. It is plausible that these environmental influences may work in concert with circannual clocks cued to photoperiod. However, it is expected that where timing is mostly driven by circannual rhythms, individual birds may be less equipped to respond flexibly to climate change effects through adjustments in their migration timing.

In sum, research to date indicates that endogenous, circannual routines control much of migration timing, particularly in longer-distance migrants, but that individual birds rely on photoperiod to entrain and cue migration timing. Further fine-tuning of migration may be driven by local environmental variation, particularly rainfall or moisture gradients. Migration timing programs can be highly tuned to species, population, or migration distance, and phenotypic plasticity (i.e., one genotype can exhibit more than one phenotype in response to environment) appears to vary for short- versus long-distance migrants; all of which is important to consider and further explore before we can determine how different species will respond to climate change. Changing migration timing to better align with global-scale climate change will require either phenotypic plasticity or ontogenetic change in how timing programs are formed [14, 15]. However, a rapid, evolutionary response in timing may be underway in some species [16], but may fall short of what is required to keep up with the pace of current change [17]. These possibilities and the consequences of mistiming are addressed in subsequent sections.

Migration Timing in the Context of Climate Change

Global climate change is altering the conditions surrounding bird migration. Increasing variability in the prevalence and intensity of storms, changes in moisture and temperature gradients, and advancing seasonal phenology are introducing new challenges to migratory birds, which must adequately compensate or face population declines—or even extinction [18]. Can birds match their migration timing to these rapid changes? The degree to which migration, particularly timing, may be flexible to environmental change has been much debated [5].

Spring in the north-temperate breeding areas of many Nearctic-Neotropical and Palearctic-Afrotropical songbird species is becoming earlier. Breeding-site arrival is also becoming earlier on a decadal scale in many species of migratory birds [19, 20]. How is this occurring? Individual flexibility or evolutionary change could both underlie timing shifts. Evidence from non-migratory species suggests a high degree of flexibility to changing spring conditions. For example, in Great tits (*Parus major*) individuals showed plasticity in first egg dates (a measure of breeding phenology) in response to spring temperature, and selection favoured increasing flexibility over time to match increasing variability in spring conditions [21]. An important advantage for non-migratory species is that they are present, on-site, to receive local cues to changing weather and climate conditions.

The challenge for migrants is greater, as they must begin migration at a large distance away from their destination, in the absence of any cues of phenology at their final destination. Pied flycatchers (*Ficedula hypoleuca*) for example, did not adjust their spring arrival timing to match advancing spring phenology over a 20-year period [33]. Individual wood thrushes had remarkably similar spring migration timing in different years, suggesting low flexibility in migration timing [22]. Similarly, purple martins (*Progne subis*) did not speed up the pace of migration in response to a highly advanced spring driven by climate change and thus did not arrive earlier at breeding sites to match the conditions in that year [8]. Such mismatches between arrival timing and local conditions may be common and on the rise in long-distance migratory songbirds. In a study of 48 Nearctic-Neotropical migratory bird species, 19% did not advance their timing sufficiently to match earlier spring green-up [23].

The consequences of timing mismatches may be severe; when birds are misaligned in time with peak resources for nesting, they may produce fewer young, which could ultimately influence population trends. Pied flycatchers suffered greater population declines in areas where peak abundance of caterpillars (their preferred prey) had advanced the most in time [24]. Such timing mismatches could explain why long-distance migrants are declining more steeply than shorter-distance migrants [25], but more research is needed to determine causal mechanisms (i.e., why is their timing not advancing?) and to explore patterns in other species and systems.

A key, unanswered question is whether birds can flexibly change their timing to suit new conditions or whether an evolved response is required; which may take much longer—perhaps too long to avoid population declines or avert extinction. In the study mentioned above, great tits have made a plastic, two-week advance in the timing of their nesting in response to climate change over 47 years, but it was estimated that a comparable shift via microevolutionary change would require 200 years [21]. However, a recent study provides new and rare evidence for a rapid, microevolutionary shift in migration timing in a long-distance migrant. In a careful replication of a lab experiment conducted 21-years ago, pied flycatchers were collected from European breeding sites and raised under identical conditions as in the earlier experiment. Any differences found in circannual timing between the earlier and present study would have to reflect inherited evolutionary change in timing programs [16]. Under these similar and constant conditions, researchers measured onset of breeding and migratory restlessness and found that spring migration and reproductive timing had advanced by 9 days, demonstrating remarkable microevolutionary change on circannual timing programs in this population over a short time frame.

Strong selection events may also drive rapid timing shifts in cohorts of migratory populations. In a field-based study of cliff swallows (*Petrochelidon pyrrhonota*), the timing of nesting was significantly later in the year after an early spring, mass mortality event [26]. When spring temperatures were well below average, all early-arriving swallows were killed, thus selecting against early phenotypes. The later nesting in the subsequent year indicated that early genotypes were lost from this population for the short term [26].

My own research with purple martins has revealed that individual arrival dates at breeding sites can vary by as much as 24 days earlier, or later, from year to year. This amount of within-individual variation in migration timing was sufficient to explain the 100-year advance of spring arrival date reported for this species [27]. Whether timing shifts occur through microevolution [16, 26], or individuals flexibly adjust their timing [27], the conclusion to date is that changes in timing are either not fast or great enough to meet new conditions with climate change [17, 23]. All three species mentioned above—pied flycatchers, cliff swallows, and purple martins—are exhibiting steep rates of population decline, despite evidence of adaptive responses to environmental change. Future work is needed in other migratory species, to determine where changes in timing may not be keeping up with the pace of change.

Impacts from anthropogenic change are also expected for the portion of variation in migratory timing that is responsive to climate, weather, and habitat quality. For example, a drying trend is predicted for the Caribbean and Central America [28], which will affect habitat quality for many Nearctic-Neotropical species that overwinter in this region. Drier habitats have been found to delay spring migration timing of species like wood thrush and American redstart (see Migration Timing Background above); therefore a drying trend could result in population-wide delays in spring. This may also carry-over to influence fitness at breeding sites. In redstarts, for example, individuals originating from drier, poorer quality overwintering habitats in the Caribbean had lower reproductive success at a North American breeding site [29] and spring migration timing has a strong influence on subsequent fitness [30].

Other impacts on migration timing

Ecotoxicological effects have also been reported to impact migration activity and departure timing. Two different projects combining an experimental, lab approach with subsequent field tracking demonstrated impacts on fall migration timing. Yellow-rumped warblers (*Setophaga coronota*) dosed with mercury had earlier fall migration departure than controls [31]. White-throated sparrows (*Zonotrichia albicollis*) were experimentally dosed with neonicotinoids, a pesticide used to coat seeds of corn and other crops. Those at higher dose levels had reduced weight and delayed fall migration timing by up to ~2 days as compared to controls [32]. While such differences in timing may seem subtle, if they are not compensated for during migration, they could have significant impacts on fitness [87]. Further, delays in departure could lead to continued delays along the rest of the migratory journey. Although, there have been no ecotoxicological studies examining the effects of toxins on the timing of start-to-finish migration.

Conservation and Management with Migration Timing

Without a doubt there will be winners and losers with climate change. It remains to be determined which migratory songbird species, or populations, are in each of these categories. While the overarching action for climate change must be to continue to combat greenhouse gas emissions and reduce global warming by as much as possible, at the finer-scale, ongoing research on migratory birds could help to better identify which species or populations may be most at-risk with current and forecasted changes. In a triage approach, these most at-risk groups could become the focus for more targeted conservation and management programs. Further, we could better prepare for captive-reintroduction programs, and possibly assisted evolution approaches, where most needed and where we anticipate success.

In migratory birds, those that travel the longest distances appear to be declining the most steeply [25]. Nearctic-Neotropical and Palearctic-Afrotropical species possess some of the longest journeys, which, as noted above, are driven mostly by circannual routines that are fine-tuned by photoperiod and environmental conditions. In these species local weather at the start of migration does not generally correlate with weather at their destination, and individuals are primarily relying on photoperiod as a cue to initiate migration. This may explain the emerging pattern of timing mismatches between migration and spring green-up at their breeding destinations [23]. Further, if the timing of arrival at breeding sites does not advance swiftly enough to match advances in spring phenology [17], then birds may simply run out of time to nest while conditions are still favourable [33], which will result in population-level consequences [24]. With shorter growing seasons, this threat could be greater for birds breeding further north. For example, in purple martins, my lab's research shows that while birds seem to have the flexibility to nest earlier in warmer springs, and later in cooler ones, more northern breeders have a much shorter interval between arrival date and first egg date than more southern breeders [10], meaning they have less time with which to adjust the timing of their nests to local conditions. Evidence to date suggests that more northern breeding songbirds will experience greater losses as a result of climate change, a pattern that has already been documented in some groups of migratory birds (e.g., aerial insectivores; [34]). This scenario also suggests that more targeted conservation action toward northern breeders, in light of climate change, would be justified.

With 3 billion birds already lost in North America and steep, long-term declines in European migratory systems [35], captive reintroduction programs for migratory songbirds might become valuable management options in the near future. For example, purple martins are nearly extirpated from their north-eastern breeding areas in the Maritime provinces of Canada. In the most recent Maritime Breeding Bird Atlas [36], martins were only detected in 9% of the areas where they were present 20 years earlier in the 1986–90 atlas. Could martins be re-introduced to these areas they inhabited in the past? For martins and other species, we may first need to understand what factors contributed to the eastern-population declines. Reintroduction efforts into subpar habitats could result in the creation of population sinks. But reintroductions can also be highly successful. A reintroduction program for loggerhead shrikes (*Lanius ludovicianus*), is already underway in Canada and has achieved much success repopulating this species into formerly occupied areas in Ontario (https://wildlifepreservation.ca/eastern-loggerhead-shrike-program/). A similar approach may counter losses seen in other migratory songbird species.

A more experimental approach could also be taken to remedy the problem of migratory timing lagging behind environmental change. The idea of 'assisted evolution', where captively-reared individuals are experimentally pre-adapted to current and future rates of change before release [37], could help species to keep up with the pace of change. Assisted evolution has been employed in coral reef systems to preadapt individuals to future ocean warming scenarios and to increase resistance to coral bleaching [37]. There may be potential applications to migratory timing, if adaptive (or shifted) timing could be instilled in captive-raised songbirds. Both laboratory and field studies suggest that migration timing in young birds may be determined by the photoperiod experienced in the nest [14, 15], which lends support to the feasibility of such an approach. But, whether timing formed in the nest carries into adult life remains unknown. Setting the starting gun for the first migration would be useless if birds simply reset to the local photoperiod in subsequent years and return to maladaptive timing. In sum, the principles of assisted evolution have not been considered or applied to the problem of mismatched migratory songbirds, but may be of value for helping species or populations to keep up with the pace of change.

In summary, mismatched timing and resulting impacts on fitness are a known contributor to population declines in migratory birds, particularly those migrating the longest distances and breeding the furthest north (where nesting seasons are shortest). Further research is required to identify species or populations that are most at-risk for targeted conservation and management. While the effects of mismatched timing are difficult to address through conservation, strategies that prioritize and support the most at-risk groups would be one method of addressing climate change impacts on migratory birds. Captive-reintroduction programs for songbirds could be more widely employed, being targeted towards species and locations where success is most anticipated. Where the pace of adaptation in migratory timing is too slow to keep up with the rate of environmental change, assisted evolution could be considered as an additional approach to mitigate mismatches between migration timing and environmental phenology due to climate change.

Box 1   Migration timing and climate change

Background behaviour

- Migration timing in songbirds is mostly directed by endogenous circannual routines, cued to photoperiod
- Birds travelling further distances may rely more heavily on endogenous factors
- Timing is highly tuned to species, location, or migration distance
- Evidence suggests flexibility to photoperiod en-route and to local environmental conditions

Migration timing and climate change

- The timing of spring green-up is advancing but many species are not adjusting fast enough to keep pace
- Responses in timing may reflect phenotypic plasticity or an evolved response
- Mismatches between migration timing and peak resources could impact survival and/or fitness and contribute to population declines

Migration timing and conservation and management

- Identify most at-risk species and populations for targeted conservation and management (e.g., long-distance migrants and more northern breeding populations)
- Apply captive-reintroduction programs for lost populations more widely
- Explore assisted evolution approaches to address mismatched timing

Figure 1   Some long-distance migrants, like purple martins (*Progne subis*) pictured here at a pre-migratory roost in Brazil, must cue migration timing thousands of kilometres away from much of their migratory routes and their breeding sites.
Photo: Marcus Amend.

Migratory Routes and Stopovers in the Context of Habitat Loss and Change

Migration Direction, Distance, and Stopover Background

Migratory routes of songbirds are largely driven by endogenous programs, which direct birds to fly in a given direction for a set period of time (Box 2). Laboratory studies with garden warblers demonstrate that, in the absence of external cues, birds continue to 'fly' for the same duration and in the same direction as their wild counterparts, including repeating directional movements at barrier crossings [38]. Birds experimentally transported to their overwintering range, to a point along their migration route, or even outside of their normal range, continued to exhibit migratory restlessness and directional preferences for the appropriate time period and direction for their migration [1]. In some cases, changes in daylength can also act as a trigger for seasonally appropriate flights north or south [1]. The mostly-endogenous control of migration routes is what enables juvenile songbirds, which are non-social migrants and do not learn their routes from others, to reach their overwintering destinations independently.

Some birds gain navigational abilities with experience, after the completion of their first full migration. Displacement experiments in the field provide evidence for this. Birds were captured during migration and moved to a new location to simulate wind drift, and were subsequently tracked to determine whether they corrected back to their appropriate route. Juvenile European starlings (*Sturnus vulgaris*) and White-crowned Sparrows (*Zonotrichia leucophrys gambelii*) were not as good at re-orienting as adults [39, 40], suggesting that migration experience provides additional navigational ability. However, a meta-analysis of laboratory studies, providing a cross-species perspective, suggests even juveniles may have this ability [41] and a recent field study demonstrates that juvenile Common cuckoos (*Cuculus canorus*) were able to re-orient after displacement [42]. Generally, these studies indicate that juveniles in some species are more at risk of navigational errors, but adult birds can compensate for drift or changes in migratory route through added navigational abilities, modifying their 'built-in' migration programs. With an array of new tracking technologies available, exploring the degree to which these patterns vary across songbird species and migratory pathways is an area ripe for new research.

Most migratory journeys require periods of rest and refueling (stopovers) interspersed with migratory flights. For diurnal songbirds that migrate at night, a stopover 'day' would be a night without flight. For diurnal birds that also migrate during the day (e.g., swallows), a stopover would be pause in the journey during daylight hours. Theoretical predictions for the ratio of flights to rest for small birds using flapping flights is approximately 7 rest days for every 1 day of flying [43]. Stopovers may be more frequent before or after crossing a large migration barrier (like an ocean or desert), during inclement weather, or before making a change in migratory direction [1]. Stopover patterns can vary largely between and even within species, and between seasons (spring versus fall migration). A general pattern among songbirds, is that with reduced time-pressure during fall migration, stopovers are of longer duration than in spring [44, 45]. Thus, stopover areas may need to provide suitable habitat for a longer duration in fall than in spring, and may be in different locations for individuals, populations, or species.

Direct tracking has revealed that several species of migratory songbird in both Nearctic-Neotropical and Palearctic-Afrotropical systems have extra-long stopovers that may be several weeks long and therefore greater than predicted by energetic migration models [45–47]. Such long stops have also been documented in some moult-migrating songbird species, that pause for moult during fall migration before continuing on to their overwintering destinations [48]. These previously unknown periods of residency are an important priority for research and a conservation target for some songbird species.

The duration as well as the consistency of use for stopover sites are important metrics for prioritizing conservation actions. Site-fidelity to stopover areas for individual songbirds is poorly understood, although return rates may be low. Where fidelity has been studied using band-recaptures, usually no more than 10% of birds were observed in the following year at the same stopover site [1]. However, mark-recapture methods are not ideally suited to determine stopover site fidelity, as returning birds are only likely to be recaptured at very small spatial scales. Direct tracking of individual songbirds over multiple migrations offers new opportunities to assess site-fidelity at stopover sites, and at different spatial scales – although this remains to be formally tested in any species. Recently-developed, miniaturized GPS tracking technology, provides a particularly exciting opportunity to test ideas regarding stopover site-fidelity at a fine-scale (<1 km) [46].

While individual fidelity patterns would yield important insight into how repeatable site and habitat use is for migrants, it is also important to determine whether populations use consistent routes, stopover sites, and exhibit stopover-site fidelity. Here, direct-tracking has provided new insight. In wood thrushes, over multiple years, individuals funneled through a narrow band of longitude (88–93°W) during spring migration after crossing the Gulf of Mexico [49]. Similarly, Canada warblers from wide-ranging breeding populations circumnavigated the gulf over multiple years [50] and purple martins from more northern breeding populations consistently took stopovers in the northern Yucatan [47]. In general, it is expected that migration barriers like gulfs or deserts will constrain routes and stopovers to similar areas, making them more likely to be chosen, and/or increasing site fidelity. Whatever the mechanisms, identifying these locations for different migratory species, or populations, remains an important goal for conservation of habitats used during migration.

Migration Routes in the Context of Habitat Loss and Change

One of the largest black boxes in the study and conservation of long-distance migratory songbirds is the impact of migratory stopover habitat quality, availability, and behaviour on population dynamics. While theoretical patterns of behaviour at stopovers have been verified and studied with birds in the wild, it is not known how global change is impacting the overall availability or the quality of stopover habitats for migrants.

There are few studies on how stopover areas and conditions experienced during stopover carry-over to impact the rest of migration, survival, and fitness; however, there is substantial evidence for how conditions at stopovers impact short-term behaviour. From this we can infer how environmental change may impact birds. For example:

1) Swainson's Thrushes (*Catharus ustulatus*) used double the amount of energy at northern U.S. stopover sites than during migration flights [51]. This has relevance for predicting climate change impacts on birds at stopovers as cooling or warming temperatures would impact stopover energetics, particularly for insectivorous migrants where temperature may strongly impact prey availability.

2) Fat gained at a spring stopover site in northwestern Colombia contributed up to a 30-day difference in subsequent migratory travel time between individual grey-cheeked thrushes (*Catharus minimus*) [52]. In a global change context, this

study demonstrated that the protection of high-quality stopover habitat can have important impacts that carry across much of migration. Based on direct-tracking data, migration stopover duration was the most important predictor of overall migration speed in 49 migrants [9], underlying a key role in stopover behaviour in the overall timing of migration. Good quality stopover sites in the right location may therefore have a large effect on subsequent migration speed, and thus on the overall duration of migration and arrival timing, which can have important fitness consequences [30]. Purple martins with more stopover days on spring migration fledged fewer young, suggesting that delays on migration through stopover dynamics may have a direct impact on fitness (Fraser unpublished data). Determining how environmental factors at migratory stopovers influence behaviour across the rest of migration and fitness is an important area for further study, with implications for range-wide conservation strategies for songbirds.

3) Urban stopover sites in the middle of New York city provided similar amounts of insect prey as compared to nearby rural sites, leading to comparable refueling rates for three species of migratory songbird [53]. Despite a high level of anthropogenic disturbance, urban sites can still provide high quality stopover sites. Many species pass through a gauntlet of urban development during their migration, which introduces many novel threats at stopover sites; however, conservation and management must consider the important role these (growing) urban habitats play in migratory stopover.

4) Declining populations of purple martins from the more northerly extents of their breeding range took longer than average stopovers at the Yucatan Peninsula after crossing in fall, demonstrating the conservation value of this stopover region for these populations [47]. Stopover sites on either side of migration barriers, like open water or deserts, can contribute importantly to successful crossing and refueling afterwards for many species. Areas around the gulf coast are recognized as critical 'fire escape' sites for migrants that are also critical to conserve with the increasing prevalence of storms with climate change, that may impose additional danger to birds on migration [54].

5) In a repeat-tracking study of wood thrushes, migration timing was highly repeatable, but routes taken at the Gulf of Mexico by the same individuals in different years were variable [22]. Similarly, red-eyed vireos (*Vireo olivaceus*) either crossed the open water of the gulf, or directed flight around the gulf, depending on their individual fuel (fat) loads [55]. Taken together, these results suggest that individual songbirds alter route selection in response to local conditions and/ or their own physiological condition. Thus, birds are expected to adapt more quickly to spatial changes to sites along their routes, as compared to adjusting their migration timing.

6) Lastly, based on the displacement experiments described above, it is predicted that juvenile birds should be the most susceptible to the negative impacts of change to their migration route. Habitat loss or the impacts of climate change along routes could require re-routing, and evidence suggests that juveniles may not be able to re-route successfully and may thus fail to complete their migration or end up in the wrong place. More experienced adult birds appear to have more flexibility in their routes and stopover site choice, and can respond to the vagaries of conditions they experience en-route. However, just as breeding site fidelity may be promoted in adults that have bred successfully, stopovers that have 'worked' in the past may continue to be selected by adults despite declines in potential resources.

Conservation and Management of Migratory Routes and Stopovers

The knowledge that anthropogenic change will impact migratory routes leads to large-scale management challenges. Can we safely facilitate the movement of billions of songbirds across hundreds to thousands of kilometres in the midst of rapidly growing urban areas, climate change, and other anthropogenically-driven change? Clearly, the adequate protection of migratory routes and stopovers will be one of the biggest challenges for bird conservation in the Anthropocene. What do we need to know to be able to do this effectively?

Independent of new challenges facing migrants with rapid global change, it is expected that migration is a survival bottleneck for many songbird species. While mark-recapture data across seasons are rare, a few studies to date indicate that survival is the lowest during migratory periods. For example, black-throated blue warblers (*Setophaga caerulescens*), journeying between breeding areas in the northeastern United States and overwintering sites in the Caribbean experienced fifteen times greater mortality (totaling 85% of mortality) during spring and fall migration as compared to breeding and overwintering periods [56]. Juvenile songbirds on their first migration have lower survival rates than adults. Further work with additional species over the annual cycle is required to identify where survival is lowest, and thus where species, or specific populations, are most limited. However, this effort must begin with the critical first step of further identifying connections between breeding areas, routes, and nonbreeding sites for different populations and species.

Determining migratory connectivity (i.e., the strength of the connection between breeding and overwintering areas and or stopover areas [57]) can reveal important spatial links and interactions across migration for songbirds. Establishing these connections is a first step for the development of population network models [58–60], that can be used to determine how habitat loss, or other impacts at different points in the annual cycle, may drive population dynamics, and where best to invest in conservation to enable the greatest impact [61]. Yet, there have been a few range-wide connectivity studies to date with songbirds, and even fewer that have used connectivity data to investigate sources of impact on populations. In wood thrushes and golden-winged warblers (*Vermivora chystopera*), variation in habitat loss at the overwintering grounds in Central and South America was associated with patterns of population decline at northern breeding regions [49, 59, 62]. My lab's migratory connectivity research with purple martins revealed that steeply declining and more stable breeding populations had overlapping overwintering ranges [63], and at a finer-scale, even shared night roost locations [64]. This pattern suggests that habitat use during the nonbreeding season is likely not driving varying rates of population decline at breeding sites [34]. Incorporating migration routes and stopovers into analyses of migratory connectivity patterns has been much more rare [46], but would advance our ability to identify survival bottlenecks and prioritize the conservation of migratory habitats.

The ingrained behaviour of migratory songbirds will interact with the success of conservation and management of migratory routes in several ways. First, given the fact that migratory routes and destinations are not socially learned in songbirds and have an important endogenous component, we may not anticipate that songbirds will be as flexible to habitat loss or other changes along their routes as other migrants that have a social component to the formation of their migration behaviour. In flocking migratory waterbirds, socially-learned migration behaviour has been used to create successful, captive re-introduction programs where geese and cranes trained by humans used new routes and overwintering destinations [e.g., 88]. This is not an option for migratory songbirds and it is hard to picture a strategy by which songbirds could be reintroduced to former routes, or 'trained' to use new, safer migratory routes. Second, we do not currently know how repeatable migratory routes are in time and space for more than a handful of species or populations. The answer will be important if conservation and management is targeted to the population or species level. For example, if a steeply declining population consistently uses the same migratory routes and main stopover locations, the conservation of suitable habitat across their flyway may be simpler to implement, as compared to groups that use very different routes and stopover sites in different seasons or years. Currently, we do not have the data to know how most species, and particularly populations, differ in their routes and stopovers between years, thus limiting a conservation strategy for connected routes at the population level.

Conserving full migratory routes for songbirds will require a coordinated effort that spans geopolitical boundaries. Some networks already exist and are active at this scale of partnership, including Partners in Flight (partnersinflight.org) and the Southern Wings program [65]. Demonstrating migratory connectivity can, in some cases justify the leveraging of federal funds from breeding areas to be applied to the protection or restoration of migratory songbird nonbreeding habitat in other countries; similar to what was done for Nearctic-Neotropical species overwintering in the highlands of Nicaragua once it was demonstrated that populations were connected with breeding areas in the northeastern U.S. using mark-recapture data [65].

While breeding to overwintering habitat conservation projects are developing, there are fewer precedents for the protection of connected, en-route habitat for songbirds. However, the conservation of migratory routes for other taxa can serve as models. For example, the successful conservation of migratory waterfowl routes utilized decades of connectivity data derived through mark-recapture (banding) programs and benefited from a diversity of support from government agencies, to duck hunting organizations, to NGOs. This enabled the protection and restoration of wetland migratory corridors across the U.S. and Canada [66]. The benefit here has been demonstrated; waterfowl in the Americas are one of the few migratory groups where there is actually an overall increase in populations [66]. This is not to state that waterfowl species are 'safe' (many are still declining) but it does point to the success of concentrated conservation and management efforts across a broad spatial scale. In another example of broad-scale conservation, the Yellowstone to the Yukon Conservation Initiative aims to conserve and manage connected upland and mountain corridor habitat to facilitate the movement, migration, and dispersal of large mammals and other taxa (y2y.net); demonstrating that a coordinated broad-scale effort can conserve migratory routes that span geo-political boundaries.

To improve conservation along migratory routes, we may also consider making 'islands' of good quality habitat in our urban areas. Like migration barriers of open water and deserts, extensive urban areas with little green space may be, and will increasingly become, like gulfs or deserts of unsuitable habitat that birds must cross during migration. We could work to make safe havens, stepping stones for migration, across urban areas. Already, songbirds typically use highly density urban areas for stopover. For example, even in New York City, one of the most densely populated urban areas in the world, migratory songbirds will typically stopover over in great number in the urban 'oasis' of Central Park. I have even seen boreal migrants such as blackpoll warbler (*Setophaga striata*) using this park as stopover, and as noted above, these sites may serve as suitable refueling sites for migration [53]. There is still much that we can do to make these habitats safer for migratory songbirds, considering the risk of cat predation and window strikes that birds currently face (see other chapters in this volume). However, effort to increase the quality of urban habitat for migratory songbirds is likely to provide benefit.

In summary, the conservation of migratory routes for songbirds has been hampered by our limited understanding of which habitats along routes are the most important for different species or populations, and deciphering where conservation will have the biggest impact on population dynamics. Further research to better elucidate migratory routes to identify bottlenecks and risk, and to inform population dynamic models, is needed. Connections should be mapped at small to large scales to suit a diversity of conservation management approaches. Further, the success of conservation strategies will benefit from an enhanced understanding of the behaviour of how, when, and where migratory songbirds use migratory habitat and where this changes in time and space for different species. While we continue to develop this body of knowledge, effective conservation strategies should currently focus on the conservation of migratory habitats anticipated to be important for refueling or safety, or areas that may be migratory bottlenecks for different species or populations [54]. Efforts to conserve other migratory corridors for waterfowl or alpine mammals may serve as good models for developing similar strategies for migratory songbirds. The future of this area of research will determine if we can help to facilitate the safe passage of migrant populations across the vast distances they travel between breeding and overwintering areas.

Box 2 Migration routes and stopover in the context of anthropogenic change

Background behaviour

- Endogenous programs tell songbirds which direction to fly and for how long
- Experience en route may improve navigation and the ability to correct for wind drift as birds age, with juvenile birds being more at risk of navigational errors
- Most migratory journeys require periods of rest and refueling (stopovers) interspersed with migratory flights. The duration of some 'stopovers' may greatly exceed the time required for refueling, and could be considered alternate periods of residency
- The degree to which populations or individuals use the same stopover sites in subsequent migrations is poorly known

Migration routes and stopover with habitat change

- Temperature impacts the amount of energy birds use during stopover and is therefore expected to be influenced by climate change
- The impact of stopover habitat quality on refueling is likely to have large impacts on migration speed and carry-over effects to arrival timing and fitness at breeding sites
- Migration routes appear to be flexible at the individual level, suggesting birds may be able to adjust to habitat change en route to some degree

Migration routes and stopover; conservation and management

- Determining how breeding, migratory routes, and overwintering sites are connected (migratory connectivity), particularly for the most at-risk populations, will be critical for successful conservation
- Songbirds may not be as flexible to habitat loss or other changes along their routes as for species that learn their migrations socially

---

Box 2 (*Contd...*)

- Conserving full migratory routes for songbirds will require a coordinated effort that spans geopolitical boundaries, but there are no examples for songbirds to date
- Efforts to improve stopover sites in urban areas could provide 'islands' of good quality habitat that support subsequent migration

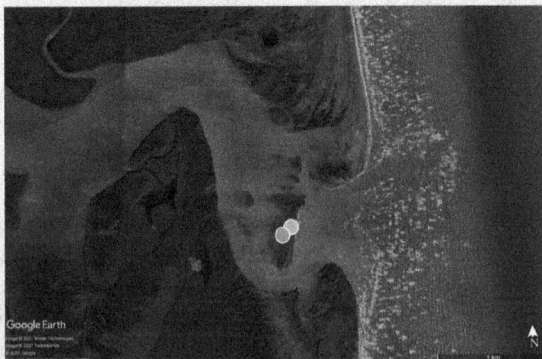

Figure 2 Stopover locations on the north east coast of Nicaragua for two purple martins originating from breeding colonies more than 2000 km apart in Texas and Florida. Whether individuals from the same, or different, populations share stopover sites or regions will be critical to determine to support the conservation of migratory routes for long-distance migrants (adapted from Fraser et al. 2017, Journal of Avian Biology 48:001–007).

Migration in the Context of Anthropogenic Light at Night and the Threat of Collisions with Structures and Windows

Migration Navigation Background

Songbirds have a migration tool-kit that enables an active response to signals and signs along migratory routes (Box 3). In addition to a basic clock and compass, many songbirds use the earth's magnetic fields, the sun and its azimuth, the rotation of earth with stars as an indicator, and direction of polarized light at sunset to cue direction [1]. Experimental evidence shows that birds use cues individually, or in combination, to adjust their routes along the way to displacement, when crossing the equator, or when generally following their innate directional routes [67].

Songbird navigation evolved to make use of long-standing, reliable navigation cues. For example, the stellar navigation system works by using stars as an indicator of the directional rotation of the earth, which enables birds to align on a north-south axis. As the stars turn, birds use their internal clocks to adjust and keep on course [68, 69]. There is some evidence that young birds need a period of exposure to starlight in order to be able to navigate using these cues [70]. Anthropogenic light at night (hereafter ALAN), or other changes in the atmosphere, such as smoke from fires, or smog, may obscure starlight, and thus impact the development of navigational abilities and navigation itself. It can also result in attracting birds into areas where collisions with human-built structures become more likely.

Migration Navigation in the Context of Anthropogenic Light at Night and Built Urban Structures

Most cues that birds use for navigation could be impacted by anthropogenic change to the environment (with the exception of the perception of the Earth's magnetic fields, which is probably unaffected). Light pollution affects birds' ability to use stars or polarized light as a navigational aid. Smoke from landscape-scale fires, or urban smog could limit or obscure a songbird's view of polarized light or stars. Even the natural light of the moon may impact navigation and we may expect ALAN to similarly obscure view of the stars. Birds have a diverse tool kit to be able to navigate by other means when starlight or polarized light are obscured, such as the more natural situation when cloud cover obscures these features, but it is unknown whether an increase in the obscuring of these light features with new anthropogenic-induced change will have a negative impact.

There is ample evidence that birds are drawn to artificial light with fatal consequences. This movement toward light, or 'phototaxis' could be the result of how birds inherently respond to visual cues that aid navigation. Movement toward artificial light, such as lighthouses or industrial gas flares, may correspond to the behaviour of orienting toward light broken through the clouds, which provides a visual cue (sun-azimuth) that birds can use to re-orient [1]. There is a veritable gauntlet of lit areas across songbird migratory bird routes; this is particularly the case for eastern North America and major flyways may overlap broadly with areas where ALAN is concentrated [71]. Some of the most dramatic and large-scale mortalities of migrants have been caused by collisions with illuminated structures. There were approximately 50,000 casualties in one night at a ceilometer (a laser device used to determine the height of clouds) in Georgia [72]. A poorly-timed gas flare during peak fall migration in Saint John New Brunswick, Canada resulted in the mortality of more than 8,000 individuals from at least 26 species, including species at risk [73]. Cruise ships may be an increasing problem, when their lit decks intersect with open-water crossings of nocturnally migrating songbirds [74, 75]. The Tribute in Light monument in New York City commemorating 9–11 victims attracted an estimated 1.1 million birds from their fall migratory routes during just 7 nights [76]. In this case, collisions were largely avoided by on-site observers that turned off the monument lights when numbers of birds 'trapped' in the lights were highest, leading to the subsequent dispersion of birds away from the site, as evidenced by radar. However, this attraction behaviour makes it possible to picture the effect of permanently lit urban structures, where there are no such on-the-ground supports to shut off the lights when too many birds are present. Indeed, lit high-rise structures most commonly associated with downtown urban areas may have the highest 'kill rates' per building of any human-made structure, with estimates of kill rates averaging about 10 birds per year, per high rise building [77]. Although houses, while killing fewer birds per building, may have the biggest impact overall due to a much higher number of houses in the landscape. Impacts with buildings are likely associated with attraction to light first, then eventual collisions as birds fly near them [77]. The threats of lit structures at night when birds are drawn toward them are exacerbated by threats of collision with windows during the day.

Although it has not been quantified, lit structures may draw more birds into stopover areas near windows. Birds may also be selecting suitable stopover habitat in developed areas, which puts them at higher risk with window collisions. The loss of birds to windows has been estimated for various regions. In Canada, it is estimated that 25 million bird mortalities a year are attributed to window collisions; and up to

998 million in the continental U.S. [77, 78]. Most of these collisions occur during migration, when birds are more concentrated along flyways and active at night.

While many nocturnally migrating songbirds may be drawn to light, some other birds may seek to avoid it entirely. Birds that are highly nocturnal and inactive during daylight may be particularly averse to ALAN along migratory routes. For example, many owls and nightjars are inactive during natural daylight and several species migrate great distances between breeding and overwintering areas. Eastern whip-poor-wills (*Antrostomus vociferus*) are inactive during daylight and must forage and migrate during dark hours. Recent tracking confirms that they do not migrate during daylight and may even go around the Gulf of Mexico, rather than across as many migrants routinely do, possibly because a crossing could not be completed during darkness [79]. Highly nocturnal birds like these may be strongly impacted by ALAN encountered along migratory routes. Indeed, eastern whip-poor-will tend to deviate around brightly lit areas during migration and select darker stopover areas as compared to what is available (Korpach et al. *under review*). Similar responses would be expected for other species restricted to the dark but it is not known if any songbird species avoid lit areas during their migrations.

Considering the importance of daylength and light intensity (as described in Migration Timing Background above) on migration timing, light pollution could have an impact on migration time-keeping mechanisms, even for birds that are migrating during the day. Experimentally, birds exposed to longer or more intense light developed breeding readiness early [1]. In a study of urban European blackbirds (*Turdus merula*), even a low level of ALAN was sufficient to advance the timing of breeding and moult by up to a month, as compared to rural conspecifics with darker skies [80]. The inference here was that ALAN influenced the perception of daylength, which cued physiological mechanisms to induce an early start to breeding. Given this, light pollution may impact migration timing in species that rely on photoperiod as a cue to time their migration, which is nearly all songbird species. Purple martins exposed to ALAN at night during overwintering in Brazil advanced spring migration by up to two weeks, compared to individuals that experienced darker skies prior to their spring migration departure [81]. Perceived night length could also alter migration departure decisions on a short timescale [82]. While the effects here would be more subtle, on the scale of hours, it is not known what the long-term consequences for migration might be.

Lastly, given that light intensity (brightness) plays a role along with photoperiod in predicting seasonal timing [11], there is the potential for light pollution to interact with onboard endogenous mechanisms to alter timing. Recent experimental work with great tits (*Parus major*) increased the intensity of light in nest-boxes. Brighter light significantly advanced the timing of daily activities of great tits [83] but it is unknown whether the intensity of ALAN can modulate the timing of migration in songbirds. Light pollution is on the increase, but also, changes in weather have the potential to impact the experience of light intensity. For example, where climate change results in more dry or wet conditions this could impact how light filters through the atmosphere to birds, with an effect on its intensity. Smoke in the atmosphere from forest fires, such as are common in the Boreal forest and have also recently been experienced in the Brazilian Amazon and Australia, could change light intensity. Smog in cities could also change light intensity. While the effects of daytime alterations of light intensity are subtler than ALAN, further research is required before the potential impact of changes in daytime sources of light can be estimated.

Conservation and Management with ALAN and Window/Building Collisions

Many songbird migrants must run a gauntlet of urban development and associated ALAN during their spring and fall migration. This is particularly the case for species that journey across eastern North America and through central Europe during their migrations. While the impacts of site-specific light sources and built structures have received a fair amount of research attention, how ALAN impacts migration timing, migratory routes, and impacts fitness has rarely been studied. Many Nearctic-Neotropical and Palearctic-Afrotropical migrants overwinter in areas with relatively less ALAN as compared to breeding and migration areas; however, with the swift global rise in ALAN, it is expected that overwintering temperate migrants will experience greater light pollution in future.

The handful of research studies to date on the impact of light on timing suggest that cities should be more 'green', and not just in the usual sense of the word. In a study of resident great tits, green lights in nest boxes had a lower impact on diel activity, as compared to white and red lights [83]. However, at higher light intensities green and white light had comparable effects. This may be because of songbird perceptual physiology, where lower wavelength lights (green or blue) are perceived as a lower intensity, even at the same lux level, as red or white lights [83]. We might expect similar impacts of light on the diel activity and timing in migratory species. While studies along these lines are in their early days, the important message is that all lights are not equal in how they impact birds. Ongoing research may reveal which type and intensity are truly the 'greenest' (i.e., as in have the least negative effects on birds and other wildlife).

There are also opportunities to turn off lights, particularly at times when illumination is expected to have the most impact. The Tribute in Light experiment indicates that many birds leave the source of the light and continue on with their migration when it is extinguished [76]. The Fatal Light Awareness Program (flap.org) and other NGOs have advocated for decades for urban buildings, particularly large office buildings in the urban core, to turn off their lights at night to reduce attraction, and ultimately fatal collisions [77]. The bonus of these types of lights-off campaigns is an energy savings that can further contribute to 'greening' urban areas.

The relatively new application of broad-scale radar tracking of migrating birds and 'forecasts' of bird migration (birdcast.info) will increase our ability to strategically manage light. When the peak risk periods for birds moving through cities during migration can be predicted, quick, short-term solutions are possible, such as turning off lights. Similar short-term monitoring and response programs have been used in conservation of non-bird wildlife, such as diverting marine traffic to reduce ship strikes of whales. It may be possible to exert collective action in urban areas to quickly implement lights out programs that would reduce the risk of collisions during peak migration periods. Indeed, some organizations send out social media messages to members during peak migration periods, to alert people to the increased need for mitigative actions. Bird forecasts could be included along with regular weather forecasts, or through social media. Lights-out strategies and window treatments are important at all times of year, as birds can collide with windows at any time, but the risk clearly increases when billions of landbirds are migrating across North America [84]. Further, strategic, short-term interventions such as closing blinds, temporarily adding window treatments, or turning lights out, might garner wider support across communities, organizations, and businesses than more permanent solutions.

Box 3 Migration navigation with anthropogenic light at night and collision risk

Background behaviour

- Songbirds use magnetic fields, the sun, the stars and rotation of the earth, and polarized light at sunset to help guide their migrations
- Navigational cues can be used independently or together
- A period of exposure to these cues in juvenile birds may assist the development of navigation

Migration navigation and anthropogenic impacts

- Light pollution may impact the use of starlight cues for navigation
- Attraction to anthropogenic light can displace millions of birds and can result in mass mortality events
- In urban areas, millions of migratory songbirds are killed each year when they collide with windows
- Anthropogenic light can impact the timing of breeding and migration, through interactions with circannual clocks cued to photoperiod

Migration navigation conservation and management

- Short-term reduction of light could be most effective during migration and enhanced by bird migration forecasting
- Window treatments can reduce collisions; these are a solvable problem with current knowledge
- Research on carry-over effects of anthropogenic light across migration is needed
- Reducing light intensity or using alternative colours of light (wavelengths) may reduce impact but research is ongoing

Figure 3 Many long-distance migrants, like this cedar waxwing (*Bombycilla cedorum*), are primarily threatened by windows during their migrations. New migration forecasting offers opportunities to enact short-term measures to reduce collisions with windows and structures. Temporary window treatments that reduce collisions, such as pictured here at the University of Manitoba, can not only reduce mortality during migration but can also be engaging public activities with important opportunities for education and outreach. Waxwing photo: Paulson des Brisay; window treatment photo: Seema Goel; Lights out alert: Houston Audubon and BirdCast).

Fatal window collisions by birds may actually become a 'curable' problem with the many window treatment solutions available, from low tech do-it-yourself options to commercially sold and/or installed treatments (see flap.org for a list). The main goal of any window treatment designed to prevent bird collisions is to break up the surface reflection or transparency of windows so they appear to birds as the solid objects that they are. There has been a move across many university campuses in North America to alert the public to this issue, to conduct research on the problem, and to investigate solutions. This has included some creative and visually pleasing ways to deal with the issue, including design competitions to create window treatments that are also artworks [85], or as a place to demonstrate window solutions to undergraduates and visiting elementary school groups during programs like Science Rendezvous (an annual Canadian event for elementary-age students across the country; https://www.sciencerendezvous.ca/). Some forward-thinking municipalities, such as Toronto, Canada, have created policies where bird-friendly windows must be a part of new building plans [86]. Clearly, policies such as these should be extended to retrofitting existing structures as well, possibly with companion public incentive programs, such as tax reduction, as has been implemented for those installing new green energy technologies at existing businesses and homes. Considering that most window-strike mortalities may happen at private residences [77] programs that promote bird-friendly windows at home could have a large impact.

Summary and Conclusions

That many migratory species are declining precipitously is startlingly clear. For example, North America alone has lost 3 billion birds since the 1960s, most of these migratory [35]. Clearly there will be winners and losers in the future, as birds must adapt to new conditions, or face possible extinction. Migratory songbirds may be in the most at-risk group, with those travelling the furthest at most risk of all. However, we need to move from documenting the losses to inventing and further implementing solutions to the problem of rapid population declines in migratory songbirds. Finding the best way to develop and incorporate knowledge of migratory bird behaviour into conservation and management will be a key factor in the success or failure of these approaches.

This chapter has summarized some of the main ways in which migratory songbird behaviour interacts with rapid anthropogenic change, often with negative consequences for fitness and resulting in population decline. Migration timing may not be keeping up with the pace of change, resulting in mismatches with resources along migratory routes and at breeding or wintering sites. While there is strong evidence that songbird migration timing is advancing through phenotypic plasticity and/or evolutionary mechanisms, it does not appear to be doing so fast enough, although further research is needed in this area. In the meantime, identifying and focussing on groups most at-risk is a viable conservation and management strategy, and assisted evolution options should be considered. Species that travel the longest distances and breed the furthest north may be the most at-risk with climate change and could be preferentially targeted. Assisted evolution approaches could be applied to addressing the well-documented mismatch between migration timing and ecological timing. There is unlikely to be a one-size fits all strategy that is effective across, or even within species, where further research could help to determine best approaches.

Much work is also needed to determine how populations of most species connect across migratory routes, and how this may vary over time, to enable the most efficient protection of migratory stopover habitat. The development and engagement of broad, international efforts to conserve connected flyways or corridors of needed habitat, as has been developed for migratory waterfowl, is needed for songbirds. There is currently no protected migratory flyway that targets any migratory songbird species or population. The success of any such strategy would need to be based on the development of connectivity data, as well as the incorporation of the best available knowledge on the repeatability of stopover use by songbird populations and consideration of the behaviour of young birds developing migration behaviour.

Migration behaviour, anthropogenic light, and our human-built environments combine with fatal consequences for songbirds. Many aspects of these negative interactions seem solvable with existing knowledge. For example, even short-term reductions in anthropogenic light during peak migration through lights-out programs could be very effective in reducing impacts on migration and window kills. This could be based on new migration forecasting to make it most effective and to limit the impact on human activity as much as possible. It is also well known that window treatments reduce collisions and lots of low-cost solutions and products are available. The challenge here may be more about public education and will than a lack of knowledge as to how to address the issues. How anthropogenic light may interact with bird timing mechanisms or influence their migratory routes as a whole requires further investigation but available data suggests ALAN does impact migration however, the colour and intensity of light could potentially be altered to reduce impacts.

Facilitating the safe passage of billions of birds across hundreds to thousands of kilometres is a major challenge in the Anthropocene. By advancing our understanding of how songbird migration works and implementing this knowledge into conservation and management, instead of behaviour interacting negatively with anthropogenic change, we will be able to work better with it to solve the problem of migratory bird declines.

▨ LITERATURE CITED

[1] Newton, I. 2008. The Migration Ecology of Birds. Oxford, UK, Elsevier.

[2] Gwinner, E. 1989. Photoperiod as a modifying and limiting factor in the expression of avian circannual rhythms. J. Biol. Rhythms. 4(2): 237–250.

[3] Gwinner, E. 1996. Circadian and circannual programmes in avian migration. J. Exp. Biol. 199(Pt 1): 39–48.

[4] Gwinner, E. and A. Scheuerlein. 1999. Photoperiodic responsiveness of equatorial and temperate-zone stonechats. The Condor. 101: 347–359.

[5] Knudsen, E., A. Linden, C. Both, N. Jonzen, F. Pulido, N. Saino, et al. 2011. Challenging claims in the study of migratory birds and climate change. Biol. Rev. Camb. Philos. Soc. 86(4): 928–946.

[6] Bojarinova, J. and O. Babushkina. 2016. Photoperiodic conditions affect the level of locomotory activity during autumn. Migration in the Long-tailed Tit *(Aegithalos c. caudatus)*. The Auk. 132(2): 370–379.

[7] Helm, B. and E. Gwinner. 2005. Carry-over effects of day length during spring migration. Journal of Ornithology. 146: 348–354.

[8] Fraser, K.C., C. Silverio, P. Kramer, N. Mickle, R. Aeppli and B.J.M. Stutchbury. 2013. A trans-hemispheric migratory songbird does not advance spring schedules or increase migration rate in response to record-setting temperatures at breeding sites. PLoS One. 8(5): e64587.

[9] Schmaljohann, H. and C. Both. 2017. The limits of modifying migration speed to adjust to climate change. Nature Climate Change. 7: 573–577.

[10] Shave, A., C.J. Garroway, J. Siegrist and K.C. Fraser. 2019. Timing to temperature: Egg-laying dates respond to temperature and are under stronger selection at northern latitudes. Ecosphere. 10(12): e02974.

[11] Gwinner, E. and B. Helm. 2003. Circannual and circadian contributions to the timing of avian migration. pp. 81–95. *In*: A. Migration, P. Berthold, E. Gwinner and E. Sonnenschein (eds). Avian Migration. Springer, Berlin Heidelberg.

[12] Studds, C.E. and P.P. Marra. 2011. Rainfall-induced changes in food availability modify the spring departure programme of a migratory bird. Proc. Biol. Sci. 278(1723): 3437–3443.

[13] McKinnon, E.A., C.Q. Stanley and B.J.M. Stutchbury. 2015. Carry-over effects of nonbreeding habitat on start-to-finish spring migration performance of a songbird. PLoS One, 10(11): e0141580.

[14] Both, C. 2010. Flexibility of timing of avian migration to climate change masked by environmental constraints en route. Curr. Biol. 20(3): 243–248.

[15] Coppack, T., F. Pulido and P. Berthold. 2001. Photoperiodic response to early hatching in a migratory bird species. Oecologia, 128(2): 181–186.

[16] Helm, B., B.M. Van Doren, D. Hoffmann and U. Hoffmann. 2019. Evolutionary response to climate change in migratory pied flycatchers. Curr. Biol. 29: 3714–3719.

[17] Visser, M.E. 2019. Evolution: adapting to a warming world. Curr. Biol. 29(22): R1189–R1191.

[18] Visser, M.E. and P. Gienapp. 2019. Evolutionary and demographic consequences of phenological mismatches. Nat. Ecol. Evol. 3(6): 879–885.

[19] Hurlbert, A.H. and Z. Liang. 2012. Spatiotemporal variation in avian migration phenology: citizen science reveals effects of climate change. PLoS One. 7(2): e31662.

[20] Rubolini, D., A.P. Møller, K. Rainio and E. Lehikoinen. 2007. Intraspecific consistency and geographic variability in temporal trends of spring migration phenology among European bird species. Clim. Res. 35: 135–146.

[21] Charmantier, A. and P. Gienapp. 2014. Climate change and timing of avian breeding and migration: evolutionary versus plastic changes. Evol. Appl. 7(1): 15–28.

[22] Stanley, C.Q., M. MacPherson, K.C. Fraser, E.A. McKinnon and B.J.M. Stutchbury. 2012. Repeat tracking of individual songbirds reveals consistent migration timing but flexibility in route. PLoS One. 7(7): e40688.

[23] Mayor, S.J., R.P. Guralnick, M.W. Tingley, J. Otegui, J.C. Withey, S.C. Elmendorf, et al. 2017. Increasing phenological asynchrony between spring green-up and arrival of migratory birds. Sci. Rep. 7(1): 1902.

[24] Both, C., S. Bouwhuis, C.M. Lessells and M.E. Visser. 2006. Climate change and population declines in a long-distance migratory bird. Nature. 441(7089): 81–83.

[25] Both, C., C.A.M. Van Turnhout, R.G. Bijlsma, H. Siepel, A.J. Van Strien and R.P.B. Foppen. 2010. Avian population consequences of climate change are most severe for long-distance migrants in seasonal habitats. Proc. Biol. Sci. 277(1685): 1259–1266.

[26] Brown, C.R. and M.B. Brown. 2000. Weather-mediated natural selection on arrival time in cliff swallows. Behav. Ecol. Sociobiol. 47(5): 339–345.

[27] Fraser, K.C., A. Shave, E. de Greef, J. Siegrist and C.J. Garroway. 2019. Individual variability in migration timing can explain long-term, population-level advances in a songbird. Front. Ecol. Evol. 6.

[28] Neelin, J.D., M. Münnich, H. Su, J.E. Meyerson and C.E. Holloway. 2006. Tropical drying trends in global warming models and observations. Proc. Natl. Acad. Sci. USA. 103(16): 6110–6115.

[29] Norris, D.R., P.P. Marra, T.K. Kyser, T.W. Sherry and L.M. Ratcliffe. 2004. Tropical winter habitat limits reproductive success on the temperate breeding grounds in a migratory bird. Proc. Biol. Sci. 271(1534): 59–64.

[30] Kokko, H. 1999. Competition for early arrival in migratory birds. J. Anim. Ecol. 68(5): 940–950.

[31] Seewagen, C.L., Y. Ma, Y.E. Morbey and C.G. Guglielmo. 2019. Stopover departure behavior and flight orientation of spring-migrant Yellow-rumped Warblers (*Setophaga coronata*) experimentally exposed to methylmercury. J. Ornithol. 160: 617–624.

[32] Eng, M.L., B.J.M. Stutchbury and C.A. Morrissey. 2017. Imidacloprid and chlorpyrifos insecticides impair migratory ability in a seed-eating songbird. Sci. Rep. 7(1): 15176.

[33] Both, C. and M.E. Visser. 2001. Adjustment to climate change is constrained by arrival date in a long-distance migrant bird. Nature. 411(6835): 296–298.

[34] Nebel, S., A. Mills, J.D. McCracken and P.D. Taylor. 2010. Declines of aerial insectivores in north america follow a geographic gradient. Avian Conserv. Ecol. 5(2).

[35] Rosenberg, K.V., A.M. Dokter, P.J. Blancher, J.R. Sauer, A.C. Smith, P.A. Smith, et al. 2019. Decline of the North American avifauna. Science. 366(6461): 120–124.

[36] Stewart, R., K.A. Bredin, A.R. Couturier, A.G. Horn, D. Lepage, S. Makepeace, et al. 2015. Second Atlas of Breeding Birds of the Maritime Provinces. Sackville, New Brunswick, Canada: Bird Studies Canada, Environment Canada, Natural History Society of Prince Edward Island, Nature New Brunswick, New Brunswick Department of Natural Resources, Nova Scotia Bird Society, Nova Scotia Department of Natural Resources, Prince Edward Island Department of Agriculture and Forestry.

[37] van Oppen, M.J., J.K. Oliver, H.M. Putnam and R.D. Gates. 2015. Building coral reef resilience through assisted evolution. Proc. Natl. Acad. Sci. USA. 112(8): 2307–2313.

[38] Gwinner, E. and W. Wiltschko. 1978. Endogenously controlled changes in migratory direction of the garden warbler, *Sylvia borin*. J. Comp. Physiol. 125: 267–273.

[39] Perdeck, A.C. 1958. Two types of orientation in migrating starlings, *Sturnus vulgaris* L. and *Chaffinches, Fringilla coelebs* L., as Revealed by Displacement Experiments. Ardea. 55: 1–37.

[40] Thorup, K., I.-A. Bisson, M.S. Bowlin, R.A. Holland, J.C. Wingfield, M. Ramenofsky, et al. 2007. Evidence for a navigational map stretching across the continental U.S. in a migratory songbird. Proc. Natl. Acad. Sci. USA. 104(46): 18115–18119.

[41] Thorup, K. and J. Rabol. 2007. Compensatory behaviour following displacement in migratory birds. A meta-analysis of cage-experiments. Behav. Ecol. Sociobiol. 61: 825–841.

[42] Thorup, K., M.L. Vega, K.R.S. Snell, R. Lubkovskaia, M. Willemoes, S. Sjöberg, et al. 2020. Flying on their own wings: young and adult cuckoos respond similarly to long-distance displacement during migration. Sci. Rep. 10(1): 7698.

[43] Hedenstrom, A. and T. Alerstam. 1998. How fast can birds migrate? J. Avian Biol. 29(4): 424–432.

[44] Nilsson, C., R.H. Klaassen and T. Alerstam. 2013. Differences in speed and duration of bird migration between spring and autumn. Am. Nat. 181(6): 837–845.

[45] McKinnon, E.A., K.C. Fraser and B.J.M. Stutchbury. 2013. New discoveries in landbird migration using geolocators, and a flight plan for the future. The Auk. 130(2): 211–222.

[46] McKinnon, E.A. and O.P. Love. 2018. Ten years tracking the migrations of small landbirds: lessons learned in the golden age of bio-logging. The Auk: Ornithol. Adv. 135: 834–856.

[47] van Loon, A., J.D. Ray, A. Savage, J. Mejeur, L. Moscar, M. Pearson, et al. 2017. Migratory stopover timing is predicted by breeding latitude, not habitat quality, in a long-distance migratory songbird. J. Ornithol. 158: 745–752.

[48] Leu, M. and C.W. Thompson. 2002. The potential importance of migratory stopover sites as flight-feather molt staging areas: a review for neotropical migrants. Biol. Conserv. 106: 45–56.

[49] Stanley, C.Q., E.A. McKinnon, K.C. Fraser, M.P Macpherson, G. Casbourn, L. Friesen, et al. 2015. Connectivity of wood thrush breeding, wintering, and migration sites based on range-wide tracking. Conserv. Biol. 29(1): 164–174.

[50] Roberto-Charron, A., J. Kennedy, L. Reitsma, J.A. Tremblay, R. Krikun, K.A. Hobson, et al. 2020. Widely distributed breeding populations of Canada warbler (*Cardellina canadensis*) converge on migration through Central America. BMC Zoology. 5. Article number 10.

[51] Wikelski, M., E.M Tarlow, A. Raim, R.H. Diehl, R.P. Larkin and G.H. Visser. 2003. Avian metabolism: costs of migration in free-flying songbirds. Nature. 423(6941): 704.

[52] Gomez, C., N.J. Bayly, D.R. Norris, S.A. Mackenzie, K.V. Rosenberg, P.D. Taylor, et al. 2017. Fuel loads acquired at a stopover site influence the pace of intercontinental migration in a boreal songbird. Sci. Rep. 7(1): 3405.

[53] Seewagen, C.L., C.D. Sheppard, E.J. Slayton and C.G. Guglielmo. 2011. Plasma metabolites and mass changes of migratory landbirds indicate adequate stopover refueling in a heavily urbanized landscape. The Condor. 113(2): 284–297.

[54] Rodewald, P.G. and K.V. Rosenberg. 2018. Bottlenecks, refueling stations, and fire escapes: 3 types of stopover sites migrants really need. *In*: Living Bird. Cornell University, Ithaca, NY. https://www.allaboutbirds.org/news/bottlenecks-refueling-stations-and-fire-escapes-3-types-of-stopover-sites-migrants-really-need/

[55] Sandberg, R. and F.R. Moore. 1996. Migratory orientation of red-eyed vireos, *Vireo olivaceus* in relation to energetic condition and ecological context. Behav. Ecol. Sociobiol. 39: 1–10.

[56] Sillett, T.S. and R.T. Holmes. 2002. Variation in survivorship of a migratory songbird throughout its annual cycle. J. Anim. Ecol. 71(2): 296–308.

[57] Webster, M.S., et al. 2002. Links between worlds: unravelling migratory connectivity. Trends Ecol. Evol. 17: 76–83.

[58] Taylor, C.M. and D.R. Norris. 2010. Population dynamics in migratory networks. Theor. Ecol. 3: 65–73.

[59] Taylor, C.M. and B.J.M. Stutchbury. 2015. Effects of breeding versus winter habitat loss and fragmentation on the population dynamics of a migratory songbird. Ecol. Appl. 26(2): 424–437.

[60] Taylor, C.M. 2019. Effects of natal dispersal and density-dependence on connectivity patterns and population dynamics in a migratory network. Front. Ecol. Evol. 7: 354.

[61] Sheehy, J., C.M. Taylor and D.R. Norris. 2011. The importance of stopover habitat for developing effective conservation strategies for migratory animals. J. Ornithol. 152: 161–168.

[62] Kramer, G.R., D.E. Andersen, D.A. Buehler, P.B. Wood, S.M. Peterson, J.A. Lehman, et al. 2018. Population trends in *Vermivora warblers* are linked to strong migratory connectivity. Proc. Natl. Acad. Sci. USA. 115(14): E3192–E3200.

[63] Fraser, K.C., B.J.M. Stutchbury, C. Silverio, P.M. Kramer, J. Barrow, D. Newstead, et al. 2012. Continent-wide tracking to determine migratory connectivity and tropical habitat associations of a declining aerial insectivore. Proc. Biol. Sci. 279(1749): 4901–4906.

[64] Fraser, K.C., A. Shave, A. Savage, A. Ritchie, K. Bell, J. Siegrist, et al. 2017. Determining fine-scale migratory connectivity and habitat selection for a migratory songbird by using new GPS technology. J. Avian Biol. 48(3): 339–345.

[65] Agencies, T.A.o.F.a.W. Southern Wings. 2019. Available from: https://www.fishwildlife.org/afwa-inspires/southern-wings.

[66] Canada, N.A.B.C.I. 2019. The State of Canada's Birds. Environment and Climate Change Canada, Ottawa, Canada. p. 12.

[67] Akesson, S. and G. Bianco. 2017. Route simulations, compass mechanisms and long-distance migration flights in birds. J. Comp. Physiol. A Neuroethol. Sens. Neural Behav. Physiol. 203(6–7): 475–490.

[68] Emlen, S.T. 1967. Migratory orientation in the Indigo Bunting, *Passerina cyanea*, Part I: Evidence for use of celestial cues. The Auk. 84(3): 309–342.

[69] Emlen, S.T. 1967. Migratory orientation in the Indigo Bunting, *Passerina cyanea*. Part II: Mechanism of celestial orientation. The Auk. 84(4): 463–489.

[70] Mouritsen, H. and O.N. Larsen. 2001. Migrating songbirds tested in computer-controlled Emlen funnels use stellar cues for a time-independent compass. J. Exp. Biol. 204(Pt 22): 3855–3865.

[71] Falchi, F., P. Cinzano1, D. Duriscoe, C.C.M. Kyba, C.D. Elvidge, K. Baugh, et al., 2016. The new world atlas of artificial night sky brightness. Sci. Adv. 2(6): e1600377.

[72] Johnston, D.W. and T.P. Haines. 1957. Analysis of mass bird mortality in october, 1954. The Auk. 74(4): 447–458.

[73] Whittam, B., D.F. McAlpine, P.-Y. Daoust, B. Rothfuss and P. Thomas. 2016. Over 8, 000 Songbirds Die at a Gas Flare on the Bay of Fundy coast, in NAOC 2016. Washington, D.C.

[74] Bocetti, C.I. 2011. Cruise ships as a source of avian mortality during fall migration. The Wilson Journal of Ornithology. 123(1): 176–178.

[75] McGlashen, A. 2020. Grisly report raises questions about the cruise industry's impact on migrating birds. *In*: Audubon Magazine. https://www.audubon.org/news/grisly-report-raises-questions-about-cruise-industrys-impact-migrating-birds

[76] Van Doren, B.M., K.G. Horton, A.M. Dokter, H. Klinck, S.B. Elbin and A. Farnsworth. 2017. High-intensity urban light installation dramatically alters nocturnal bird migration. Proc. Natl. Acad. Sci. USA. 114(42): 11175–11180.

[77] Machtans, C.S., C.H.R. Wedeles and E.M. Bayne. 2012. A first estimate for canada of the number of birds killed by colliding with building windows. Avian Conservation and Ecology. 8(2): 6.

[78] Loss, S.R., T. Will, S.S. Loss and P.P. Marra. 2014. Bird–building collisions in the United States: Estimates of annual mortality and species vulnerability. The Condor. 116(1): 8–23.

[79] Korpach, A.M., A. Mills, C. Heidenreich, C.M. Davy and K.C. Fraser. 2019. Blinded by the light? Circadian partitioning of migratory flights in a nigthjar species. J. Ornithol. 160: 835–840.

[80] Dominoni, D., M. Quetting and J. Partecke 2013. Artificial light at night advances avian reproductive physiology. Proceedings of the Royal Society B-Biological Sciences. 280: 20123017.

[81] Smith, R., M. Gagné and K.C. Fraser. 2021. Pre-migration artificial light at night advances the spring migration timing of a trans-hemispheric migratory songbird. Environ. Pollut. 269: 116136.

[82] Muller, F., G. Ruppel and H. Schmaljohann. 2018. Does the length of night affect the timing of nocturnal departures in a migratory songbird? Anim. Behav. 141: 183–194.

[83] de Jong, M., S.P. Caro, P. Gienapp, K. Spoelstra and M.E. Visser. 2017. Early birds by light at night: effects of light color and intensity on daily activity patterns in blue tits. J. Biol. Rhythms. 32(4): 323–333.

[84] Dokter, A.M., A. Farnsworth, D. Fink, V. Ruiz-Gutierrez, W.M. Hochachka, F.A. La Sorte, et al. 2018. Seasonal abundance and survival of North America's migratory avifauna determined by weather radar. Nat. Ecol. Evol. 2(10): 1603–1609.

[85] Ryan, D. 2019. Bird-friendly windows reduce collision deaths at UBC. *In*: Vancouver Sun. https://vancouversun.com/news/local-news/bird-friendly-windows-reduce-collision-deaths-at-ubc

[86] Toronto, C.o. 2020. Bird-friendly Guidelines. Toronto. https://www.toronto.ca/city-government/planning-development/official-plan-guidelines/design-guidelines/bird-friendly-guidelines/

[87] Kokko, H. 1999. Competition for early arrival in migratory birds. J. Anim. Ecol. 68(5): 940–950. JSTOR, www.jstor.org/stable/2647239.

[88] Urbanek, R.P., E.K. Szyszkoski and S.E. Zimorski. 2014. Winter distribution dynamics and implications to a reintroduced population of migratory whooping cranes. J. Fish Wildl. Manage. 5(2): 340–362. https://doi.org/10.3996/092012-JFWM-088

Chapter 4

Conspecific and Heterospecific Interactions

Michael P. Ward[1], Valerie L. Buxton[1],
Janice K. Enos[1] and Jinelle H. Sperry[2]

Introduction

Interactions among individuals make up the critical glue holding ecological communities together. Songbird communities in particular are of longstanding interest to ecologists and conservation biologists alike, with entire textbooks dedicated to the subject of interactions among songbird species [1, 2, 3]. Likewise, foundational topics in community ecology, such as community assembly and "ghosts of competition past" [4] are frequently exemplified with classic studies in avian ecology [5]. Studies of songbird communities

[1]University of Illinois, Urbana-Champaign.
[2]Engineer Research and Development Center, United States Army.

demonstrate that at the core of all interactions, both within [conspecific) and between (heterospecific) species, are individual decisions, thus embedding the topic of songbird interactions deeply into animal behavior. That is, the fitness costs and benefits accrued by individuals are what shape interactions in communities.

This chapter provides a broad overview of the many types of interactions songbirds experience, bringing attention to how the fitness benefits of individual decisions influence both conspecific and heterospecific interactions. Interactions are categorized by their "direction" in community ecology: *positive interactions*, such as mutualism and commensalism (addressed in Section I), and *negative interactions*, such as predation, competition and parasitism (addressed in Section II). In both categories, the benefits of information exchange between individuals (both conspecific and heterospecific) can shape and hold communities together. This information exchange can occur directly between senders and receivers through communication networks [6]. Or, information exchange can occur indirectly through social information use, where songbirds make decisions based on observations of other individuals interacting with their environment [7]. Anthropogenic changes can disrupt information exchange in songbird communities, which in turn impacts the broader ecological community (addressed in Section III). The chapter concludes with case studies demonstrating how species interactions, both conspecific and heterospecific, can be manipulated in the context of conservation and management.

Section I: **Positive Conspceific and Heterospecific Interactions**

Many songbird communities exist because of fitness benefits associated with interactions between individuals in space and time. These interactions ultimately occur as [1] mutualism, whereby both individuals receive a net fitness gain by interacting with one another, or [2] commensalism, with one individual accruing benefits and the other netting a zero sum. For songbirds, social information use is a common mechanism promoting these beneficial species interactions. Social information use is well-documented among many songbird species [7–9]. Songbirds are continually gathering information from others to make decisions with important fitness implications. When there is a net benefit to making decisions based on information collected from conspecifics or heterospecifics, mutualistic or commensalistic relationships are expected to follow. Social information does not require direct interaction among individuals, as even the mere presence of others can provide valuable information (see case studies).

Social Information Promoting Habitat Selection

Many songbirds use social information to find and select habitat, leading to a phenomenon known as conspecific attraction [8]. Conspecific attraction is the preferential aggregation of individuals of the same species in a particular location [10]. Questions related to what types of information are used, when information is obtained and used, who uses information, and the fitness consequences of using information have been productive areas of research for nearly two decades. Here we review each in the context of habitat selection.

Social information may be derived from the presence of an individual (referred to as location cues) or from the performance of an individual (referred to as performance-based cues; [7, 11]). In songbirds, location cues generally take the form of conspecific song, while performance-based cues go beyond the simple presence of a conspecific and are often more closely correlated with fitness (e.g., clutch size, number of nestlings, fledgling calls; [11]). Performance-based cues provide more detailed information on habitat quality, while location cues may only indicate habitat presence. Songbirds will use both types of cues depending on when the cue is available. For example, performance-based cues may not be available to migrants arriving to breeding areas if breeding has not yet initiated. When performance-based cues become available, individuals collecting information on habitat quality (termed prospectors) often use these cues to make settlement decisions in the current or subsequent year. Veeries (*Catharus fuscescens* Stephens 1817), for example, preferentially settle in areas that contained fledgling calls in the previous year [12]. Both male and female yellow-headed blackbirds (*Xanthocephalus xanthocephalus* Bonaparte 1826), prospect and in the subsequent year move to sites that produced more young per nest than the site in which they bred [13]. Prospecting collared flycatchers (*Ficedula albicollis* Temminck 1815) visit more successful nests and, in the following year, settle at sites close to where they prospected [14]. Further, collared flycatchers will preferentially settle in nests boxes that contained a conspecific in the previous year [15]. In some species, males switch to singing a certain type of song after pairing, which may also be used as a correlate of performance. Yellow warblers (*Setophaga petechial* Linnaeus 1776), for example, are more abundant in areas where males are singing a paired song [16].

Location cues, on the other hand, are generally available pre-settlement and throughout the breeding period. Multiple studies have experimentally broadcast conspecific song of species in the pre-settlement period and documented positive responses (e.g., higher territory density, higher occupancy rates) in plots with conspecific cues compared to silent control plots. This has been documented in a wide range of species including golden-cheeked warblers (*Setophaga chrysoparia* Sclater and Salvin 1861; [17]), Cape Sable seaside sparrows (*Ammodramus maritimus mirabilis* Howell 1919; [18]), black-throated blue warblers (*Setophaga caerulescens* Gmelin 1789; [19]), black-capped vireos (*Vireo atricapilla* Woodhouse 1852; [20]), least flycatchers (*Empidonax minimus* Baird 1843; [21]), Kirtland's warblers (*Setophaga kirtlandii* Baird 1852; [22]) and American redstarts (*Setophaga ruticilla* Linnaeus 1758; [21]).

In addition to conspecific attraction, individuals may use information from heterospecifics to find and select habitat. Heterospecific attraction tends to be more common when species are ecologically similar. In Minnesota, for example, migrant arboreal insectivores were attracted to areas with higher densities of resident parids (Family Paridae) in the same foraging guild, likely because the presence of parids contained information on food resources related to habitat quality [23]. In Finland, migrant songbird densities increased with experimentally increased resident parid densities, similarly indicating that migrants use residents to indicate high-quality habitat [24]. Migrants may also use cues of other migrants when selecting habitat if one species arrives before the other. Long-distance migrating wood warblers (*Phylloscopus sibilatrix* Bechstein 1793) settled more quickly and more often on experimental plots where short-distance migrating chiffchaff (*Phylloscopus collybita* Vieillot 1817) cues had been broadcast [25]. In addition to location cues, heterospecifics will also base settlement decisions on performance-based cues. Migrant European pied flycatchers (*Ficedula hypoleuca* Pallus 1764), for example, are more likely to copy nest site preferences of resident blue tits (*Cyanistes caeruleus* Linnaeus 1758) as the number of offspring in a tit nest increases [26].

While there are many examples of social information resulting in conspecific and heterospecific attraction, the fitness benefits of using this strategy are surprisingly less clear. For individuals using conspecific cues, we should expect to see some benefit in terms of higher survival or increased reproduction. Few studies have quantified these benefits, but for those that have, the results collectively are equivocal [27]. It is possible that only certain individuals may benefit from using social information to select habitat. For example, juveniles appear to use social information more frequently than adults, and failed breeders more frequently than successful breeders [8].

Reduced Predation and Parasitism Risk: Benefits of Having Neighbors

Perhaps the most important benefit to interacting with others is access to information about shared threats. A well-studied behavior in songbirds is alarm calling, where one or many individuals vocalize with call notes in the presence of a threat [28, 29]. This communication system is well-studied in relation to nest predation risk, as nest predation is the leading cause of reproductive failure for many songbird species [30]. For example, nest survival improves when at least one parent alarm calls in the presence of a nest predator [31]. Increased nest survival is likely due to changes in nestling behavior in response to alarm calls, such as reduced begging or movements that would otherwise make the nest more conspicuous (reviewed in [31, 32]).

Sharing information about threats extends to neighboring conspecifics, as well. For colonially-nesting songbirds, such as swallows (Family Hirundinidae), alarm calling elicits a "mobbing" response, where members of the group collectively attack and harass the shared threat [33–35]. Mobbing in response to alarm calls is a highly effective nest defense system. For many colonial songbirds, per capita nest survival increases with colony size [36]. In some territorial species, such as European pied flycatchers [37] and Eurasian reed warblers (*Acrocephalus scirpaceus* Hermann 1804; [38]) increased alarm calling rates recruit even more conspecifics to mob, further enhancing nest defense and reproductive success for the group. Within conspecific flocks, the physical location of an individual relative to others can also provide a benefit. Oftentimes dominance hierarchies dictate this relationship: in red-winged blackbirds (*Agelaius phoeniceus* Linnaeus 1766) older males occupy more central locations in a roost, which provides a predation risk benefit [39]. Young males still receive a benefit, though less than older males, by roosting together.

Information about threats is not limited to communication with conspecific receivers. Unintended receivers, called "eavesdroppers," can usurp information from alarm calls and stay informed about shared threats [6]. Heterospecific eavesdropping, where different species listen in on each other's alarm calls, is fairly common [29]. For example, young superb fairy-wrens (*Malarus cyaneus* Ellis 1782), with little to no experience with predators, will reduce their begging when heterospecific brown thornbills (*Acanthiza pusilla* Shaw, 1790) alarm call near their nests [40]. Even non-avian community members pay attention to songbird alarm calls: eastern chipmunks (*Tamias striatus* Linnaeus 1758) stay more vigilant while foraging when tufted titmice (*Baelophus bicolor* Linnaeus, 1766) are alarm calling [41]. Heterospecific groups benefit from alarm calling, as mortality risk is greatly reduced for all when there is information available about predator presence [29]. This is certainly the case for mixed species foraging flocks, where members often share predators due to similarity in niche and body size. Here, alarm calls are specifically referred to as "mobbing calls," which recruit members of the mixed species flock to drive off a shared predator [42].

Interestingly, a relatively small number of species tend to provide social information about predator presence. In the temperate zone during the boreal winter, species from the family Paridae typically elicit the mobbing calls in mixed foraging species flocks [43–46]. For example, in North America tufted titmice signal to black-capped chickadees (*Poecile atricapillus* Linnaeus 1766) and red-breasted nuthatches (*Sitta canadensis* Linnaeus 1766) when predators of adults are near [47]. Carolina chickadees (*Poecile carolinensis* Audubon 1834) also respond to mobbing calls produced by tufted titmice. Notably, playback experiments demonstrate the great tit (*Parus major* Linnaeus, 1758) of Europe also responds to mobbing calls of black-capped chickadees of North America [48]. Mobbing call structure could be a phylogenetically conserved trait in the family Paridae, which could be why its members often find themselves playing the role of "sentinels."

Heterospecific responses to alarm calls are also well-documented in the Neotropics. In the Amazon, many species occur in flocks that tend to form around just 1–2 species from the family Thamnophilidae (antbirds and antshrikes). Recent studies indicate that reduced predation risk, rather than increased foraging success, is the primary benefit to flocking with antshrikes (genus *Thamnomanes*) [49–51]. When antshrike sentinels were removed from their mixed-species flocks, members spent more time in vegetation cover, and species occurrence in flocks decreased [52]. This would indicate that, not only does heterospecific social information use affect community composition, it also contributes to the realized niche of some songbird species.

Last, information sharing about brood parasitism risk can provide a benefit to those that use the information. Many songbirds alarm call when brood parasites are present at nests [53–56]. Some, such as superb fairy-wrens [57] and *Acrocephalus* genus warblers [38, 58], use this social information about brood parasitism risk from neighboring pairs to improve their own nest defense. But only the yellow warbler signals the presence of its brood parasite, the brown-headed cowbird (*Molothrus ater* Boddaert, 1783), with a unique referential alarm call (called "seet calls;" [59, 60]). Instead of mobbing, the female returns to her nest in response to seet calls and "sits tightly," therefore reducing the risk of receiving a brood parasitic egg [59, 60]. At least one heterospecific, the red-winged blackbird, eavesdrops on yellow warbler seet calls and becomes more defensive at their own nests in response [61]. Moreover, red-winged blackbirds nesting close to yellow warblers respond more aggressively to seet calls compared to those nesting further away, suggesting a "neighborhood watch" effect between these two species [61].

Improved Foraging: Mutualistic Effects of Information Sharing

Species interactions that occur while foraging have long been of interest to behavioral ecologists, particularly the mutual benefits that are potentially realized for each species. A classic example of mutualism occurs between oxpeckers (Family Buphagidae) and large African mammals, where oxpeckers forage on ectoparasites from large mammals, and in turn mammals benefit from parasite removal [62]. Although often referenced, this type of heterospecific interaction is rare among songbirds as only two species have been documented exhibiting this behavior (both genus *Buphagus*; [62]). There is also likely little social information passed between the large mammals and oxpeckers. More commonly, the benefit of foraging with others (conspecific or heterospecific) takes one of two forms. First, group foraging reduces mortality risk by means of dilution effects [63] or having "many eyes" that improves predator detection [64]. Second, groups have access to more information about food resources that can improve foraging efficiency for all members [65].

This second benefit is rooted in the Optimal Foraging Theory, which uses mathematical models to predict foraging decisions of individuals based on various inputs of information [66]. This fundamental topic in animal behavior often considers personal information, such as time spent in a foraging patch and diminishing returns [67]. There is growing evidence, however, that individuals use social information to make optimal foraging decisions [68–70]. Theoretically, individuals that use social information can estimate the quality of a foraging patch more quickly and accurately compared to a solitary forager with no information (referred to as "public information" in the literature, [71]). Laboratory experiments on conspecific flocks support these foraging benefits for many species, including house sparrows (*Passer domesticus* Linnaeus 1758; [72]), house finches (*Haemorhous mexicanus* Müller 1776; [73]), European starlings (*Sturnus vulgaris* Linnaeus 1758; [74]), and red crossbills (*Loxia curvirostra* Linnaeus 1758; [75]). Improved foraging via social information use occurs in wild songbird populations, as well. For example, field studies on several tit species (Family Paridae; [76–78]) indicate that individuals assess group members' foraging behaviors to decide when and where to forage.

Heterospecifics can also benefit from foraging information. A major benefit to foraging in mixed species flocks is access to information about food availability. Foraging with other species provides two key benefits over foraging with conspecifics. First, even slight niche partitioning among ecologically similar species will greatly reduce competition for overlapping food resources [79]. Second, foraging in mixed species flocks enhances decision-making, as each species brings a different "skill set" that diversifies foraging strategies [80–82]. Studies at feeders in Europe demonstrate that coal tits (*Periparus ater* Linnaeus 1758), blue tits, marsh tits (*Poecile palustris* Linnaeus 1758), great tits, and Eurasian nuthatches (*Sitta europaea* Linnaeus 1758) all use heterospecific social information to make foraging patch (i.e., feeder) decisions [69, 76, 83]. Recent playback experiments suggest that tit species recruit one another to high quality and/or novel foraging patches by vocalizing at feeders [84]. Such "recruitment" calls have been observed in other mixed species flocks [85] as well as conspecific flocks [86]. In these cases, it seems that food availability is actively communicated to other group members (conspecific and/or heterospecific) although the benefit of sharing such information likely diminishes with group size and depletion of the foraging patch [84].

Collecting social information about foraging locations has been documented in tropical systems as well. Playback experiments on a tropical avian community in China show that frugivores are attracted to playbacks of other frugivore species, but not playbacks of sympatric insectivorous species [87]. Likewise, playback experiments conducted in Panama demonstrate that ant-following birds species are more attracted to vocalizations of species from the family Thamnophilidae (species that are primary obligate ant followers) than sympatric control species that do not follow army ants [52, 88]. Social information may thus play a critical role in community assembly through key species interactions, at least for those in the Neotropics with unique foraging niches.

Box 1   Positive conspecific and heterospecific interactions

Background

- Interactions between conspecifics and heterospecifics can be positive or negative.
- Positive interactions in birds often involve the transfer of information.
- The transfer of information can be direct, with a sender and receiver, or indirect where inadvertent information is used.

Information from conspecifics or heterospecifics can be used to improve decision-making, an important behavior, in many contexts:

- Habitat selection
  - Location cue – the simple presence of a conspecific or heterospecific can provide information at a location contains habitat suitable enough to reside in.
  - Performance cue – the presence of successfully-breeding individuals can provide information that not only is the habitat suitable for residency, but also high-quality because reproductive success is possible.
- Presence of parasites or predators
  - Conspecific and heterospecifics can share information about the presence of predators and brood parasites.
  - Conspecifics and heterospecifics can engage in mobbing behavior to attempt to drive the shared predator or brood parasite away.
- Foraging locations
  - Conspecifics and heterospecifics can provide both direct and indirect information on the location and quality of forage items.
  - Although having more individuals in a foraging flock can result in increased competition, individuals in these flocks can benefit via information from conspecifics on where to find food and to avoid predators.

Figure 1   Cliff Swallows breed in colonies where they benefit from sharing information about the location of food (information center hypothesis [89] and mob or attack predators in force as a group [34]. Photo credit: Mary Kay Rubey.

Finally, it is worth noting that social information has been credited in the evolution of coloniality and/or communal roosting for some species. Known as the "information center hypothesis," it has been suggested that individuals aggregate in large conspecific groups to share and learn information about foraging locations [89]. However, this hypothesis has received considerable criticism [90, 91], and almost exclusively applies to communal/colonial non-passerines [92, 93]. Recent improvements on the information center hypothesis relaxes some assumptions and provides testable predictions more applicable to songbird aggregations [91]. Bijleveld and colleagues [91] propose expanding the definition beyond foraging to learning information about predation risk and potential mates, and expanding beyond the active transfer of information to include inadvertent transfer of information. The information center hypothesis could therefore be applied songbird species, but generally has only been used with members of the family Corvidae. Common ravens (*Corvus corax* Linnaeus 1758; [94]) and American crows [95], for example, have been found to depart for foraging locations with or following knowledgeable conspecifics (i.e., active transfer of information). Likewise, hooded crows (*Corvus cornix* Linnaeus 1758) appear to collect information about food resources at communal roosts [96]. The information center hypothesis is compelling and intuitively logical, but still in need of rigorous testing.

Section II: Negative Conspecific and Heterospecific Interactions

Negative interactions that reduce fitness for at least one individual can place strong selective pressures on adaptive behaviors. Individuals can dampen costs associated with competition, predation risk, and parasitism risk by collecting information about their environment [97]. Songbirds can, and often do, use social information to help mitigate the effect of negative interactions. In turn, these decisions of who to interact with in space and time can ultimately shape songbird communities, such as conspecific abundances and species richness in particular. Here, we review important topics in competition, predation, and parasitism, bringing attention to how songbirds use information to help avoid these costly interactions.

Competition

Conspecific and heterospecific competition are dominant forces shaping songbird communities. This is particularly true during the breeding season, when competition for space, resources, and access to mates is strong and pronounced. Avian ecology has a long history studying both forms of competition and their roles in shaping communities. Charles Darwin, for example, hypothesized that competition for food led to species diversification in Galapagos finch communities (otherwise known as "adaptive radiation"; [98]). This hypothesis is well-supported, with studies demonstrating rapid character displacement among *Geospiza* finches in the face of strong competition for food [99]. Similarly, niche partitioning among North American boreal warblers (specifically foraging and nesting sites) likely resulted from intense competition over evolutionary time [5].

Competing species, however, may not always have a negative effect on one another as predicted by traditional competition theory. In recent years, behavioral ecologists have suggested that social information use strategies may increase niche overlap among competing species. Indeed, a species may use the presence of another ecologically similar species to indicate the location and quality of resources or the level of predation risk. Using heterospecifics with some ecological overlap as information sources, rather than conspecifics, may even be preferable if conspecific competition is greater than heterospecific competition [7]. Additionally, conspecifics may not provide information that is more useful, because conspecific individuals operate under similar ecological constraints. For example, conspecific migrants arrive synchronously at breeding grounds and may not yet have access to conspecific performance-based cues for quality assessment (e.g., number of offspring); in contrast, heterospecifics operating on a different timeline could provide such information about quality [7]. In Sweden, the density of foliant-gleaning migrant songbirds was positively correlated with the density of resident titmice species (genus *Parus*), likely because migrants use titmice to collect information about shared food resources [100]. However, a trade-off occurs between competition and the value of information. As more individuals settle near each other and/or use the same resources, fitness costs of competition can be incurred such that at some threshold, density may exceed fitness benefits derived from using information. When this threshold is reached, negative density dependence would result in heterospecific avoidance rather than attraction [100]. At intermediate densities, social information should be most beneficial, whereas low densities might indicate poor habitat or high mortality risk [101] and high densities indicate high costs of competition [102]. Experimental evidence from migrant collared flycatchers and resident tits in Sweden demonstrate this unimodal relationship, with flycatchers preferentially settling in areas with intermediate tit densities and avoiding areas of low and high tit densities [101].

In some cases, heterospecific information use may be dependent on dominance or competitor success. Regarding the former, some authors have proposed that subordinate species benefit more by collecting information directly from the environment, while dominant species benefit more by collecting social information [103]. Certainly subordinate species likely suffer higher costs if using cues from dominants to find resources. If these costs are greater than the benefits derived from social information, then heterospecific avoidance should occur. Indeed, when dominant least flycatcher cues were broadcast in experimental plots in Montana, colonization rates of small-bodied migrants were significantly reduced whereas subordinate American redstart cues had no effect on colonization [104]. Competitor success also appears to influence information use, with European pied flycatchers selectively copying nest site preferences of heterospecific great tits with experimentally manipulated high clutch sizes [105]. This finding has interesting implications for niche segregation: avoidance of the behaviors associated with unsuccessful individuals can increase niche partitioning while the converse can increase niche overlap [105].

While we have primarily discussed heterospecific competition, the same ideas are applicable to conspecific competition. At a certain threshold density, the benefits of using conspecific cues and social information may no longer outweigh the costs of competition. At an individual level, the benefits of using information from a perceived high-quality conspecific may not outweigh the competitive costs incurred from settling near that individual. Recent research has revealed that some songbirds select habitat based on the quality of nearby conspecifics, suggesting that individuals assess the tradeoff between competition and information. For example, wood warblers in Poland preferentially settle in areas with poor-quality conspecifics, likely because the costs of settling near high-quality conspecifics (e.g., decreased mating success, loss of paternity, poorer quality territory) are deemed too high [106].

Predation and Brood Parasitism Risk

Predation, given its lethal effects, is an influential force shaping avian behaviors. Indeed, how predators influence prey population sizes and life-history traits have been well-studied in avian ecology, dating as far back as Lack [107] and Skutch's [108] classic studies comparing clutch sizes of temperate and tropical birds. More recently there has been growing interest in "nonlethal effects" of predators: individuals change their behavior when perceived predation risk is high, resulting in tradeoffs in activity budgets [109, 110]. For songbirds this tradeoff is often between gaining energy and being vigilant, known as the widely accepted "predation-starvation trade-off" [111, 112]. Dark-eyed juncos (*Junco hyemalis* Linnaeus, 1758), for example, forage less when predation risk is perceived to be high, despite there being no predator present [113]. For great tits, perceived predation risk can negatively affect weight gain and fat reserves, providing particularly compelling evidence for the predation-starvation tradeoff in songbirds [114, 115]. Nonlethal effects can also alter mixed species flock composition: in winter foraging flocks of tit species (Family Paridae), perceived predation risk causes high turnover at foraging sites and in turn disrupts foraging success and stability of the group [116]. In combination, these nonlethal effects can result in overall population declines due to reduced survival and/or reproduction [109].

For songbirds, perceived nest predation risk has equally strong and negative effects on parental investment into offspring, which in turn compromises nest survival and reproductive success [110]. Compelling evidence comes from studies that experimentally manipulate perceived predation risk. For example, Fontaine and Martin [117] found that on experimental plots where nest predators were removed, females laid larger eggs and spent less time incubating. Females could thus invest more in themselves and their young when perceived nest predation risk was low. Investment carried over to the nestling stage, as well: parents increased feeding rates to nestlings on plots where nest predators were removed compared to control plots [117]. This study included several coexisting songbird species across many nesting guilds (cavity, ground-nesting, subcanopy), suggesting reproductive costs of perceived predation risk are widespread. Another similar experiment broadcasted predator vocalizations at song sparrow (*Melospiza melodia* Wilson, 1810) nests, while also controlling for direct nest predation with predator exclosures at nests [118]. Annual offspring production reduced by 40% during this experiment [118], providing strong evidence that perceived predation risk alone can indeed affect songbird population dynamics (as originally hypothesized in Cresswell [109]).

Brood parasitism also has large, negative fitness consequences on many songbird species. For some species, brood parasitism has an even greater impact on reproductive failure than nest predation [119–121]. The mechanisms driving reproductive failure are diverse among brood parasites. The brown-headed cowbird, a well-studied North American brood parasite, often removes a host egg while laying [122, 123]. Brown-headed cowbird nestlings also have a "competitive edge" over host nestlings for food, reducing nestling survival rate [120, 124]. Likewise, the common cuckoo (*Cuckoo canorus* Linnaeus 1758), a well-studied brood parasite of Europe and Asia, has equally strong and negative effects on nest and nestling survival. Compared to other brood parasites, cuckoos are especially virulent to hosts (genus *Cuckoo*) as egg mimicry makes it difficult to reject parasitic eggs [122, 125], and once hatched parasitic nestlings actively evict nest mates [126].

There is evidence that songbirds can reduce brood parasitism risk by using social information. For example, both superb fairy-wrens [56, 57] and Eurasian reed warblers [58], actively guard their nests when social information indicates brood parasitism risk is high. Additionally, a playback experiment in Finland demonstrates that several host species avoid settling in habitat where perceived common cuckoo abundance is high (made apparent by playback of vocalizations; [127]). Similarly, Eurasian magpies (*Pica* Linnaeus, 1758) prefer to place their nests where perceived brood parasitism risk is low (specifically the great spotted cuckoo, *Clamator glandorius* Linnaeus, 1758; [128]). Notably, informed breeding habitat selection has been well studied in the context of perceived predation risk, as many songbirds avoid habitat where perceived predation risk is high [129–131]. In contrast, the effects of perceived brood parasitism risk is a topic in need of research, with only a handful of studies assessing if songbirds incorporate social information about brood parasitism risk into their settlement decisions [127, 128].

Section III: Changing Interactions and Altered Communities

In today's world many songbird species face a barrage of threats that change how they interact, communicate, and ultimately behave. How species respond to these changing interactions can have population-level consequences, fundamentally alter community structure, and pose conservation challenges. Some species will alter their behavior in response to changes in species interactions, whereas species that cannot modify their behavior will likely suffer consequences. Multiple anthropogenic activities change the environments songbird communities reside in both directly (e.g., habitat destruction, introduction of exotic species) and indirectly (e.g., climate change). The multitude of anthropogenic changes is beyond the scope of this chapter. Instead, this chapter focuses on two major anthropogenically-sourced changes, climate change and habitat destruction, and how songbird behavior could be harnessed to mitigate the effect of these changes on populations and communities alike.

To date much of the focus on conspecific and heterospecific interactions is concerned with negative consequences due to individuals being unable to cope with environmental changes from anthropogenic activity. A new, innovative approach could be to take advantage of the behavior of songbirds in order to conserve and manage target species or communities. This chapter closes with two case studies: the first illustrates how social cues may be used to extend a species geographic range, and the second illustrates how behavioral cues can be used to elicit an ecological service from introduced species.

A Changing Climate Results in Changing Interactions

Climate change can differentially affect interactions within and between species through multiple mechanisms, including shifts in the phenology of life history events and influences on survival or reproduction [132]. The effects on one species and their responses can have subsequent implications for coexisting species and potentially increase or decrease competitive interactions between them. How species respond to climate-related changes can also influence information exchange between species, including the type, quality, and value of information available. Research on migratory European pied flycatchers and resident tits (genus *Parus*) in Europe illustrates how climate change may directly influence competitive interactions and disrupt social information exchange. As in the case of other resident songbirds, tits appear more able to closely track phenological shifts in food availability in response to warming and adjust their breeding phenology accordingly [133], while flycatchers are less responsive to these changes [134]. In years with warm winters and springs, tit population density increases, likely due to enhanced overwinter survival. In turn, increased tit density results in greater occupancy of nest boxes and increased heterospecific competition [135]. In years where the mean arrival date of female flycatchers coincides with the mean laying date of tits, male flycatchers are more likely to be killed by great tits [136]. Earlier breeding by tits, however, may also increase the information available to flycatchers. If flycatchers arrive when nestlings or fledglings are present, this social cue may indicate high quality habitat. Correspondingly, when blue and great tit hatchings were experimentally advanced and delayed, female flycatchers were far more likely to settle in plots with advanced tit hatching [136].

In addition to temporal mismatches, climate change may also shift the spatial distributions of species' geographic range and create novel heterospecific interactions. For example, fox sparrows (*Passerella iliaca* Merrem 1786), native to North America, once only migrated through Madarte Island in British Columbia, Canada, but now reside and breed in this region [137]. Fox sparrows now outcompete song sparrows where they are novel competitors, and contribute considerably to population declines [137]. Similarly, collared flycatchers have shifted their range, displacing European pied flycatchers and potentially disrupting community-level alarm calling systems in the process [138]. These novel heterospecific interactions have the potential to lead to the extirpation of a species in an area, the increase in hybridization, and large changes in the behavior of a species to reduce competition.

Climate change may also directly influence survival and reproductive success of some species more than others, giving an "edge" to certain species. In the breeding season following a record-breaking drought in Texas, body condition of female brown-headed cowbirds was significantly lower. Additionally, fewer females developed pre-ovulatory follicles compared to the previous year [139]. Correspondingly, rates of nest parasitism were lower and nest success of their hosts, black-capped vireos, was significantly higher. In a region where drought is expected to increase, this finding may have implications for population dynamics of these two species. While more research is needed on the effect of climate change on songbird interactions, it is clear that extreme climate events such as drought can have species-specific effects [140]. Understanding how these effects will impact the long-term population dynamics of different species is a crucial area of future research.

Climate change may also impact interactions between songbirds and their predators through both lethal and non-lethal effects. For predators such as snakes, higher temperatures are positively correlated with movement and activity, which in turn is correlated with predation rates [141]. A simulation by DeGregorio and colleagues [142] found that a 2°C increase in temperature resulted in higher nest predation rates by rat snakes (*Pantherophis* spp. James 1823), with snakes becoming active earlier in the season and depredating early nesting species, and also depredating more nests at night when adult birds are less successful at defending their nests. Temperature may also indirectly alter the behavior of songbird parents at the nest and subsequently influence nest predation rates. In wetland habitats, for example, increasing spring temperatures are predicted to increase food availability [143]. When wetland-breeding Eurasian reed warblers were presented with nest predator models, warblers that were provided with a supplemental food source returned more quickly to their nest to defend against the predator than birds that were not supplemented [143]. Further, supplemented birds spent less time off the nest, suggesting that a nearby food source allowed birds to be more vigilant and respond more quickly to predators [143]. However, higher temperatures could also increase energetic demands of nestlings, requiring parents to increase visitation rates to the nest to provide for the nestlings [144].

Changing Habitats, Changing Communities, and Changing Behaviors

One of the most prevalent changes to the world is the loss of natural habitats, as ecosystems around the world are being destroyed, degraded, and altered [145, 146]. These changes have a myriad of impacts on songbird populations, and in some cases change interactions between community members. Urbanization, for example, tends to convert native communities into homogenized communities of invasive and exotic species [147], favoring certain species. For example, a study addressing how songbird communities in Illinois changed from the early 1900s to the early 2000s showed that most of the species that expanded their distribution did so by colonizing and using urban habitats [148]. Once established, successful urban adapters have great potential to influence competition and predation dynamics. House sparrows (*Passer domesticus* Linnaeus 1758), for example, with their high aggression levels and pre-adaptation to anthropogenically disturbed landscapes, quickly outcompete native species once they are introduced. In the United States, house sparrows contributed considerably to the nationwide decline of eastern bluebirds specifically (*Sialia sialis* Linnaeus 1758; [149]) and, in general, overall biodiversity of native songbird communities [150]. Similarly, invasive ring-necked parakeets (*Psittacula krameri* Scopoli 1769) outcompete some native species for nesting cavities in urban environments in Europe [151].

All songbirds, native or invasive, that persist in human-modified landscapes face a variety of novel threats. Predation by domestic cats is a major cause of juvenile and adult bird mortality, particularly in urban environments [152, 153]. In the United States alone, domestic cats are estimated to kill ~1 to 4 billion songbirds annually [154]. Songbirds of Canada face similar decimation, with domestic cats responsible for the loss of 100 to 350 million birds per year [155]. Other factors, such as chronic anthropogenic noise have been shown to impact bird song, by causing males to sing at a higher acoustic frequencies, likely to avoid inference from low-frequency anthropogenic noise [156]. In non-urbanized habitats, noise from oil production has been found to lower the abundance of some species [157]. Alterations or masking of song due to anthropogenic noise can have a variety of negative impacts including increased risk of predation, altered energy budgets and loss of access to social information (reviewed in [158]).

Songbird species in human-modified landscapes are not the only ones facing novel threats. In forested habitats, fragmentation can lead to situations where novel threats may require behavioral changes. One of the most studied threats is forest fragmentation leading to the increase in brown-headed cowbirds and subsequent increases in brood parasitism among forest species [159]. Changes such as fragmentation lead to species that have historically not been exposed to brown-headed cowbirds becoming common hosts (e.g., wood thrush, *Hyocichla mustelina* Baird 1864). Species that do not have adaptive defenses to combat brood parasitism will experience a reduction in reproductive success as brown-headed cowbird abundances increase with fragmentation. The fragmentation of forests therefore leads to new heterospecific interactions, a situation in which some species may "win" but others may "lose." From a conservation perspective, the "loser" species either have to change their behaviors (e.g., eject the cowbird eggs), or managers will have to actively control brown-headed cowbird populations by reducing habitat fragmentation or implementing lethal control.

Box 2 Changing communities, changing interactions

Background

- Anthropogenic changes can produce novel songbird communities that disrupt conspecific and heterospecific interactions, often with negative consequences.
- Conservation efforts could take advantage of key songbird behaviors, such as habitat selection and foraging choices, to help manage target species or communities.

Conspecific and heterospecific interactions are most impacted by the following three factors associated with anthropogenic change (none are mutually exclusive):

- Habitat destruction:
  - Habitat fragmentation can rapidly change species interaction by bringing together species without an evolutionary history together, such as brood parasitic brown-headed cowbirds and forest songbird species in North America.
  - Urbanization often goes alongside with habitat fragmentation, which favors few species and can disrupt communication networks within remnant communities due to heavy anthropogenic noise.
- Climate change:
  - Climate change "shuffles" songbird communities and produces novel competition between heterospecifics, often with major negative impacts on the "loser" species.
  - Predators changing their behavior and habitat use with rising temperatures could increase predation rates in some songbird communities.
- Invasive species:
  - Non-native competitors often outcompete native species for food resources and nesting sites, and quickly dominate avian communities.
  - Non-native predators associated with anthropogenic change, such as the domestic house cats, can decimate songbird populations and negatively impact several species.

Figure 2 Rat snake eating a nest (warmer weather = more active predators)

Figure 3 Urban Geese

In combination, habitat loss and climate change are leading to the formation of "novel communities", comprised of species that have not interacted in their evolutionary history. A study in California, for example, suggests that climate change will result in novel "no-analog" communities [160], thereby drastically influencing species interactions. Similarly, winter bird community structure is shifting with an observable increase in warm-adapted species represented in bird communities of eastern North America over a 22 year period [161]. Changes in habitat and climate are likely to continue, as will alterations in songbird communities. It will be imperative to understand how the "reshuffling" of songbird communities impacts species behaviors, and whether these changes lead to population declines or community collapses.

Using Behavior to Manage and Conserve Species

Traditional management and conservation efforts are focused on managing suitable habitat for target species. For example, to combat the decline of grassland bird communities, the management solution has been to create or restore more grassland [162]. While this approach has been successful, it may be more effective to both create grassland and understand the habitat selection behavior of grassland birds, thus ensuring that efforts to create habitat are successful in attracting the focal species [163]. Beyond creating or restoring habitat, understanding behavior can help mitigate threats and be applied to manipulate songbirds to improve conservation. In the context of conspecific and heterospecific interactions, managers may be able to use social information to influence the behavior of individuals, and in turn restore populations and communities. Below are two cases, one in which conspecific social cues are used to invoke conspecific attraction and increase the geographic range of the Kirtland's warbler [22]. In the other example, heterospecific social cues are used to recruit Hawaiian birds to target locations and restore critical seed dispersal ecosystem services [164]. These two examples highlight how understanding interactions within and between species can lead to creative behavioral tools that can assist in conservation and management.

Case Study 1:   Recruiting Endangered Songbirds to Habitat with Conspecific Attraction

Kirtland's warblers have one of the smallest geographic ranges of any breeding bird in North America [165]. At one point there were as few as 400 individuals, but via conservation and management actions, the species has recently been removed from the federal endangered species list. The Kirtland's warbler is rare but occasionally occurs in northern Wisconsin in seemingly suitable habitat. Area in northern Wisconsin contained the jack pine habitat that is required for the species and on occasion a lone bird was observed at these locations. Notably, these locations were hundreds of kilometers from the core of the species' range. With the use of conspecific vocalizations, researchers were able to attract Kirtland's Warblers to sites 225 km from the closest breeding location and

550 km from the core of the species' range [22]. Once attracted to the location, individuals paired and successfully nested (Fig. 4).

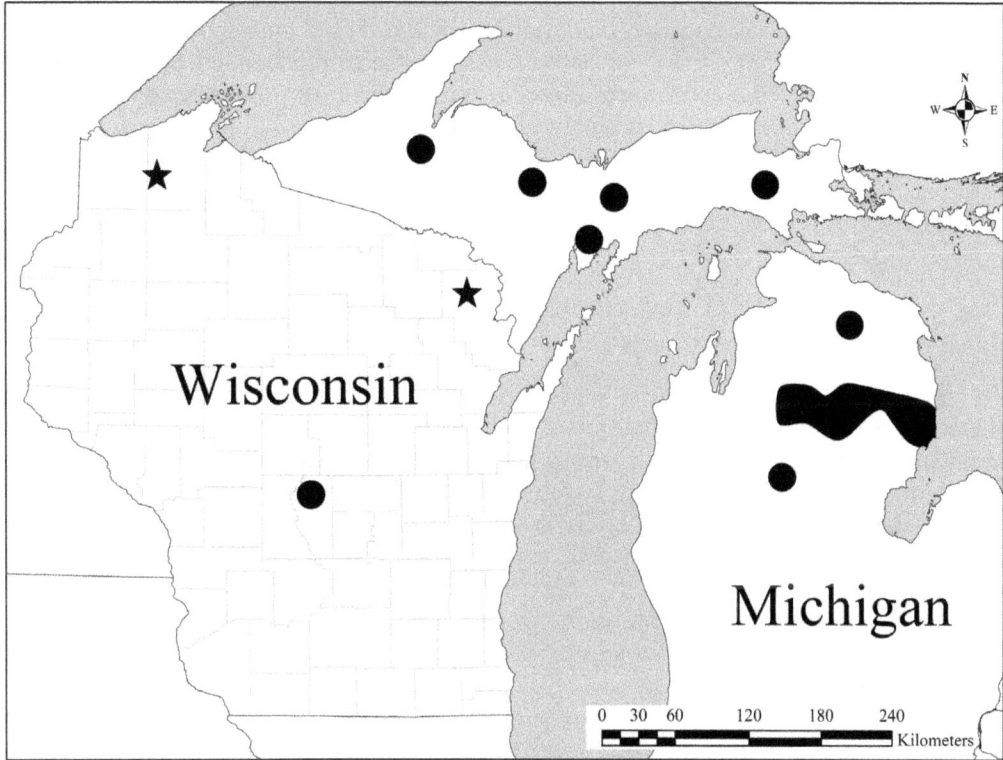

Figure 4 The distribution of Kirtland's Warblers. The circles and large area in Michigan are the historic/current distribution the stars are the study sites where Kirtland's Warblers were attracted using social cues. The information was retrieved from Anich and Ward [22].

With changes in climate and habitat a species' current distribution may not completely overlap their habitat or climate niche [166]. The Kirtland's warbler example highlights how the addition of social cues can influence habitat selection and ultimately result in the creation of a population. This approach has the potential to help expand or alter the geographic range of other bird species and provide a means to help species track habitat changes (potentially due to climate change; [167]). Approaches such as this provide insight into the behavior and the mechanisms that may establish a species' geographic range, and shed light on how behavior can be used to potentially manage and conserve a species. There are, however, many important questions to research before using this behavioral approach in a responsible and effective way. In the case of Kirtland's warblers, where were these attracted individuals dispersing from? If these are "vagrants" that wandered beyond the species' geographic range, it may be more effective to use behavioral approaches to keep the species closer to the core instead of the edge of its range. Likewise, which demographic groups are more likely to wander and settle outside the core of the species' range? A study on Kirtland's warbler in the core of the species range found that males move around (i.e., prospect) as much as 77 km during the breeding season, likely collecting information needed for future habitat selection decisions [168]. In previous research on another species, younger individuals were more likely to exhibit conspecific attraction [20], but little is known about whether males or females are more likely to settle in response to conspecific cues. In conclusion, there may be great potential to use behavioral cues to attract individuals to new locations, but more basic research on habitat selection is needed. Species that may be good candidates for this type of research/management are migratory species with a restricted geographic range and a patchy distribution. Species that breed in successional or ephemeral habitat and engage in prospecting behavior would be good candidates. However, for nearly all species that meet these criteria, more basic research is need before the use of social cues to expand or alter a species geographic range is attempted.

Case Study 2:   Social Information use Facilitates Seed Dispersal in Hawaiian Bird Communities

The majority of studies examining the use of conspecific attraction or social information, as a management tool, have focused on avian conservation goals [8]. Recent work, however, has also demonstrated the effectiveness of broadcasting calls to facilitate critical mutualisms between frugivorous birds and fruiting plants. The majority of tropical plants are adapted for vertebrate dispersal [169]. Yet in many regions around the globe, the extinction of frugivorous animals has resulted in seed dispersal limitation [170, 171] and, ultimately, plant species extinctions [172, 173].

The Hawaiian archipelago presents a particularly dramatic example of this situation with nearly all native frugivorous birds extinct across the islands [174] and numerous native plant species left without native dispersers [170]. Non-native birds may, at least partially, replace the role of native birds by dispersing native plant seeds, though they have been shown to disperse non-native plant species in far greater numbers than native species [175]. Increasing frugivory and seed dispersal of native plants, therefore, presents a key challenge for native plant restoration and ecosystem functioning (Fig. 5. A and B). MacDonald et al. [164] broadcast the calls of several non-native frugivorous birds at native fruiting plants and compared avian visitation and frugivory between the periods of playback calls and silent control periods. They found that visitation increased 4-fold and frugivory increased 10-fold during playback, suggesting that broadcasting calls may be an effective tool for increasing frugivory and seed dispersal of at-risk plant species. This work demonstrated that strength of response varied among species, highlighting the need to take life history factors into account when considering adoption of this technique for conservation goals. This research is also an example of how behavior can be used to manipulate a bird species to interact with another species (in this case a plant); and ultimately the researchers were able to manipulate the behavior of exotic species to provide an environmental service. This management approach is likely to be most appropriate and tractable on islands. Many island ecosystems are highly altered and often contain many introduced bird species. We suggest that researchers and/or managers may develop creative ways to manipulate non-native species on islands to assist in valuable ecological services. In addition, there may be other approaches where behavioral cues could be used to "introduce" both native and non-native species to non-native pest, such as a novel insect that bird predation may be able to control. In all situations basic research would be needed to fully investigate the effect of such approaches.

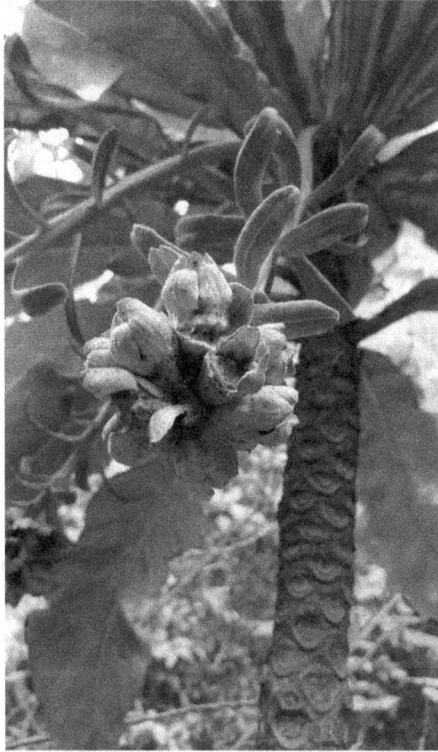

Figure 5A *Cyanea superba* a native fruit that was eaten by introduced bird species when social cues were broadcast from the plant.

Figure 5B The landscape of the area and the pink flow is *Schefflera actinophylla* an introduced plant that is commonly fed upon by introduced bird species [164]. By broadcasting social cues near introduce plants with fruit, that introduced birds are accustomed to feeding on managers might be able to attract these bird species to feed on native fruits and potential help spread the seeds of native plants.

Conclusion

A wide range of conspecific and heterospecific interactions occur that can have both positive and negative effects on individuals. Much research is needed to further understand the scope and scale of these interactions. This research is needed to not only advance our understanding of songbird behavior, but also to address pressing conservation issues. Whether the challenge is climate change or habitat alteration, these changes will likely result in new interactions and creation of novel communities. Traditional habitat management has the potential to mitigate some conservation issues, but by integrating species-specific behaviors, the scientific community may be able to develop "novel" behavior-based management approaches to address conservation issues. We encourage researchers to further our understanding of conspecific and heterospecific interactions, particularly in the context of species conservation and management. While there are plenty of conservation issues that could be addressed, working with species of decline where extensive previous research is available (e.g., Kirtland's warbler) can be a fruitful approach. Working with species in novel assemblages, such as in ecosystems with multiple non-native birds and plants, may provide the opportunity to both advance science and develop new conservation and management approaches.

▓ LITERATURE CITED

[1]   Wiens, J.A. 1989. The Ecology of Bird Communities, Vol. 2: Foundations and patterns. Cambridge University Press, Cambridge, UK.

[2]   Wiens, J.A. 1989. The Ecology of Bird Communities, Vol. 2: Processes and variation. Cambridge University Press, Cambridge, UK.

[3]   Newton, I. 1998. Population Limitation in Birds. Academic Press, Cambridge. MA. USA.

[4]   Connell, J. 1980. Diversity and the coevolution of competitors, or the ghost of competition past. Oikos. 35: 131–138.

[5]   MacArthur, R.H. 1958. Population ecology of some warblers of Northeastern coniferous forests. Ecology. 39: 599–619.

[6]   Bradbury, J. and S.L. Vehrencamp. 2011. Principles of Animal Communication, 2nd ed. Oxford University Press, Oxford, UK.

[7]   Seppänen, J., J.T. Forsman, M. Mönkkönen and R.L. Thomson. 2007. Social information use is a process across time, space, and ecology, reaching heterospecifics. Ecology. 88: 1622–1633.

[8]   Ahlering, M.A., D. Arlt, M.G. Betts, R.J. Fletcher Jr., J.J. Nocera and M.P. Ward. 2010. Research needs and recommendations for the use of conspecific-attraction methods in the conservation of migratory songbirds. Condor. 112: 252–264.

[9]   Szymkowiak, J. 2013. Facing uncertainty: how small songbirds acquire and use social information in habitat selection process? Springer Science Reviews. 1: 115–131.

[10] Reed, J.M. and A.P. Dobson. 1993. Behavioural constraints and conservation biology: conspecific attraction and recruitment. TREE. 8: 253–256.

[11] Danchin É., L-A. Giraldeau, T.J. Valone and R.H. Wagner. 2004. Public information: from nosy neighbors to cultural evolution. Science. 305: 487–491.

[12] Kelly, J.K. and K.A. Schmidt. 2017. Fledgling calls are a source of social information for conspecific, but not heterospecific, songbird territory selection. Ecosphere. 8: 1–9.

[13] Ward, M.P. 2005. Habitat selection by dispersing yellow-headed blackbirds: evidence of prospecting and the use of public information. Oecologia. 145: 650–657.

[14] Pärt, T. and B. Doligez. 2003. Gathering public information for habitat selection: prospecting birds cue on parental activity. Proc. R. Soc. B 270: 1809–1813.

[15] Kivela, S.M., J-T. Seppänen, O. Ovaskainen, B. Doligez, L. Gustaffson, M. Mönkkönen, et al. 2014. The past and the present in decision-making: the use of conspecific and heterospecific cues in nest site selection. Ecology. 95: 3428–3439.

[16] Kelly, J.K. and M.P. Ward. 2017. Do songbirds attend to song categories when selecting breeding habitat? A case study with a wood warbler. Behaviour. 154: 1123–1144.

[17] Farrell, S.L., M.L. Morrison, A.J. Campomizzi and R.N. Wilkins. 2012. Conspecific cues and breeding habitat selection in an endangered woodland warbler. J. Anim. Ecol. 81: 1056–1064.

[18] Virzi T., R.L. Boulton, M.J. Davis, J.J. Gilroy and J.L. Lockwood. 2012. Effectiveness of artificial song playback on influencing the settlement decisions of an endangered resident grassland passerine. Condor. 114: 846–855.

[19] Betts, M.G., A.S. Hadley, N. Rodenhouse and J.J. Nocera. 2008. Social information trumps vegetation structure in breeding-site selection by a migrant songbird. P. R. Soc. B. 275: 2257–2263.

[20] Ward, M.P. and S. Schlossberg. 2004. Conspecific attraction and the conservation of territorial songbirds. Conserv. Biol. 18: 519–525.

[21] Fletcher, R.J. Jr. 2007. Species interactions and population density mediate the use of social cues for habitat selection. J. Anim. Ecol. 76: 598–606.

[22] Anich, N.M. and M.P. Ward. 2017. Using audio playback to expand the geographic breeding range of an endangered species. Divers Distrib. 23: 1499–1508.

[23] Mönkkönen, M., P. Helle, G.J. Nieme and K. Montgomery. 1997. Heterospecific attraction affects community structure and migrant abundances in northern breeding bird communities. Can. J. Zool. 75: 2077–2083.

[24] Thomson, R.L., J.T. Forsman and M. Mönkkönen. 2003. Positive interactions between migrant and resident birds: testing the heterospecific attraction hypothesis. Oecologia. 134: 431–438.

[25] Szymkowiak, J., R.L. Thomson and L. Kuczyński. 2017. Interspecific social information use in habitat selection decisions among migrant songbirds. Behav. Ecol. 28: 767–775.

[26] Seppänen, J.-T., J.T. Forsman, M. Mönkkönen, I. Krams and T. Salmi. 2011. New behavioural trait adopted or rejected by observing heterospecific tutor fitness. Proc. R. Soc. B. 278: 1736–1741.

[27] Buxton, V.L., J. Enos, J. Sperry and M.P. Ward. 2020. A review of conspecific attraction for habitat selection across taxa. Ecol. Evol. 10(23): 12690–12699

[28] Curio, E., U. Erns and W. Vieth. 1978. The adaptive significance of avian mobbing. Z. Tierpsychol. 48: 184–202.

[29] Magrath, R.D., T.M. Haff, P.M. Fallow and A.N. Radford. 2015. Eavesdropping on heterospecific alarm calls: from mechanisms to consequences. Biol. Rev. 90: 560–586.

[30] Martin, T. 1993. Nest predation and nest sites—new perspectives on old patterns. Bioscience. 43: 523–532.

[31] Ibáñez-Álamo, J.D., R.D. Magrath, J.C. Oteyza, A.D. Chalfoun, T.M. Haff, K.A. Schmidt, et al. 2015. Nest predation research: recent findings and future perspectives. J. Ornithol. 156: 247–262.

[32] Magrath, R.D., T.M. Haff, A. Horn and M.L. Leonard. 2010. Calling in the face of danger: predation risk and acoustic communication by parent birds and their offspring. Adv. Stud. Behav. 41: 187–253.

[33] Winkler, D. 1994. Anti-predator defence by neighbors as a responsive amplifier of parental defence in tree swallows. Anim. Behav. 47: 595–605.

[34] Brown, C.R. and M.B. Brown. 1996. Coloniality in the Cliff Swallow: The Effect of Group Size on Social Behavior. University of Chicago Press, Chicago.

[35] Wang, J.S. and C.M. Hung. 2019. Barn swallow nest predation by a recent urban invader, the Taiwan whistling thrush: implications for the evolution of urban avian communities. Zool. Stud. 58: 1.

[36] Brown, C.R. 2016. The ecology and evolution of colony-size variation. Behav. Ecol. Sociobiol. 70: 1613–1632.

[37] Krams, I., A. Bērziņš, T. Krama, D. Wheatcroft, K. Igaune and M.J. Rantala. 2010. The increased risk of predation enhances cooperation. P. R. Soc. B. 277: 513–518.

[38] Thorogood, R. and N.B. Davies. 2016. Combining personal with social information facilitates host defences and explains why cuckoos should be secretive. Scientific Reports. 6: 19872.

[39] Weatherhead, P.J. and D.J. Hoyak. 1984. Dominance structuring of a Red-winged Blackbird Roost. Auk. 101: 551–555.

[40] Haff, T.M. and D.M. Magrath. 2012. Learning to listen? Nestling response to heterospecific alarm calls. Anim. Behav. 84: 1401–1410.

[41] Schmidt, K.A., E. Lee, R.S. Ostfeld and K. Sieving. 2008. Eastern chipmunks increase their perception of predation risk in response to titmouse alarm calls, Behav. Ecol. 19: 759–763.

[42] Caro, T. 2005. Anti-predator Defenses in Birds and Mammals. University of Chicago Press, Chicago, USA.

[43] Nolan, M.T. and J.R. Lucas. 2009. Asymmetries in mobbing behaviour and correlated intensity during predator mobbing by nuthatches, chickadees and titmice. Anim. Behav. 77: 1137–1146.

[44] Courtier, J.R. and G. Ritchison. 2010. Alarm calls of tufted titmice convey information about predator size and threat. Behav. Ecol. 21: 936–942.

[45] Hetrick, S.A. and K.E. Sieving. 2012. Antipredator calls of tufted titmice and interspecific transfer of encoded threat information. Behav. Ecol. 23: 83–92.

[46] Carlson, N.V., E. Greene and C.N. Templeton. 2020. Nuthatches vary their alarm calls based upon the source of the eavesdropped signals. Nat. Commun. 11: 526.

[47] Templeton, C.N. and E. Greene. 2007. Nuthatches eavesdrop on variations in heterospecific chickadee mobbing alarm calls. PNAS. 104: 5479–5482.

[48] Randler, C. 2012. A possible phylogenetically conserved urgency response of great tits (*Parus major*) toward allopatric mobbing calls. Behav. Ecol. Sociobiol. 66: 675–681.

[49] Martínez, A.E., J.P. Gomez, J.M. Ponciano and S.K. Robinson. 2016. Functional traits, sociality and predation risk in an amazonian understory bird community. Am. Nat. 187: 607–619.

[50] Martínez, A.E., E. Parra, L.F. Collado and V.T. Vredenburg. 2017. Deconstructing the landscape of fear in stable multi-species. Ecology. 98: 2447–2455.

[51] Martínez, A.E., E. Parra, O. Muellerklein and V.T. Vredenenburg. 2018. Fear-based niche shifts in neotropical birds. Ecology. 99: 1–8.

[52] Martínez, A.E., H.S. Pollock, J.P. Kelley and C.E. Tarwater. 2018. Social information cascades influence the formation of mixed-species flocks of ant-following birds in the Neotropics. Anim. Behav. 135: 25–35.

[53] Moksnes, A., E. Røskaft, A.T. Bra, L. Korsnes, H.M. Lampe and H.C. Pedersen HC. 1990. Behavioural responses of potential hosts towards artificial cuckoo eggs and dummies. Behaviour 116: 64–89.

[54] Sealy, S.G., D.L. Neudorf, K.A. Hobson and S.A. Gill. 1998. Nest defense by potential hosts of the Brown-headed Cowbird. In: S.I. Rothstein and S.K. Robinson. [eds]. Parasitic birds and their hosts: studies in coevolution. Oxford University Press, New York, USA.

[55] Soler, J.J., M. Soler, T. Pérez-Contreras, S. Aragón and A.P. Møller. 1999. Antagonistic antiparasite defenses: nest defense and egg rejection in the magpie host of the great spotted cuckoo. Behav. Ecol. 10: 707–713.

[56] Feeney, W.E., J.A. Welbergen and N.E. Langmore. 2012. The frontline of avian brood parasite–host coevolution. Anim. Behav. 84: 3–12.

[57] Feeney, W.E. and N.E. Langmore. 2015. Superb fairy-wrens (*Malurus cyaneus*) increase vigilance near their nest with the perceived risk of brood parasitism. Auk. 132: 359–364.

[58] Campobello, D. and S.G. Sealy. 2011. Use of social over personal information enhances nest defense against avian brood parasitism. Behav. Ecol. 22: 422–428.

[59] Gill, S.A. and S.G. Sealy. 2004. Functional reference in an alarm signal given during nest defence: Seet calls of yellow warblers denote brood-parasitic brown-headed cowbirds. Behav. Ecol. Sociobiol. 56: 71–80.

[60] Gill, S.A. and S.G. Sealy. 2003. Tests of two functions of alarm calls given by yellow warblers during nest defence. Can. J. Zool. 81: 1685–1690.

[61] Lawson, S.L., J.K. Enos, N.C. Mendes, S.A. Gill and M.E. Hauber. 2020. Heterospecific eavesdropping on an anti-parasitic referential alarm call. Comm. Biol. 3: 143.

[62] Mikula, P. and J. Hadrava, T. Albrecht and P. Tryjanowski. 2018. Large-scale assessment of commensalistic-mutualistic associations between African birds and herbivorous mammals using internet photos. PeerJ. 6: e4520.

[63] Lehtonen, J. and K. Jaatinen. 2016. Safety in numbers: the dilution effect and other drivers of group life in the face of danger. Behav. Ecol. Sociobiol. 70: 449–458.

[64] Olsen, R.S. P.B. Haley, F.C. Dyer and C. Adami. 2015. Exploring the evolution of a trade-off between vigilance and foraging in group-living organisms. Proc. R. Soc. 2: 150135.

[65] Valone, T.J. 2007. From eavesdropping on performance to copying the behavior of others: a review of public information use. Behav. Ecol. Sociobiol. 62: 1–14.

[66] Stephens, D.W. and J.R. Krebs. 1986. Foraging Theory. Monographs in Behavior and Ecology. Princeton University Press, New Jersey, USA.

[67] Charnov, E.L. 1976. Optimal foraging: the marginal value theorem. Theor. Popul. Biol. 9: 129–136.

[68] Valone, T. 1989. Group foraging, public information and patch estimation. Oikos. 56: 357–363.

[69] Aplin, L.M., D.R. Farine, J. Morand-Ferron and B.C. Sheldon. 2012. Social networks predict patch discovery in a wild population of songbirds. P. R. Soc. B. 279: 4199–4205.

[70] Jones, T.B., L.M. Aplin, I. Devost and J. Morand-Ferron. 2017. Individual and ecological determinants of social information transmission in the wild. Anim. Behav. 129: 93–101.

[71] Valone, T. and J. Templeton. 2002. Public information for the assessment of quality: a widespread social phenomenon. Philos. T. R. Soc. B. 357: 1549–1557.

[72] Barnard, C.J. and R.M. Sibly. 1981. Producers and scroungers: a general model and its application to captive flocks of house sparrows. Anim. Behav. 29: 543–550.

[73] Giraldeau, L-A., C. Soos and G. Beauchamp. 1994. A test of the producer-scrounger foraging game in captive flocks of spice finches *Lonchura punctulate*. Behav. Ecol. Sociobiol. 34: 251–256.

[74] Fernández-Juiricic, E., S. Siller and A. Kacelnik. 2003. Flock density, social foraging, and scanning: an experiment with starlings. Behav. Ecol. 15: 371–379.

[75] Smith, J.W., C.W. Benkman and K. Coffey. 1999. The use and misuse of public information by foraging red crossbills. Behav. Ecol. 10: 54–62.

[76] Firth, J.A. and B.C. Sheldon and D.R. Farine. 2016. Pathways of information transmission among wild songbirds follow experimentally imposed changes in social foraging structure. Biol. Lett. 12: 20160144.

[77] Aplin, L.M. and J. Morand-Ferron. 2017. Stable producer-scrounger dynamics in wild birds: sociability and learning speed covary with scrounging behaviour. P. R. Soc. B. 284: 20162872.

[78] Hämäläinen, L., H. Rowland, J. Mappes and R. Thorogood. 2017. Can video playback provide social information for foraging blue tits? PeerJ. 5: e3062.

[79] Sridar, H., G. Beauchamp and K. Shanker. 2009. Why do birds participate in mixed-species foraging flocks? A large-scale synthesis. Anim. Behav. 78: 337–347.

[80] Hogstad, O. 1978. Differentiation of foraging niche among tits, *Parus* spp., in Norway during winter. Ibis. 120: 139–146.

[81] Krause, J. and G.D. Ruxton and S. Krause. 2010. Swarm intelligence in animals and humans. Trends Ecol. Evol. 25: 28–34.

[82] Freeberg, T., S. Eppert, K. Sieving and J.R. Lucas. 2017. Diversity in mixed species groups improves success in a novel feeder test in a wild songbird community. Sci. Rep. 7: 43014.

[83] Farine, D.R., L.M. Aplin, B. Sheldon and W. Hoppitt. 2015. Interspecific social networks promote information transmission in wild songbirds. P. R. Soc. B. 282: 20142804.

[84] Hillemann, F., E.F. Cole, S.C. Keen, B.C. Sheldon and D.R. Farine 2019. Diurnal variation in the production of vocal information about food supports a model of social adjustment in wild songbirds. P. R. Soc. B. 286: 20182740.

[85] Suzuki, T.N. 2012. Long-distance calling by the willow tit, *Poecile montanus*, facilitates formation of mixed species foraging flocks. Ethology. 11: 10–16.

[86] Mahurin, E.J. and T.M. Freeberg. 2009. Chick-a-dee call variation in Carolina chickadees and recruiting flockmates to food. Behav. Ecol. Sociobiol. 20: 111–116.

[87] Gu, H.J. Chen, J., H. Ewing, L. Xiaohu, Z. Jiangbo and E. Goodale. 2017. Heterospecific attraction to the vocalizations of birds in mass-fruiting trees. Behav. Ecol. Sociobiol. 71: 82.

[88] Pollock, H.S., A.E. Martínez, J.P. Kelley, J.M. Touchton and C.E. Tarwater. 2017. Heterospecific eavesdropping in ant-following birds of the Neotropics is a learned behavior. Proc. R. Soc. B. 274: 20171785.

[89] Ward, P. and A. Zahavi. 1973. The importance of certain assemblages of birds as "information-centres" for Food-finding". Ibis. 115: 517–534.

[90] Mock, D.W., T.C. Lamey and D.B.A. Thompson. 1988. Falsifiability and the information centre hypothesis. Ornis Scandinavica. 19: 231–248.

[91] Bijleveld, A.I., M. Egas, J.A. van Gils and T. Piersma. 2010. Beyond the information centre hypothesis: communal roosting for information on food, predators, travel companions and mates? Oikos. 119: 277–285.

[92] Beauchamp, G. 1999. The evolution of communal roosting in birds: origin and secondary losses. Behav. Ecol. 10: 675–687.

[93] Harel, R., O. Spiegel, W.M. Getz and R. Nathan. 2017. Social foraging and individual consistency in following behaviour: testing the information centre hypothesis in free-ranging vultures. P. R. Soc. B. 284: 20162654.

[94] Wright, J., R.E. Stone and N. Brown, N. 2003. Communal roosts as structured information centres in the raven, *Corvus corax*. J. Anim. Ecol. 72: 1003–1014.

[95] Moore, J.E. and Switzer, P.V. 1998. Preroosting aggregations in the American crow, *Corvus brachyrhyncos*. Can. J. Zool. 76: 508–512.

[96] Sonerud, G.A., C.A. Smedshaug and O. Brathan. 2001. Ignorant hooded crows follow knowledgeable roost-mates to food: support for the information centre hypothesis. Proc. R. Soc. B. 268: 827–831.

[97] Koops, M.A. 2004. Reliability and the value of information. Anim. Behav. 67: 103–111.

[98] Darwin, C. 1859. The Origin of Species. John Murray, London.

[99] Grant, R.B. and P.R. Grant. 2003. What Darwin's finches can teach us about the evolutionary origin and regulation of biodiversity? BioScience. 53: 965–975.

[100] Forsman, J.T., M.B. Hjernquist and L. Gustafsson, L. 2009. Experimental evidence for the use of density based interspecific social information in forest birds. Ecography. 32: 539–545.

[101] Forsman, J.T., M.B. Hjernquist, J. Taipale and L. Gustafsson. 2008. Competitor density cues for habitat quality facilitating habitat selection and investment. Behav. Ecol. 19: 539–545.

[102] Gil, M.A., M.L. Baskett and S.J. Schreiber. 2019. Social information drives ecological outcomes among competing species. Ecology. 100: e20835.

[103] Goodale, E., G. Beauchamp, R.D. Magrath, J.C. Nieh and G.D. Ruxton. 2010. Interspecific information transfer influences animal community structure. Trends Ecol. Evol. 25: 354–361.

[104] Fletcher, R.J. Jr. 2008. Social information and community dynamics: non target effects from simulating social cues for management. Ecol. Appl. 18: 1764–1773.

[105] Loukola, O.J., J-T. Seppänen, I. Krams, S.S. Torvinen and J.T. Forsman. 2013. Observed fitness may affect niche overlap in competing species via selective social information use. Am. Nat. 182: 474–483.

[106] Szymkowiak, J., R.L. Thomson and L. Kuczyński. 2016. Wood warblers copy settlement decisions of poor quality conspecifics: support for the tradeoff between the benefit of social information use and competition avoidance. Oikos. 125: 1561–1569.

[107] Lack, D. 1948. The significance of clutch size. Part 3. Some interspecific comparisons. Ibis. 90: 25–45.

[108] Skutch, A.F. 1949. Do tropical birds rear as many young as they can nourish? Ibis. 91: 430–455.

[109] Cresswell, W. 2008. Non-lethal effects of predation in birds. Ibis. 150: 3–17.

[110] Lima, S.L. 2009. Predators and the breeding bird: behavioral and reproductive flexibility under the risk of predation. Biol. Rev. 84: 485–513.

[111] Houston, A.I., J.M. McNamara and J.M.C. Hutchinson. 1993. General results concerning the trade-off between gaining energy and avoiding predation. Philos. T. R. Soc. B. 341: 375–397.

[112] Bonter, D.N., B. Zuckerberg, C.W. Sedgwick and W.M. Hochachka. 2013. Daily foraging patterns in free-living birds: exploring the predation–starvation trade-off. P. R. Soc. B. 280: 20123087.

[113] Lima, S.L. 1988. Initiation and termination of daily feeding in dark-eyed juncos: influences of predation risk and energy reserves. Oikos. 53: 3–11

[114] Gosler, A. and J. Greenwood and C. Perrins. 1995. Predation risk and the cost of being fat. Nature. 377: 621–623.

[115] Gentle, L.K. and A.G. Gosler. 2001. Fat reserves and perceived predation risk in the great tit, *Parus major*. P. R. Soc. B. 268: 487–491.

[116] Voelkl, B., J. Firth and B. Sheldon. 2016. Nonlethal predator effects on the turn-over of wild bird flocks. Sci. Rep. 6: 33476.

[117] Fontaine, J.T. and T.E. Martin. 2006. Parent birds assess nest predation risk and adjust their reproductive strategies. Ecol. Lett. 9: 428–434.

[118] Zanette, L., A.F. White, M.C. Allen and M. Clinchy. 2011. Perceived predation risk reduces the number of offspring songbirds produce per year. Science. 334: 1398–1401.

[119] Ortega, C.P. and J.C. Ortega. 2001. Effects of brown-headed cowbirds on the nesting success of chipping sparrows in Southwest Colorado. Condor. 103: 127–133.

[120] Hauber, M.E. 2003. Hatching asynchrony, nestling competition, and the cost of interspecific brood parasitism. Behav. Ecol. 14: 224–235.

[121] Hoover, J.P. 2003. Experiments and observations of prothonotary warblers indicate a lack of adaptive responses to brood parasitism. Anim. Behav. 65: 935–944.

[122] Davies, M.B. 2010. Cuckoos, Cowbirds and other Cheats. A&C Black, New York, NY.

[123] Hoover, J.P. and S.K. Robinson. 2007. Retaliatory mafia behavior by a parasitic cowbird favors host acceptance of parasitic eggs. PNAS. 104: 4479–4483.

[124] Lichtenstein, G. and S.G. Sealy. 1998. Nestling competition, rather than supernormal stimulus, explains the success of parasitic brown-headed cowbird chicks in yellow warbler nests. P. Roy. Soc. B. 265: 249–254.

[125] Rothstein, S.I. 1990. A model system for coevolution: avian brood parasitism. Annu. Rev. Ecol. Syst. 21: 481–508.

[126] Anderson, M.G., C. Moskát, M. Bán, T. Grim, P. Cassey and M.E. Hauber. 2009. Egg eviction imposes a recoverable cost of virulence in chicks of a brood parasite. PLoS One 4: e7725.

[127] Tolvanen, J., J.T. Forsman and R.L. Thomson. 2017. Reducing cuckoo parasitism risk via informed habitat choices. Auk. 134: 553–563.

[128] Expósito-Granados, E., D. Parejo, J.G. Martínez, M. Precioso, M. Molina-Morales and J.M. Avilés. 2017. Host nest site choice depends on risk of cuckoo parasitism in magpie hosts. Behav. Ecol. 28: 1492–1497.

[129] Forstmeier, W. and I. Weiss. 2004. Adaptive plasticity in nest-site selection in response to changing predation risk. Oikos. 104: 487–499.

[130] Eggers, S., M. Griesser, M. Nystrand and J. Ekman. 2006. Predation risk induces changes in nest-site selection and clutch size in the Siberian jay. P. R. Soc. B. 273: 701–706.

[131] Chalfoun, A.D. and K.A. Schmidt. 2012. Adaptive breeding-habitat selection: Is it for the birds? Auk. 129: 589–599.

[132] Ahola, M.P., T. Laaksonen, T. Eeva and E. Lehikoinen. 2007. Climate change can alter competitive relationships between resident and migratory birds. J. Anim. Ecol. 76: 1045–1052.

[133] Charmantier A., R. McCleery, L.R. Cole, C. Perrins, L. Kruuk and B. Sheldon. 2008. Adaptive phenotypic plasticity in response to climate change in a wild bird population. Science. 320: 800–803.

[134] Ouwehand, J., C. Burger and C. Both. 2017. Shifts in hatch dates do not provide pied flycatchers with a rapid ontogenetic route to adjust offspring time schedules to climate change. Functional Ecology. 31: 2087–2097.

[135] Samplonius, J.M. and L. Bartosova, M.D. Burgess, A.V. Bushuev, T. Eva, E.V. Ivankina, et al. 2018. Phenological sensitivity to climate change is higher in resident than in migrant bird populations among European cavity breeders. Global. Change Biol. 24: 3780–3790.

[136] Samplonius, J.M. and C. Both. 2019. Climate change may affect fatal competition between two bird species. Curr. Biol. 29: 327–331.

[137] Johnson, K.M., R.R. Germain, C.E. Tarwater, J.M. Reid and P. Arcese. 2018. Demographic consequences of invasion by a native, controphic competitor to an insular bird population. Oecologia. 187: 155–165.

[138] Wheatcroft, D., M. Gallego-Abenza and A. Qvarnström. 2016. Species replacement reduces community participation in avian antipredator groups, Behav. Ecol. 27: 1499–1506.

[139] Buxton, V.L., W.M. Schelsky, T.J. Boves, S. Summers, P.J. Weatherhead and J.H. Sperry. 2017. Effects of drought on brood parasite body condition, follicle development, and parasitism: Implications for host-parasite dynamics. Auk. 135: 908:918.

[140] Martin, K., S. Wilson, E.C. MacDonald, A.F. Canfield, M. Martin and S.A. Trefry. 2017. Effects of severe weather on reproduction for sympatric songbirds in an alpine environment: Interactions of climate extremes influence nesting success. Auk: Ornithol. Adv. 134: 696–709.

[141] Sperry, J.H. R.G. Peak, D.A. Cimprich and P.J. Weatherhead. 2008. Snake activity affects seasonal variation in nest predation risk for birds. J. Avian Biol. 39: 379–383.

[142] DeGregorio, B.A., J.D. Westervelt, P.J. Weatherhead and J.H. Sperry. 2015. Indirect effect of climate change: shifts in ratsnake behavior alter intensity and timing of avian nest predation. Ecol. Modell. 312: 239–246.

[143] Vafidis, J.O., R.J. Facey, D. Leech and R.J. Thomas. 2018. Supplemental food alters nest defense and incubation behavior of an open-nesting wetland songbird. J. Avian Biol. 49: e01672.

[144] Cox, A.W., F.R. Thompson and J.L. Reidy. 2013. The effects of temperature on nest predation by mammals, birds, and snakes. Auk. 130: 784–790.

[145] Vitousek, P.M., H.A. Mooney, J. Lubchenco and J.M. Melillo. 1997. Human domination of Earth's ecosystems. Science. 277: 494–499.

[146] Hoekstra, J.M., T.M. Boucher, T.H. Ricketts and C. Roberts. 2004. Confronting a biome crisis: global disparities of habitat loss and protection. Ecol. Lett. 8: 23–29.

[147] McKinney, M.L. 2006. Urbanization as a major cause of biotic homogenization. Biol. Conserv. 127: 247–260.

[148] Ward, M.P., K.W. Stodola, J.W. Walk, T.J. Benson, J.L. Deppe and J.D. Brawn. 2018. Changes in bird distributions in Illinois, USA over the 20th century were driven by use of alternative rather than primary habitats. Condor: Ornithol. Appl. 120: 622–631.

[149] Sauer, J.R. and S. Droege. 1990. Recent population trends of the Eastern Bluebird. Wilson Bull. 102: 239–252.

[150] MacGregor-Fors, I.L. Morales-Pérez, J. Quesada and J.E. Schondube. 2010. Relationship between the presence of House Sparrows (*Passer domesticus*) and Neotropical bird community structure and diversity. Biol. Invasions. 12: 87.

[151] Strubbe, D. and E. Matthysen. 2009. Establishment success of invasive ring-necked and monk parakeets in Europe. J. Biogeogr. 36: 2264–2278.

[152] Baker, P.J., S.E. Molony, E. Stone, I.C. Cuthill and S. Harris. 2008. Cats about town: is predation by free-ranging pet cats *Felis catus* likely to affect urban bird populations? Ibis. 150: 86–99.

[153] Van Heezik, Y., A. Smyth, A. Adams and J. Gordon. 2010. Do domestic cats impose an unsustainable harvest on urban bird populations? Biol. Conserv. 143: 121–130.

[154] Loss, S., T. Will and P. Marra. 2013. The impact of free-ranging domestic cats on wildlife of the United States. Nat. Commun. 4: 1396.

[155] Blancher, P. 2013. Estimated number of birds killed by house cats (*Felis catus*) in Canada. Avian Conserv. Ecol. 8: 3.

[156] Wood, W.E. and S.M. Yezerinic. 2006. Song sparrow (*Melospiza melodia*) song varies with urban noise. Auk 123: 650–659.

[157] Bayne, E.M., L. Habib and S. Boutin. 2008. Impacts of chronic anthropogenic noise from energy-sector activity on abundance of songbirds in the Boreal forest. Cons. Biol. 22: 1186–1193.

[158] Read, J., G. Jones and A.N. Radford. 2014. Fitness costs as well as benefits are important when considering responses to anthropogenic noise. Behav. Ecol. 25: 4–7.

[159] Robinson, SK., F.R. Thompson, T.M. Donovan, D.R. Whitehead and J. Faaborg. 1995. Regional forest fragmentation and the nesting success of migratory birds. Science. 267: 1987–1990.

[160] Stralberg, D, D. Jongsomjit, C.A. Howell, M.A. Snyder, J.D. Alexander, J.A. Wiens, et al. 2009. Re-shuffling of species with climate disruption? A no-analog future for California birds? PLoS One. 4(9): e6825. https://doi.org/10.1371/journal.pone.0006825

[161] Prince, K. and B. Zuckerberg. 2014. Climate change in our backyards: the reshuffling of North America's winter bird communities. Global Change Biol. 21: 572–585.

[162] Brennan, L.A. and W.P. Kuvlesky Jr. 2005. North American grassland birds: an unfolding conservation crisis? J. Wildl. Manage. 69: 1–13.

[163] Andrews, J.E., J.D. Brawn and M.P. Ward. 2015. When to use social cues: Conspecifc attraction in newly created grasslands. Condor. 117: 297–305.

[164] MacDonald, S.E., M.P. Ward and J.H. Sperry. 2019. Manipulating social information to promote frugivory by birds on a Hawaiian Island. Ecol. Appl. 29: e01963.

[165] Bocetti, C.I., Donner, D.M. and H.F. Mayfield. 2014. Kirtland's Warbler (*Setophaga kirtlandii*). *In*: Rodewald, P.G. (ed.). The Birds of North America. Ithaca, NY, Cornell Lab of Ornithology.

[166] Tingley, M.W., W.B. Monahan, S.R. Beissinger and C. Moritz. 2009. Birds track their *Grinnellian niche* through a century of climate change. PNAS. 106: 19637–19643.

[167] Stodola, K.W and M.P. Ward. 2017. The emergent properties of conspecific attraction can limit a species' ability to track environmental change. Am. Nat. 189: 726–733.

[168] Cooper, N.W. and P.P. Marra. 2020. Hidden Long-Distance Movements by a Migratory Bird. Curr. Biol. 30(20): 4056–4062.

[169] Howe, H.F. and J. Smallwood. 1982. Ecology of seed dispersal. Annu. Rev. of Ecol. Evol. Syst. 13: 201–228.

[170] Chimera, C.G. and D.R. Drake. 2010. Patterns of seed dispersal and dispersal failure in a Hawaiian dry forest having only introduced birds. Biotropica. 42: 493–502.

[171] Markl, J.S., M. Schleuning, P.M. Forget, P. Jordano, J.E. Lamber, A. Traveset, et al. 2012. Meta-analysis of the effects of human disturbance on seed dispersal by animals. Conserv. Biol. 26: 1072–1081.

[172] Temple, S.A. 1977. Plant-animal mutualism: coevolution with dodo leads to near extinction of plant. Science. 197: 885–886.

[173] Kirika, J.M., N. Farwig and K. Bohning-Gaese. 2008. Effects of local disturbance of tropical forests on frugivores and seed removal of a small-seeded afrotropical tree. Conserv. Biol. 22: 318–328.

[174] Walther, M. and J.P. Hume. 2016. Extinct Birds of Hawai'i. Mutual Publishing, Honolulu, Hawaii.

[175] Vizentin-Bugoni, J., C.E. Tarwater, J.T. Foster, D.R. Drake, J.M. Gleditsch, A.M. Hruska, et al. 2019. Structure, spatial dynamics and stability of novel seed dispersal mutualistic networks in Hawai'i. Science. 364: 78–82.

Chapter 5

Sexual Selection and Mating Systems under Anthropogenic Disturbance

Ken A. Otter[1], Matthew W. Reudink[2],
Jennifer R. Foote[3], Ann E. McKellar[4] and Nancy J. Flood[2]

Introduction

Conservation plans often focus on habitat availability and quality, population sizes, and population interconnectivity, but include little or no consideration of the sensory ecology of the animals they are managing [1, 2]. Yet, how animals perceive the world

[1]Natural Resources and Environmental Studies, University of Northern British Columbia
[2]Department of Biological Sciences, Thompson Rivers University
[3]Department of Biology, Algoma University
[4]Canadian Wildlife Service, Environment and Climate Change Canada

around them influences all aspects of their life history. Among these is the need to find mates, which relies upon the ability of one sex ("choosers") to assess the suitability of potential mating partners; this in turn relies upon the ability to perceive and accurately process sexual signals displayed by the other sex ("courters") [3, 4]. Sexual selection theory is predicated on the ability of choosers (typically females) to compare the displays of courters (typically males), using them, via intersexual selection to discriminate among available suitors. Such displays can convey accurate information on the relative condition or resource holding potential of courters [5], which may benefit choosers in their selection of social mates or sires. As a result, any anthropogenic disturbance that alters either the production of displays or the ability of receivers to perceive them has the potential to disrupt evolved life history traits.

Much of the work on anthropogenic impacts on signaling in birds has focused on noise pollution and its capacity to mask vocalizations (see [6, 7] for recent reviews). As this is a focus of Chapter 8 (Communication), we will spend little time on that topic here. Rather, we will consider how anthropogenic effects might alter other aspects of acoustic and visual signals, such as measures of signaler condition embedded within their song output and/or appearance and its impact on the ability of both courters to signal, and choosers to assess, their quality. Because many visual and auditory signals in birds used in sexual displays are condition-dependent, the honesty of the signal is maintained by a tight link between the relative condition of the signaler and the level of trait expression. Given this linkage between sexual signals and individual condition, habitat quality can indirectly influence signal expression through its effect on the condition of individuals. Here, we review literature on changes in sexual signals and the impacts on reproduction in anthropogenically disturbed landscapes, and argue that these might reflect habitat-induced impacts on individual condition. Secondly, we consider the potential for anthropogenic disturbance to affect the settlement of individuals, which could disrupt communal display arenas (leks) and/or limit the number of signalers a receiver can assess as potential mates.

Sexual Selection and Condition-dependent Signaling

A fundamental tenet of sexual selection theory is that there is variation among courters in mating success. This could arise through intrasexual competition to secure resources that choosers desire (e.g., direct benefits, such as nesting sites and territorial resources) or through intersexual mate choice, whereby choosers directly assess courters based on the expression of secondary sexual traits [4]. While recent reviews are correct in that choosing a particular courter based on the level of expression of a trait is not conclusive proof of 'good genes' benefits (i.e., increased genetic fitness of offspring) [4], this does not mean that the level of trait expression doesn't convey accurate information about the condition of the signaler [5]. In fact, many signals used in courting by male birds are known to tightly co-vary with the somatic condition and resource access of the signaler. Bird song is one such trait; it serves the dual function of territorial defense and mate attraction [8] and many studies have been conducted on how males structure their signaling to interact with rivals during territorial disputes [9–12]. Female birds also appear to use condition-dependent aspects of male song in mate-choice decisions, such as song output [13–18] or the performance complexity of signals themselves [19–21]. This suggests choosers are attempting to differentiate courters on some measure of individual condition, regardless of the final benefit choosers might derive. This, however, creates a conundrum—if choosers gain an advantage by basing decisions on differences in courter "quality" (either through direct or some form of indirect benefits), how does the chooser assess underlying "quality" of another individual? Grafen [22] suggested that choosers cannot assess courter quality directly, but have to rely on assessing overt signals that honestly co-vary with courter condition. By the use of such signals, choosers can potentially assess other qualities that will benefit either themselves (resources available within the courter's territory, e.g., [23] or their offspring (heritable components of fitness, e.g., [24]). A key determinant in this argument is that the traits used by choosers to discriminate between courters must honestly reflect the courter's condition or ability to acquire/provide resources of interest to the chooser, and that only individuals in good condition can afford (metabolically) to have high trait expression.

Such a relationship has been shown for male song in birds. The link between male condition and song output, for example, has been well established. Song output correlates with several independent measures of condition, such as dominance rank [14], parental abilities [15], age or size [25], survival [15] and parasite loads [26, 27]. Further, numerous studies have documented the relationship between food availability [28–32] or somatic condition [33] and these same metrics of song, establishing the link between song production and honest signaling.

Song is only one signaling modality through which male birds can provide information on their relative quality: birds also signal quality with visual signals including plumage colouration. As with song, the literature linking visual signals with male condition is extensive. One of the species in which this link has been most studied is the house finch (*Haemorhous mexicanus*). Male finches express carotenoid-based red plumage that varies with condition, and females are known to have strong preferences for certain types of males [34, 35]. As carotenoids are acquired in the diet, rather than synthesized by the body (in contrast to melanin-based pigments; see below), expression of these traits in males may be more likely to convey honest signals about a male's condition or ability to acquire resources [36]. Ample evidence shows the link between condition and trait expression in this species, with reduced expression associated with decreased condition (parasite load; [37]), and increased expression associated with increased nutritional access or physical condition [38, 39]. Similarly, the intensity of yellow carotenoid-based belly colouration in great tits (*Parus major*) was negatively associated with parasite load [40]. A recent meta-analysis, however, demonstrates that though the general pattern of links between plumage coloration and quality exists, the honesty of carotenoid-based signaling is strongest for converted carotenoids (e.g., reds, oranges) rather than dietary carotenoids that do not undergo metabolic conversion [41]. In addition, the mechanism(s) maintaining signal honesty remain unclear and hotly debated [42, 43].

Regardless of the mechanisms underlying signal honesty, the relationship between colouration and individual quality/condition is not restricted to carotenoid-based pigments. The blue and iridescent plumages created by structural layers of melanin in feathers (structural colours) are also condition-dependent [44–47] and this expression can influence female choice [48–49]. Similarly, melanin-based plumage, which is synthesized *de novo* rather than acquired from the diet, had long been considered a poor candidate for condition-dependence due to its high heritability and relatively low environmental influence. However, more recent work and comparative studies have revealed compelling evidence for the condition-dependence of melanin-based plumage [50, 51].

Interconnection between Habitat Quality, Individual Condition and Signal Expression

Sexually-selected traits often evolved to be effective signals *because* of their dependence on the individual condition of the signalers that express them [5]. Assessment of an individual on the basis of traits that cannot be easily faked (because the cost associated with expressing the trait at a particular level is higher for a low-quality than high-quality individual), is a central tenet of honest signaling theory. As a result, we would predict that any environmental influences that affect individual condition, such as habitat quality, also have the potential to impair or elevate the expression of the traits [49, 52] and the ability of choosers to use them in assessment [53]. Thus, anthropogenic disturbance has the potential to disrupt evolved mating systems. Establishing a link between habitat quality, individual condition, and ultimately trait expression is the first step in this process.

The link between individual condition and habitat quality has been extensively considered in relation to intraspecific breeding success, with extensive work in Europe comparing populations of blue tits and great tits settling in evergreen vs deciduous oak forests [54]. Deciduous oak forests have a greater proportion of their leaves replaced annually, which leads to higher abundance of caterpillar larvae in these forests—the primary prey used by tits to feed young [55]. By comparison, evergreen oak forests found on the island of Corsica have lower prey availability, and birds in these forests tend to have lower clutch sizes and are in poorer physical condition [56–58]. Black-capped chickadees (*Poecile atricapillus*) occupying forests of similar composition but varying in age (mature forests vs young forests) in western Canada showed similar habitat-based effects on reproduction and condition [59]. Compared to their mature forest counterparts, chickadees in young forests have higher nest abandonment rates, which result in lower fledging success [59]; in addition, males in young forests have poorer body condition [60]. This may be related to variation in relative food availability, which appears to differ between the two habitat types. Female chickadees solicit food from males during the egg-laying period with a particular call (*broken dees*). Otter et al. [61] used supplemental feeding studies to show that call rates were positively associated with relative hunger levels, and that females in young forests called at higher rates than those in mature forests. Combined, these studies suggest that variation in habitat quality can greatly impact individual condition and reproductive success; but does it also affect the expression of sexually-selected signals?

Studies of chickadee song would imply that it does. Song output during the dawn chorus appears to be an honest signal of condition in male chickadees, with dominant males having higher song output than their subordinate flockmates [14]. Supplemental feeding studies show that this metric of song output is directly related to relative food availability [32]. Females in this species have been shown to use singing behavior directly in mate choice decisions [16, 17]. However, van Oort et al. [62] found that dominant males occupying young forests (low-quality habitats) had reduced song output during the dawn chorus compared with their counterparts in mature forests (high-quality habitats); dominant males in young forests sang at similar rates to subordinate males. This reduction of song output in otherwise dominant males would potentially put them at a disadvantage in a mating system where neighbouring females are assessing both social mates [63, 64] and extra-pair partners [14, 16, 17] based on perceived male rank, which females are partially assessing through song performance. Further, male chickadees appear to not only signal relative condition through song output, but also with how consistent successive songs are in internal frequency ratios within and between notes [65]. Grava et al. [66] found that male chickadees occupying young forest, regardless of social rank, were less able to maintain this internal consistency in their songs than were mature-forest males. Playback studies using stimuli of dominant male songs recorded in either young or mature forest further showed that mature-forest stimuli were perceived as a greater threat than young-forest stimuli, regardless of the fact that the males from which these songs were recorded were of equivalent, relative social rank. These studies strongly suggest that in this system, habitat can influence both the expression of sexually-selected traits, and how the individuals that express these are perceived by others [67].

A similar relationship between habitat quality and individual condition, and the effects these might have on female perception of relative male quality, can be found in visual signals of plumage. In an interesting early study in this area, Gustafsson et al. [68] manipulated parental effort of collared flycatchers (*Ficedula albicollis*) by increasing or decreasing clutch size. This resulted in reduced food per nestling at enlarged nests, mimicking a poor-quality environment, but increased food per nestling at reduced nests, mimicking a higher-quality habitat. The following year, juvenile males reared in enlarged nests had smaller forehead patches (a sexually-selected plumage signal in the species; [69]) compared with males reared in reduced nests. This suggests a direct link between food availability experienced as nestlings and development of secondary sexual characters. Similar links between habitat quality and expression of plumage signals occur in carotenoid-dependent plumage traits. Hill [70] found that the extent of red plumage in house finches varied among populations, but this was not due to genetic differences between regions—birds from the different populations converged on similar plumage expression when brought into aviaries and fed on an overlapping diet. This suggests that variation in plumage colour was likely due to environmental differences in accessibility to carotenoids in the diet. These studies suggest that relationships between individual condition and expression of sexually-selected traits can be influenced by the linkage between individual condition and habitat quality. In poor-quality habitat, the ability of signalers to express condition-dependent traits may be so generally depressed that these traits no longer allow discrimination amongst individuals, even if there are underlying differences in individual signaler quality. As a result, choosers in lower-quality habitats may be less able to discriminate among available courters. This could result, at best, in the loss or relaxation of mate choice decisions in these habitats or, at worst, in maladaptive decisions. The above studies have focused on variation in condition/signal expression in habitats that are still largely natural. However, anthropogenic disturbances and their effects on habitat quality might exacerbate these effects.

Anthropogenic Influences on Signal Expression and Condition

There are two ways that we envision anthropogenic disturbance might affect sexually selected signals; (1) it may affect the signaling environment, resulting in altered signal transmission; or (2) it may impact the condition of the signalers and so alter their ability to fully express the signals. Transmission issues in acoustic signals have been heavily studied in relation to noise pollution (for additional detail see Chapter 8). We will focus on anthropogenic disturbances on acoustic and visual signals and how these disturbances may impact the ability to fully express secondary sexual characters.

Light Environment and Visual Signals

Delhey and Peters [2] make a strong case that alteration of habitats may impact the light environments in which signals evolved. This occurs either through altering the vegetative structure of the environment or via the introduction of artificial light that differs in emphasized wavelengths from natural light. Much of the work in this field has focused on alterations of aquatic environments and the consequent impacts this has had on visual signaling in fish. However, similar impacts may be seen in birds. Endler and Théry [71] found that Cock-of-the-Rock males prefer to display in forest patches that have sunlight beams breaking through the canopy, particularly when display perches allow them to move in and out of light beams that emphasize their bright orange feathers. Although other studies on lekking manakin species have found less evidence for the generality of this filtered-light effect [72], it may still influence locations of leks in some species, particularly those with greater emphasis on long-wavelength orange/red signals, like Cock-of-the-Rocks, compared to the blue-black plumage of many manakin species. Disruption to vegetation that alters the frequency/nature of these natural light gaps could in turn disrupt the suitability of the site for leks.

More generally, artificial light and air pollution have the potential to change the nature of light [73], which in turn could alter the perception of colour. Much of the work in disruption of visual signaling in fish centers on alteration of the light environment, which in turn may alter colour perception in receivers (reviewed in [2]). Such alteration of the light environment can decrease the information content of visual cues to receivers and lead to increased interspecific matings and hybridization, as species-specific colour markers are lost. While this likely would not be the case in birds, which also rely upon species-specific auditory signals in recognition, disruption of perceived colours could easily occur through changes in light environments, but the relative importance of perceived color in promoting hybridization is unknown. Some artificial lights have restricted wavelengths, and can alter the perception of colour in plumage. This could decrease the reflectance of visual cues, including those presented by ultraviolet plumage, which are important in the sexual displays of some species [74]. Similarly, increased amounts of fine particles in the air, such as those arising from air pollution, can refract light and alter the perceived reflectance of plumage. This may even dampen perceived colour via simple accumulation of industrial grime on plumage [75]. Griggio et al. [76] exposed throat feathers from taxidermied starling specimens to atmospheric pollutants (dust/smoke, pollution, etc.) in one of the most polluted valleys in Italy and compared them to feathers kept in sealed plastic bags. The exposed feathers had significantly decreased transmission, particularly in the UV light range, over 3–6 weeks. Control (bagged) feathers had no such reduction in light transmission. Light transmittance continued to decrease over time, and was most strongly affected in the UV and shorter wavelengths, which correspond to known sexual signals in starling plumage [74]. Birds, however, are likely to use multiple signals in mate assessment [53], and transmission of one signal may simply shift reliance onto other signals. Further, some forms of this kind of anthropogenic impact, such as accumulated grime on feathers, may even be mitigated by the birds themselves, by increased preening behavior [76].

Urbanization and Signaling

For most species, urban environments are generally considered poorer habitat than rural environments. A meta-analysis of studies comparing urban vs rural breeding populations of multiple species [77] found a general pattern of earlier laying, but with smaller clutch sizes, reduced fledging success and smaller nestling mass in urban landscapes. Similar patterns continue to emerge, suggesting that many species considered "urban adapters" [78] have lower productivity when breeding in urban environments [79, 80]. There are multiple explanations for these types of patterns [81, 82], but perhaps two of the best explored are the effects of altered food availability and industrial toxicity.

Winter food supplementation, as well as slightly warmer temperatures offered by the heat-island effect in cities, may explain advanced laying dates [77, 79, 82], but the food types typically used in winter are not those used to feed nestlings; indeed, insect prey limitation may be one source of reduced nestling condition in cities [77, 79–81]. A recent detailed study showed that the density of caterpillars, preferred for nestling provisioning, was lower in urban versus forested areas, and resulted in reduced nestling provisioning and survival in blue tits [83]. Further, supplementation of insect prey during nestling provisioning alleviated some of the negative effects on nestlings [80], pointing to resource availability as one of the major factors affecting differences in habitat quality in urban vs rural environments.

An additional consequence of living in highly industrialized landscapes is exposure to toxic chemicals. Eens et al. [84] found that levels of heavy metals in the feathers of both blue and great tits were correlated with proximity to industrial pollution sources. Although neither Dauwe et al. [85] or Snoeijs et al. [86] found differences in the morphology or haematocrit levels of great tits relative to the source of pollution, Snoeijs et al. [86] did find that humoral response to an immune challenge was highest among birds living farthest from contamination sources. This suggests that chemical pollutants may decrease the immunological condition of birds, and could exacerbate differences across habitats already created by differences in relative prey availability.

How, then, do these consequences of living in urban environments impact trait expression? Gorissen et al. [87] found that great tits living near pollution sources had smaller repertoires and lower song output than those living farther from contaminants. Other work has shown that song structure and/or brain areas associated with song complexity are related to levels of specific pollutants including polychlorinated biphenyls (PCBs) [88], mercury ([89]; but see also [90]), and brominated flame retardants [91]. Similarly, Geens et al. [92] found that the chroma (purity of colour) and hue (colour) of carotenoid-based breast plumage was reduced in more polluted sites among both adults and nestlings. Pérez et al. [93] concluded that contaminants found in oil pollution correlated with a reduction in the size of the sexually-selected, carotenoid-based red bill spot in gulls. Recently, Grunst et al. [94] found similar effects, with UV chroma lower in birds closer to pollution sources, and carotenoid chroma lowest in birds settling closest to roads; in this study both the proximity to roads and the pollution source (smelter) were associated with higher heavy metal levels in feathers. There is also evidence of colour disruption in species with structural colouration; for example, mercury contamination affects plumage on belted kingfishers [95]. In this case, the effect – brightening of blue back feathers and dulling of white breast feathers – may be due to interference by mercury in the melanin production pathway. Melanin organization is essential for the colouration of structural blue feathers [46], and the chemical pathway for the body to produce melanin requires the activation of the enzyme tyrosinase to catalyze the reaction; tyrosinase is activated when the element copper (a co-factor to the enzyme) binds to tyrosinase. Because of its similarity in chemical structure, the element mercury can bind competitively to the activation site on tyrosinase instead of copper; mercury, though, does not activate tyrosinase and so the enzyme cannot catalyze its normal reaction when mercury binds to it, thus impairing the melanin production pathway. The result is that contaminants can disrupt the normal production of melanin, which in turn can disrupt the microstructure of the feathers, altering the scattering of white light and resulting in changes to the brightness and hues of feathers. How much these colours affect mate choice in kingfishers is currently untested but colour dimorphism between males and females may suggest that the structural colours are sexually selected.

These changes in song and colours may point to toxin-induced differences in condition but could also be compounded by general differences in prey availability in urban landscapes. Carotenoid-based plumage colouration relies on acquisition of these pigments through diet, and availability may be reduced in urban sites either through lower overall food availability or lower carotenoid-bearing food. Hõrak et al. [96] found that great tits reared in urban sites had lower yellow carotenoid levels in breast feathers than those in rural nests. Moreover, while rural nestlings transplanted to urban nests showed a reduction in feather colour compared to their siblings reared in the original home nest, the reciprocal was not the case – urban birds transplanted to rural nests remained similarly depressed in expressing yellow plumage as their non-transplanted siblings. Biard et al. [97] studied the colour of nestling great tits in two rural and two urban locations in France. While the brightness of nestlings' yellow feathers did not vary between the types of sites, chroma was significantly higher in woodlands; Biard et al. [97] suggest that this was due to a reduced ability on the part of parents to deposit carotenoids in eggs and/or feed chicks carotenoid rich food in urban environments. Similarly, Giraudeau et al. [98] measured carotenoid based plumage in house finches at multiple sites that differed in a continuum from urban center to desert in Arizona. Males were reddest and heaviest in the less industrialized sites. These findings suggest that birds may be in poorer overall condition in urban areas, and this is expressed in their plumage. This effect is sufficiently widespread that some have suggested using trait expression as a proxy for assessing the relative quality of different habitats [36, 52]. However, as with most aspects of urbanization studies, there may be subtleties to such generalizations.

Some species may benefit from resources available in urban environments; for example, abundance of carotenoid-containing berry bushes may increase with ornamental plants in suburban landscapes [99]. The red plumage of northern cardinals (*Cardinalis cardinalis*) is tightly correlated with body condition, but while city birds had overall duller plumages, the relationship between condition and trait expression was depressed [99]. This suggests that general availability of carotenoid-bearing plants may allow males in poorer condition to express traits at similar levels to those in better condition, making the signal less reliable in urban areas. Indeed, Rodewald and Arcese [100] found female selection pressure seemed to be relaxed in urban habitats, likely due to greater homogeneity among males/territories in such environments. Similarly, some urban populations of house finches may not be as greatly hampered by the overall pattern of reduced carotenoids in plumage as others. Giraudeau et al. [98] tested females from three populations for preferences for male colours. They used carotenoid diet-deprived males, who were artificially coloured with marker pens to manipulate colour independent of male condition; females from two of the three populations tested preferred the reddest males, even if males in their population were not on average this red. One population of females, however, preferred yellower males; interestingly, males in their population were much yellower on average. This indicates that sexually preferred traits may also shift in some habitats, depending on what females experience. This adaptation, however, would require low levels of gene flow among populations.

Such restricted movement and differential selection within different environments has been suggested in house finches, where Badyaev et al. [101] compared the bill morphology of desert-dwelling vs urban-dwelling house finches, finding that harder seeds in urban sites appear to select for larger bills. However, larger bills can also constrain the motor ability of birds to sing rapidly while simultaneously spanning broad frequency bandwidths [19], which is considered a performance constraint upon which female birds may assess males [20, 21]. Indeed, Badyaev et al. [101] also found that urban birds had slower trill rates and fewer notes in their songs than rural birds. This could be viewed as compromised song performance. This examination of change in song structure was coupled with genetic analysis that showed high differentiation between the urban and rural populations, suggesting that reproductive. isolation may be occurring and allowing divergence. Job et al. [102] found a similar compromise associated with urban song; house wrens in more urban habitats reduced the overall frequency bandwidth of their songs in response to urban noise but did not correspondingly increase their trill rates. This could lower perception of the urban males' song performance relative to rural males. Indeed, when noise broadcasts were played to unpaired vs paired male house wrens, paired males shifted their songs up to a higher, but narrower, frequency bandwidth to compensate. Unpaired males did not shift their songs, suggesting that unpaired males may be constrained to sing songs preferred by females even if these may have reduced transmission [103]. Similar changes in song structure in cardinals [104] and white-crowned sparrows [105, 106] suggest that attempting to compensate for urban noise may compromise perceived song performance. In contrast, male European starlings exposed to endocrine disrupting chemicals (EDCs) had enlarged HVC volumes (*High Vocal Center*; a brain region associated with song complexity), and longer and more complex songs, which were preferred by females [107]. However, exposure to EDCs also caused reduced immune response in these males. Thus, exposure to pollutants in this case made a sexually selected signal of male quality less honest, which could have population-level consequences if mating with immunocompromised males decreased female fitness.

As mentioned previously, nutrition is another contributor to singing performance, particularly to song rates or output over extended periods. As urban areas are typically associated with lower food availability, one might expect reduced song output in urban vs rural settling birds. However, this has not necessarily been the case. Kempenaers et al. [108] reported higher song output rates of male blue tits living in close proximity to artificial light sources. While they equate this to the direct impact of light inducing increased dawn singing, our own studies have found that when controlling for the dominance ranks of male chickadees, males singing near street lamps begin chorusing earlier, but do not sing for longer or at higher rates (K.A. Otter, unpublished data). This suggests that light pollution is more likely to clock-shift the chorus than it is to increase overall song output, the latter of which may instead be more limited by food availability. Blue tits that sing near street lights are also likely to be in closer proximity to artificial food sources, which may not only advance breeding in urban-dwelling birds, but could also influence the relative condition of males early in the breeding season when song output is typically highest. In our own studies on mountain chickadees, we found an increased song output during the early-season dawn chorus in urban vs rural birds; this was not only associated with singing earlier, but also with singing for longer and at higher rates of songs/minute [109]. In this context, the song output of urban-dwelling males may help compensate for potential differences in other sexual signals (e.g., plumage), and allow some degree of phenotypic plasticity for mate choice decisions among birds settling in urban landscapes. Despite being considered a habitat specialist, mountain chickadees breeding in urban sites have comparable fledging rates and nestling growth rates to those characteristic of rural birds [110], suggesting that it is possible to adjust to some of the challenges of urban living.

An interesting facet of anthropogenic impacts on signaling is that living in proximity to humans may facilitate aspects of signals that do not occur in more natural settings. Various bird species make use of anthropogenic features in their signals, including the many vocal mimics that incorporate car alarms and other city sounds into their repertoires [111]. For species where the diversity of sounds produced is under directional sexual selection, the addition of novel stimuli to integrate into their repertoire may enhance attractiveness as a potential mate. Similarly, the extended phenotype that males of some species use to attract females can also be affected by humans. For example, male bower birds decorate with both natural and artificial objects (e.g., bits of glass and plastic) collected from their environments, the colour of the materials selected varying by species [112]. Satin bowerbirds (*Ptilonorhynchus violaceus*) prefer blue. The number of blue objects in a bower is positively related to female visitation rate, while the availability of such objects in the environment is negatively related to theft of decorations by other males [113]. Rosenthal and Stuart-Fox [114] note that if human disturbance greatly increased availability of blue objects in the environment, this might reduce variation among males in their ability to collect suitable decorations, compromising the reliability of this as an honest signal and leading to reduced sexual selection on this trait. In contrast, integration of human-made items into displays may enhance perception of quality in other species; black kites (*Milvus migrans*), which use nest decoration as an honest signal of territory quality and social status, preferentially collect white, especially artificial (e.g., plastic) objects. Birds with more of these objects in their nests are in better condition, and although they suffer from fewer territorial intrusions, birds with more decorations were more successful at repelling intruders than birds with few decorations [115]. This stresses the complexity of determining whether anthropogenic disturbance will enhance or detract from the perceived signal value of traits, as this is highly context-dependent and specific to how the life-history of individual species interacts with the changes associated with anthropogenic disturbance.

Disruption to Reproduction and Mating Systems from Anthropogenic Disturbance

The expression-level of signals may be impacted by anthropogenic disturbance, but this is only one aspect of signal assessment in sexual selection. Any land change that impacts density and spacing of birds may also disrupt mate choice. For choosers, the ability to directly compare the quality of potential mates is an important feature of most mating systems, but is perhaps most acute in lek-based mating systems. Changes in habitat quality due to anthropogenic disturbance, including oil and gas or wind development, forestry and agricultural practices, and urban development, can influence the spacing of leks and the density of lek attendants [116–120]. While these effects are often considered from a population monitoring standpoint, their impact on mating systems is often overlooked, despite mating systems and mate choice being clearly linked to density (reviewed by [121]). In lek-based mating systems, reproductive success is often limited to one or several males (e.g., [122]), which provide females with only indirect benefits [121]. Females visit leks to assess males, mating with males that have the best visual displays or plumage traits [123, 124] and in some cases, females are more likely to visit (and mate at) larger leks (reviewed in [125]). When density is reduced due to anthropogenic disturbance, how does this influence male mating skew in leks and the intensity of sexual selection? Models suggest that as the availability of mates declines, so too may female selectivity, and indiscriminate mating may result [121]. However, it is also possible that mate choice errors may be reduced at lower density; for example, the frequency of "spillover copulations" that occur when less suitable males crowd near dominant males in a lek may be lower at lower densities [125–127]. At very low density, selection could be relaxed leading to the reduction of ornaments [121]. Thus, the impacts of land change on lek mating systems is difficult to predict a priori, and requires study of the specific system under investigation.

In territorial species with social monogamy, anthropogenic changes to habitat can also influence density and, as a consequence, sexual selection through both primary mate choice and extra-pair mate choice [128, 129]. For example, in gray catbirds (*Dumatella carolinensis*), eastern bluebirds (*Sialia sialis*), blue tits, American redstarts (*Setophaga ruticilla*) and reed buntings (*Emberiza schoeniclus*), rates of extra-pair paternity were higher at higher breeding densities [130–134]. In tree swallows, an interaction between density and colouration on determining extra-pair mating success suggests that successful males had different colour attributes when they occurred in occurred high vs. low density [135]. In addition to altering breeding density, variation in habitat quality can also affect territorial behavior; Fort and Otter [136] found chickadees that occupy young forest (lower-quality habitat) had much greater territorial overlap and lower levels of defense of territorial boundaries than males in mature forests (high-quality habitats). Could this in turn suggest there are more opportunities for extra-pair copulations (EPCs) in such habitats due to increased contact rates between non-mated males and females? Where anthropogenic alteration of the landscape influences breeding density or intersexual contact rates, there may be potential for changing the strength and direction of, or opportunity for, sexual selection.

There are few studies that have directly compared mating tactics, such as extra-pair paternity rates, between urban and rural populations within the same species. The few that have find contrasting patterns among the different species tested. In some species, urbanization may increase mating opportunities due to increased resources and/or breeding density. Urban Cooper's hawks (*Accipiter cooperi*) had high rates of extra-pair young in nests (19.3%) compared to the low rates typically found in raptors (0–11%) [137]. Rosenfeld et al. [137] suggest that because Cooper's hawks tend to copulate when the male returns with food during pre-nesting, that females might accept or solicit EPCs from other males in this food-rich, high breeding-density setting. Smith et al. [138] found higher extra-pair paternity rates among spotted towhee (*Piilo maculatus*) nests on the edges of urban parks; the authors proposed that anthropogenic food sources available near these edge habitats may attract individuals, increasing encounter rates between females and potential extra-pair sires. Blue tits in areas lit by artificial street lights (which may also have been closer to urban food sources) experience elevated success in extra-pair matings [108]. However, other studies have found little evidence that paternity is disrupted in anthropogenic habitats.

Bonderud et al. [139] found no difference in extra-pair paternity rates between urban and rural nesting mountain chickadees. Similarly, despite urban populations having higher nesting densities, Rodriguez-Martínez et al. [140] did not find this affected either the extra-pair paternity rates or intra-specific brood parasitism rates of burrowing owls (*Athene cunicularia*). Perhaps the differences in the studies above reflect how disturbance affects the habitat quality as perceived by different species. If anthropogenic disturbance provides increased resource availability and alters condition-dependent signaling, it may affect not only density but the propensity of females to assess males worthy of extra-pair matings. In other species, anthropogenic landscapes do not appear to promote these alternate mating tactics. It is interesting to note that, despite limited research on the topic, none of the species studied experienced reduced extra-pair behavior in urban areas. This is an area of research that is worthy of further investigation.

Finally, it is also possible that divergent sexual selection due to urbanization could facilitate speciation events, although there is a dearth of research on the subject. Most relevant is a study on dark-eyed juncos (*Junco hyemalis*), which found a decrease in the amount of white in tail feathers—a sexually selected signaling trait—in a recently established urban population compared to nonurban populations [141]. Changes in tail whiteness were suggested to be caused by an increase in length of the breeding season in the urban environment [142]. Such a divergence could potentially contribute to premating isolation between urban and nonurban populations, but more research is needed into this subject [143].

Recommendations for Management

Making general recommendations for management of anthropogenic disruption to sexual selection is difficult. Unlike noise pollution, which has a somewhat general masking impact on organisms that signal within a particular band-width, how the expression of sexually selected signals is impacted by anthropogenic disturbance is dependent on how disturbance affects the quality of the habitat as it is experienced by the birds. If disturbance reduces food availability, as appears to be the case with many insectivorous species occupying urban habitats, there may be reduced capacity for courters to produce condition-dependent signals and this could reduce the quality of mating displays. This in turn can disrupt mate choice, and lead to overall reductions in mating success. Supplemental provisioning of food may alleviate some of these impacts, but typical supplementation via seed feeders could have undesired effects. For example, song sparrow (*Melospiza melodia*) pairs that were provided with extra food produced sons with repertoires that were smaller than those of their fathers and smaller than those of unfed pairs nesting in the same area, due to fed pairs producing more, but smaller, offspring [144]. Winter feeding may elevate male condition and increase condition-dependent signalling early in the season, but it may also fail to provide resources necessary for successful nestling provisioning later in the season, leaving urban landscapes as potential ecological traps [145]. Perhaps a better long-term strategy for creating positive urban landscapes is to increase the density of deciduous trees that provide foraging habitat, but we recommend these plans focus on tree species typical of the region rather than non-native ornamentals. Retention or restoration of insect-producing shrub/grasslands (again focusing on native vegetation) may also supplement food supplies, particularly for edge-associated or grassland species occupying urban landscapes. Carotenoid-rich exotics can result in courters expressing elevated signals in urban landscapes, but which do not necessarily reflect the courter's true condition or the suitability of their territory for breeding [146]. However, urban landscapes can also result in depressed expression of carotenoid-dependent plumage in other species, so even increasing green spaces containing natural food sources within cities may only partially alleviate negative impacts. Reducing pollution (e.g., particulates) and toxin exposure may not only increase individual condition, but may also help alleviate the impact these have on obscuring light transmission or soiling plumage. Thus, the best means of creating urban environments that allow for evolved systems of sexual selection is to simply focus on making our urban landscapes as naturalistic as possible.

A recent study has highlighted the potential that altering our landscapes to alleviate pressures on sexual signaling can have, as well as the resilience of birds to respond to such changes. Derryberry et al. [147] recorded white-crowned sparrows in the reduced noise-scapes of San Francisco during the COVID-19 shutdown. The reduction in traffic during spring 2020 shutdowns resulted in a reduction to noise levels not seen since the 1950s. This not only made birds more audible, but males responded to the noise relief by decreasing the minimum frequencies of the trills at the end of their songs. The result was that the terminal trills of males had broader bandwidths in urban areas than they had had pre-COVID lockdown. As noted previously, performing rapid trills with broad bandwidth is considered a performance constraint on song [20, 21]. Previous studies on this species [105] suggested that urban males constrained to produce lower bandwidths to avoid being masked by noise pollution may, as perceived by choosing females, have potentially compromised signals. Derryberry et al. [147] showed that within a single year, the alleviation of this masking noise pollution resulted in urban males producing songs with lower minimum frequencies and larger bandwidths than males in these same areas in preceding years. More importantly, the songs of urban males in 2020 also did not differ from those of rural males in the years both preceding and during COVID. This suggests that urban planning that reduces noise (or other impediments to signaling), could result in rapid rebound effects.

Even if noise and other sources of pollution are addressed, the variability in spacing of birds in urban areas relative to wild spaces may impact mating systems. Low densities and irregular inter-individual spacing could decrease the ability of choosers to assess the signals of multiple courters, thus decreasing discriminating mate choice. However, concentrated settlement into small patches of suitable, available habitat may have the alternate effect of promoting increased density for some species, and make assessment of multiple courters easier. The ideal solution may be to plan cities with retention of sufficient native vegetation to create regular and even settlement of species, rather than either clumped or scattered settlement. Even if the settlement densities are lower than native habitat, spacing may allow better opportunities for maintaining signal transmission and assessment in a network-like fashion than either clumping or scattering populations. This in turn may alter mating tactics, such as the rate of extra-pair matings. Thus, understanding how urbanization impacts urban-settling species will require research into how changes to urban landscapes interact with individual species' life history traits, and whether these allows them to adjust to the challenges of city life. This is what makes this field interesting.

Box 1   Consideration of signalling and sexual selection in urban management plans

Any conservation plan needs to account not only for how disturbance affects presence/absence of species, but how it affects a species' ability to produce and perceive signals.

All aspects of a species' natural history rely upon individuals being able to advertise for, or assess the advertisements of, potential mates.

Anthropogenic disturbance that impedes visual (plumage) or auditory (song) sexual signals can potentially disrupt avian mating systems, due to altering either detection (Chapter 8) or signal expression (this chapter).

Figure 1   Carotenoid-base visual signals (Pine Grosbeak) and auditory signals (Mountain Chickadee) in urban-dwelling birds.
(photo credits Mountain Chickadee - Cara Snell; Pine Grosbeak - Glen Dreger)

Honest Signalling and Sexual Selection

- In most bird species, *choosers* (typically females) select among *courters* (typically males) based on assessing the expression of secondary sexual traits, such as song or plumage.

- Selection of particular courters by choosers often relies on a strong correlation between the level of trait expression (e.g., redness of plumage, song output) and the courter's physical condition or ability to gain access to resources, such as food/territories.

  ○ This association between courter condition and trait expression makes the advertisement an 'honest signal' to the chooser.

- This association works so long as anthropogenic disturbance doesn't disrupt the expression of courter signals, or the chooser's ability to perceive them.

Anthropogenic impacts

- In many environments, the level of trait expression is strongly linked to the resource quality of the environment; the more food in the territory a courter can secure, the more energy they can invest in high song rates and/or elaborate plumage colours.

- Habitat disturbance can impact signals if it alters habitat quality.

  ○ This may arise through decreased food availability (e.g., insect larvae) in urban landscapes, or through industrial pollutants affecting avian health.

    ▪ Both of these might lower the ability of courters to produce sexually selected signals, and thus reduce their attractiveness to choosers.

Box 1 (Contd....)

- ○ In other urban landscapes, supplementation of food (bird feeders), presence of novel food sources (non-native plants), addition of additional nest decorations, and/or increased temperatures (heat-island effects) may actually increase the expression of sexual signals.
  - ▪ Addition of resources becomes problematic if it allows all courters high trait expression, irrespective of their age, condition or quality of the territory for nesting; if there is diminished difference between level of trait expression in courters, it reduces the utility of these signals for choosers. This in turn can affect mate choice and mating success.
- Finally, disturbance may also impact avian mating systems by affecting spacing.
  - ○ Lowered habitat quality could reduce the sizes or spacing of leks, disrupting female mate choice.
  - ○ Changes in density can also alter inter-individual contact rates, potentially disrupting extra-pair mating opportunities.
- Because the impact of urbanization may affect individual species in different ways, detailed monitoring of the health/signaling of species of specific management concern may be required.

LITERATURE CITED

[1] Lim, M.L.M., N.S. Sodhi and J.A. Endler, J.A. 2008. Conservation with sense. Science. 319: 281.

[2] Delhey, K. and A. Peters. 2016. Conservation implications of anthropogenic impacts on visual communication and camouflage. Conserv. Biol. 31: 30–39.

[3] Andersson, M. 1994 Sexual Selection. Princeton University Press.

[4] Achorn, A.M. and G.G. Rosenthal. 2020. It's not about him: mismeasuring 'good genes' in sexual selection. TREE. 35: 206–218.

[5] Maynard-Smith, J. and D. Harper. 2003. Animal Signals. Oxford University Press.

[6] Ortega, C.P. 2012. Effects of noise pollution on birds: a brief review of our knowledge. Ornithol. Monogr. 74: 6–22.

[7] Slabbekoorn, H. 2013. Songs of the city: noise-dependent spectral plasticity in the acoustic phenotype of urban birds. Anim. Behav. 85: 1089–1099.

[8] Catchpole, C.K. and P.J.B. Slater.. 2008. Bird Song: Biological Themes and Variations, 2nd Ed. Cambridge University Press.

[9] Stoddard, P.K., M.D. Beecher, S.E. Campbell and C.L. Horning. 1992. Song-type matching in the song sparrow. Can. J. Zool. 70: 1440–1444.

[10] Beecher, M.D., P.K. Stoddard, E.S. Campbell and C.L. Horning. 1996. Repertoire matching between neighbouring song sparrows. Anim. Behav. 51: 917–923.

[11] Todt, D. and M. Naguib, 2000. Vocal interactions in birds: the use of song as a model in communication. Adv. Study Behav. 29: 247–296.

[12] Foote, J.R., L.P. Fitzsimmons, D.J. Mennill and L.M Ratcliffe. 2008. Male chickadees match neighbors interactively at dawn: support for the social dynamics hypothesis. Behav. Ecol. 19: 1192–1199.

[13] Kempenaers, B., G.R. Verheyen, M. van den Broeck, T. Burke, C. van Broeckhoven and A.A. Dhondt. 1992. Extra-pair paternity results from female preference for high-quality males in the blue tit. Nature. 357: 494–496.

[14] Otter, K., B. Chruszcz and L. Ratcliffe. 1997. Honest advertisement and singing during the dawn chorus of Black-capped Chickadees, *Parus atricapillus*. Behav. Ecol. 8: 167–173.

[15] Welling, P.P., S.O. Rytkönen, K.T. Koivula and M.I. Orell. 1997. Song rate correlates with paternal care and survival in willow tits: advertisement of male quality? Behaviour. 134: 891–904.

[16] Mennill, D.J., L.M. Ratcliffe and P.T. Boag. 2002. Female eavesdropping on male song contests in songbirds. Science. 296: 873.

[17] Mennill, D.J., P.T. Boag and L.M. Ratcliffe. 2003. The reproductive choices of eavesdropping female black-capped chickadees, *Poecile atricapillus*. Naturwissenschaften. 90: 577–582.

[18] Poesel, A., T. Dabelsteen and S.B. Pedersen. 2004. Dawn song of male blue tits as a predictor of competitiveness in midmorning singing interactions. Acta Ethol. 6: 65–71.

[19] Podos, J. 2001. Correlated evolution of morphology and vocal signals in Darwin's finches. Nature. 409: 185–188.

[20] Ballentine, B. 2009. The ability to perform physically challenging songs predicts age and size in male swamp sparrows, *Melospiza georgiana*. Anim. Behav. 77: 973–978.

[21] Logue, D.M., J.A. Sheppard, B. Walton, B.E. Brinkman and O.J. Medina. 2019. An analysis of avian vocal performance at the note and song levels. Bioacoustics. DOI: 10.1080/09524622.2019.1674693.

[22] Grafen, A. 1990. Biological signals as handicaps. J. Theor. Biol. 144: 517–546.

[23] Manica, L.T., R. Maia, A. Dias, J. Podos and R.H. Macedo. 2014. Vocal output predicts territory quality in a Neotropical songbird. Behav. Processes. 109: 21–26.

[24] Kempenaers, B., G.R. Verheyen and A.A. Dhondi. 1997. Extrapair paternity in the blue tit (*Parus caeruleus*): female choice, male characteristics, and offspring quality. Behav. Ecol. 8: 481–492.

[25] Murphy, M.T., K. Sexton, A.C. Dolan and L.J. Redmond. 2008. Dawn song of the eastern kingbird: an honest signal of male quality? Anim. Behav. 75: 1075–1084.

[26] Møller, A.P. 1991. Parasite load reduces song output in a passerine bird. Anim. Behav. 41: 723–730.

[27] Bischoff, L.L., B. Tschirren and H. Richner. 2009. Long-term effects of early parasite exposure on song duration and singing strategy in great tits. Behav. Ecol. 20: 265–270.

[28] Alatalo, R.V., C. Glynn and A. Lundberg. 1990. Singing rate and female attraction in the Pied Flycatcher: an experiment. Anim. Behav. 39: 601–603.

[29] Cuthill, I.C. and W.A. Macdonald. 1990. Experimental manipulation of the dawn and dusk chorus in the Blackbird Turdus merula. Behav. Ecol. Sociobiol. 26: 209–216.

[30] Lucas, J.R., A. Schraeder and C. Jackson. 1999. Carolina chickadee (Aves, Paridae, *Poecile carolinensis*) vocalization rates: effects of body mass and food availability under aviary conditions. Ethology. 105: 503–520.

[31] Thomas, R.J. 1999. The effect of variability in the food supply on the daily singing routines of European Robins: a test of a stochastic dynamic programming model. Anim. Behav. 57: 365–369.

[32] Grava, T, A. Grava and K.A. Otter. 2009. Supplemental feeding and dawn singing in black-capped chickadees. Condor. 111: 560–564.

[33] Godfrey, J.D. and D.M. Bryant. 2001. State-dependent behaviour and energy expenditure: an experimental study of European robins on winter territories. J. Anim. Ecol. 69: 301–313.

[34] Hill, G.E., P.M. Nolan and A.M. Stoehr. 1999. Pairing success relative to male plumage redness and pigment symmetry in the house finch: temporal and geographic constancy. Behav. Ecol. 10: 48–53.

[35] Hill, G.E. 2002. A Red Bird in a Brown Bag: The Function and Evolution of Colourful Plumage in the House Finch. Oxford: Oxford University Press.

[36] Hill, G.E. 1995. Ornamental traits as indicators of environmental health. Bioscience. 45: 25–31.

[37] Thompson, C.W., N. Hillgarth, M. Leu and H.E. McClure. 1997. High parasite load in house finches (*Carpodacus mexicanus*) is correlated with reduced expression of sexually selected trait. Am. Nat. 149: 270–294.

[38] Hill G.E. and R. Montgomerie. 1994. Plumage colour signals nutritional condition in the house finch. Proc. Roy. Soc. Lond., B. 258: 47–52.

[39] Hill, G.E. 2000. Energetic constraints on expression of carotenoid-based plumage colouration. J. Avian Biol. 31: 559–566.

[40] Dufva R. and K. Allander. 1995. Intraspecific variation in plumage coloration reflects immune response in great tit (*Parus major*). Funct. Ecol. 9: 785–789.

[41] Weaver, R.J., E.S.A. Santos, A.M. Tucker, A.E. Wilson and G. Hill. 2018. Carotenoid metabolism strengthens the link between feather coloration and individual quality. Nature Communications. 9: 73.

[42] Garratt, M. and R.C. Brooks. 2012. Oxidative stress and condition-dependent sexual signals: more than just seeing red. Proc. Roy. Soc. Lond., B. 279: 3121–3130.

[43] Weaver, R.J., R.E. Koch and G.E. Hill. 2017. What maintains signal honesty in animal colour displays used in mate choice? Philos. Trans. R. Soc. Lond., B. Biol. Sci. 372: 20160343.

[44] Keyser, A.J. and G.E. Hill. 1999. Condition-dependent variation in the blue-ultraviolet coloration of a structurally based plumage ornament. Proc. Roy. Soc. Lond., B. 266: 771–777.

[45] Siefferman, L. and G.E. Hill. 2003. Structural and melanin coloration indicate parental effort and reproductive success in male Eastern Bluebirds. Behav. Ecol. 14: 855–861.

[46] Shawkey, M.D. and G.E. Hill. 2006. Significance of a basal melanin layer to production of non-iridescent structural plumage color: evidence from an amelanotic Steller's Jay (*Cyanocitta stelleri*). J. Exp. Biol. 209: 1245–1250.

[47] White, T.E. 2020. Structural colours reflect individual quality: a meta-analysis. Biol. Lett. 16: 20200001.

[48] Siefferman, L. and G.E. Hill. 2005. Evidence for sexual selection on structural plumage coloration in female Eastern Bluebirds (*Sialia sialis*). Evolution. 59: 1819–1828.

[49] Hill, G.E. 2006. Female choice for ornamental colouration. Bird Coloration. 2: 137–200.

[50] Guindre-Parker S. and O.P. Love. 2014. Revisiting the condition-dependence of melanin-based plumage. J. Avian Biol. 45: 29–33.

[51] Roulin, A. 2015. Condition-dependence, pleiotropy and the handicap principle of sexual selection in melanin-based colouration. Biol. Rev. 91(2): 328–348.

[52] Godfrey, J.D. 2003. Potential use of energy expenditure of individual birds to assess quality of their habitats. Sci. Conserv. 214: 11–24.

[53] Otter, K.A. and L. Ratcliffe. 2005. Enlightened decisions: female perspectives on communication networks. pp. 133-151. *In*: P.K. McGregor (ed.). Communication Networks. Cambridge University Press, Cambridge, UK.

[54] Otter, K.A., H. van Oort and K.T. Fort. 2007. Habitat quality, reproduction and behaviour in chickadees and titmice: insights on the use of habitat matrixes in conservation? pp. 277–298. *In*: K.A. Otter (ed.). Ecology and Behavior of Chickadees and Titmice: An Integrated Approach. Oxford University Press, Oxford, UK.

[55] Blondel, J. and P.C. Dias. 1994. Summergreenness, evergreenness and life history variation in Mediterranean blue tits. pp. 25–36. *In*: M. Arianoutsou and R.H. Groves (eds). Plant-Animal Interactions in Mediterranean-Type Ecosystems. Netherlands, Kluwer Academic Publishers.

[56] Blondel, J., P.C. Dias, M. Maistre and P. Perret. 1993. Habitat heterogeneity and life-history variation of mediterranean blue tits (*Parus caeruleus*). Auk. 110: 511–520.

[57] Blondel, J., M. Maistre, P. Perret, S. Hurtrez-Boussčs and M.M. Lambrechts. 1998. Is the small clutch size of a Corsican blue tit population optimal? Oecologia. 117: 80–89.

[58] Lambrechts, M.M., S. Caro, A. Charmantier, N. Gross, M.J. Galan, P. Perret, et al. 2004. Habitat quality as a predictor of spatial variation in blue tit reproductive performance: a multi-plot analysis in a heterogeneous landscape. Oecologia. 141: 555–561.

[59] Fort, K. and K.A. Otter. 2004. Effects of habitat disturbance on reproduction in black-capped chickadees (*Poecile atricapillus*) in Northern British Columbia. Auk. 121: 1070–1080.

[60] van Oort, H., K.A. Otter, K. Fort and Z. McDonell. 2007. Habitat, dominance and the phenotypic quality of male black-capped chickadees. Condor. 109: 88–96.

[61] Otter, K.A., S.E. Atherton and H. van Oort. 2007. Female food-solicitation calling, hunger levels and habitat differences in the black-capped chickadee. Animal Behaviour. 74: 847–853.

[62] van Oort, H., K.A. Otter, K. Fort and C.I. Holschuh. 2006. Habitat quality, social dominance and dawn chorus song output in black-capped chickadees. Ethology. 112: 772–778.

[63] Otter, K. and L. Ratcliffe. 1996. Female initiated divorce in a monogamous songbird: abandoning mates for males of higher quality. Proc. Roy. Soc. Lond. 263: 351–354.

[64] Ramsay, S.M., K. Otter, D.J. Mennill and L. Ratcliffe. 2000. Divorce & extrapair mating in female black-capped chickadees (*Poecile atricapillus*): separate strategies with a common target. Behav. Ecol. Sociobiol. 49: 18–23.

[65] Christie, P.J., D.J. Mennill and L.M. Ratcliffe. 2004. Pitch shifts and song structure indicate male quality in the dawn chorus of black-capped chickadees. Behav. Ecol. Sociobiol. 55: 341–348.

[66] Grava, T., A. Grava and K.A Otter. 2012. Vocal performance varies with habitat quality in the black-capped chickadee. Behaviour. 149: 35–50.

[67] Grava, T., A. Grava and K.A. Otter. 2013. Habitat-induced changes in song consistency affect perception of social status in male chickadees. Behav. Ecol. Sociobiol. 67: 1699–1707.

[68] Gustafsson, L., A. Qvarnström and B. Sheldon. 1995. Trade-offs between life-history traits and a secondary sexual character in male collared flycatchers. Nature. 375: 311–313.

[69] Sheldon, B.C. and H. Ellegren. 1999. Sexual selection resulting from extrapair paternity in collared flycatchers. Anim. Behav. 57: 285–298.

[70] Hill, G.E. 1993. Geographic variation in the carotenoid plumage pigmentation of male house finches (*Carpodacus mexicanus*). Biol. J. Linn. Soc. 49: 63–86.

[71] Endler, J.A. and M. Thery. 1996. Interacting effects of lek placement, display behavior, ambient light, and color patterns in three neotropical forest-dwelling birds. Am. Nat. 148: 421–452.

[72] Anciães, M. and R.O. Prum. 2008. Manakin display and visiting behaviour: a comparative test of sensory drive. Anim. Behav. 75: 783–790.

[73] Gaston K.J., J. Bennie, T.W. Davies and J. Hopkins. 2013. The ecological impacts of night time light pollution: a mechanistic appraisal. Biol. Rev. Biol. Proc. Camb. Philos. 88: 912–927.

[74] Bennett, A.T.D., I.C. Cuthill, J.C. Partridge and K. Lunau. 1998. Ultraviolet plumage colours predict mate preferences in starlings. PNAS. 94: 8618–8621.

[75] DuBay, S.G. and C.C. Fuldner. 2017. Bird specimens track 135 years of atmospheric black carbon and environmental policy. PNAS. 114: 11321–11326.

[76] Griggio, M., L. Serra and A. Pilastro. 2011. The possible effect of dirtiness on structurally based ultraviolet plumage. Ital. J. Zool. 78: 90–95.

[77] Chamberlain, D.E., A.R. Cannon, M.P. Toms, D.I. Leech, B.J. Hatchwell and K.J. Gaston. 2009. Avian productivity in urban landscapes: a review and meta-analysis. Ibis. 151: 1–18.

[78] Blair, R.B. 1996. Land use and avian species diversity along an urban gradient. Ecol. Appl. 6: 506–519.

[79] Bailly, J., R. Scheifler, S. Berthe, V.A. Clément-Demange, M. Leblond, B. Pasteur and B. Faivre. 2016. From eggs to fledging: negative impact of urban habitat on reproduction in two tit species. J. Ornithol. 157: 377–392.

[80] Meyrier, E., L. Jenni, Y. Bötsch, S. Strebel, B. Erne and Z. Tablado. 2017. Happy to breed in the city? Urban food resources limit reproductive output in Western Jackdaws. Ecol. Evol. 7: 1363–1374.

[81] Seress, G. and A. Liker. 2015. Habitat urbanization and its effects on birds. Acta Zool. Acad. Sci. Hung. 61: 373–408.

[82] Isaksson, C. 2018. Impact of Urbanization on Birds. pp. 235–257. *In:* D. Tietze (ed). Bird Species. Fascinating Life Sciences. Springer, Cham.

[83] Pollock, C.J., P. Capilla-Lasheras, R.A.R. McGill, B. Helm and D.M. Dominoni. 2017. Integrated behavioural and stable isotope data reveal altered diet linked to low breeding success in urban-dwelling blue tits (*Cyanistes caeruleus*). Sci. Rep. 7: 5014.

[84] Eens, M., R. Pinxten, R.F. Verheyen, R. Blust and L. Bervoets. 1999. Great and blue tits as indicators of heavy metal contamination in terrestrial ecosystems. Ecotoxicol. Environ. Saf. 44: 81–85.

[85] Dauwe, T., E. Janssens and M. Eens. 2006. Effects of heavy metal exposure on the condition and health of adult great tits (*Parus major*). Environ. Pollut. 140: 71–78.

[86] Snoeijs, T., T. Dauwe, R. Pinxten, F. Vandesande and M. Eens. 2004. Heavy metal exposure affects the humoral immune response in a free-living small songbird, the great tit (*Parus major*). Arch. Environ. Cons. Tox. 46: 399–404.

[87] Gorissen, L., T. Snoeijs, E.V. Duyse and M. Eens. 2005. Heavy metal pollution affects dawn singing behaviour in a small passerine bird. Oecologia. 145: 504–509.

[88] DeLeon, S., R. Halitschke, R.S. Hames, A. Kessler, T.J. DeVoogd and A.A. Dhondt. 2013. The effect of polychlorinated biphenyls on the song of two passerine species. PloS One. 8: e73471.

[89] Hallinger, K.K., D.J. Zabransky, K.A. Kazmer and D.A. Cristol. 2010. Birdsong differs between mercury-polluted and reference sites. Auk. 127: 156–161.

[90] Greene, V.W., J.P. Swaddle, D.L. Moseley and D.A. Cristol. 2018. Attractiveness of male Zebra Finches is not affected by exposure to an environmental stressor, dietary mercury. Condor. 120: 125–136.

[91] Eng, M.L., V. Winter, J.E. Elliott, S.A. MacDougall-Shackleton and T.D. Williams. 2018. Embryonic exposure to environmentally relevant concentrations of a brominated flame retardant reduces the size of song-control nuclei in a songbird. Dev. Neurobiol. 78: 799–806.

[92] Geens, A., T. Dauwe and M. Eens. 2009. Does anthropogenic metal pollution affect carotenoid colouration, antioxidative capacity and physiological condition of great tits (*Parus major*)? Comp. Biochem. Physiol. C. 150: 155–163.

[93] Pérez, C., I. Munilla, M. Lopex-Alonso and A. Velando. 2010. Sublethal effects on seabirds after the *Prestige* oil-spill are mirrored in sexual signals. Biol. Lett. 6: 33–35.

[94] Grunst, M.L., A.S. Grunst, R. Pinxten, L. Bervoets and M. Eens. 2020. Carotenoid- but not melanin-based plumage coloration is negatively related to metal exposure and proximity to the road in an urban songbird. Environ. Pollut. 256: 113473.

[95] White, A.E. and D.A. Cristol. 2014. Plumage coloration in belted kingfishers (*Megaceryle alcyon*) at a mercury-contaminated river. Waterbirds. 37: 144–152.

[96] Hõrak, P., H. Vellau, I. Ots and A.P Møller. 2000. Growth conditions affect carotenoid-based plumage colorationof great tit nestlings. Naturwissenschaften. 87: 460464.

[97] Biard, C., F. Brischoux, A. Meillère, B. Michaud, M. Nivière, S. Ruault, et al. 2017. Growing in cities: an urban penalty for wild birds? A study of phenotypic differences between urban and rural great tit chicks (*Parus major*). Front. Ecol. Evol. 5: 79.

[98] Giraudeau, M., M.B. Toomey, P. Hutton and K.J. McGraw. 2018. Expression of and choice for condition-dependent carotenoid-based color in an urbanizing context. Behav. Ecol. 29: 1307–1315.

[99] Jones, T., A. Rodewald and D. Shustack. 2010. Variation in plumage coloration of northern cardinals in urbanizing landscapes. Wilson J. Ornithol. 122: 326–333.

[100] Rodewald, A. and P. Arcese. 2017. Reproductive contributions of cardinals are consistent with a hypothesis of relaxed selection in urban landscapes. Front. Ecol. Evol. 5: 77.

[101] Badyaev, A.V., R.L. Young, K.P. Oh and C. Addison. 2008. Evolution on a local scale: developmental, functional, and genetic bases of divergence in bill form and associated changes in song structure between adjacent habitats. Evolution 62: 1951–1964.

[102] Job, J.R., S.L. Kohler and S.A. Gill. 2016 Song adjustments by an open habitat bird to anthropogenic noise, urban structure, and vegetation. Behav. Ecol. 27: 1734–1744.

[103] Grabarczyk E.E., M.A. Pipkin, M.J. Vonhof and S.A. Gill. 2018. When to change your tune? unpaired and paired male house wrens respond differently to anthropogenic noise. JEA. 2: #LHGRVC.

[104] Narango, D.L. and A.D. Rodewald. 2016. Urban-associated drivers of song variation along a rural–urban gradient. Behav. Ecol. 27: 608–616.

[105] Luther, D., J. Phillips and E. Derryberry. 2016. Not so sexy in the city: urban birds adjust songs to noise but compromise vocal performance. Behav. Ecol. 27: 332–340.

[106] Derryberry, E.P., R.M. Danner, J.E. Danner, G.E. Derryberry, J.N. Phillips, S.E. Lipshutz, et al. 2016. Patterns of song across natural and anthropogenic soundscapes suggest that white-crowned sparrows minimize acoustic masking and maximize signal content. PLoS One. 11: e0154456.

[107] Markman, S., S. Leitner, C. Catchpole, S. Barnsley, C.T. Müller, D. Pascoe, , K.L. Buchanan. 2008. Pollutants increase song complexity and the volume of the brain area HVC in a songbird. PloS One. 3: e1674.

[108] Kempenaers, B., P. Borgström, P. Loës, E. Schlicht and M. Valcu. 2010. Artificial night lighting affects dawn song, extra-pair siring success and lay date in songbirds. Curr. Biol. 20: 1735–1739.

[109] Marini, K.L.D., M.W. Reudink, S.E. LaZerte and K.A. Otter. 2017. Urban mountain chickadees (*Poecile gambeli*) begin vocalizing earlier, and have greater dawn chorus output than rural males. Behaviour. 154: 1197–1214.

[110] Marini, K.L.D., K.A. Otter, S.E. LaZerte and M.W. Reudink. 2017. Urban environments are associated with earlier clutches and faster nestling feather growth compared to natural habitats. Urban Ecosyst. 20: 1291–1300.

[111] Marx, V. 2018. What makes birds and bats the talk of the town. Nature Methods. 15: 485–488.

[112] Marshall, A.J. 1954. Bower-Birds: Their Displays and Breeding Cycles. Oxford, U.K. Clarendon Press.

[113] Hunter, C.P. and P.D. Dwyer. 1997. The value of objects to Satin Bowerbirds (*Ptilonorhynchus violaceus*). Emu. 97: 200–206.

[114] Rosenthal, G.G. and D. Stuart-Fox. 2012. Environmental disturbance and animal communication. pp. 16–31. *In*: U. Candolin and B. Wong (eds). Behavioural Responses to a Changing World: Mechanism and Consequences. Oxford University Press, Oxford, UK.

[115] Sergio, F., J. Blas, G. Blanco, L. López, J.A. Lemus and F. Hiraldo. 2011. Raptor nest decorations are a reliable threat against conspecifics. Science. 331: 327–330.

[116] Blickley, J.L., K.R. Word, A.H. Krakauer, J.L. Phillips, S.N. Sells, C.C. Taff, et al. 2012. Experimental chronic noise is related to elevated fecal corticosteroid metabolites in lekking male greater sage-grouse. (*Centrocerus urophasianus*). PLoS One. 7(11): e50462.

[117] Hess, J.E. and J.L. Beck. 2012. Disturbance factors influencing greater sage-grouse lek abandonment in north-central Wyoming. J. Wildl. Manage. 76: 1625–1634.

[118] Kouffeld, M.J., M.A. Larson and R.J. Gutiérrez. 2013. Selection of landscapes by male ruffed grouse during peak abundance. J. Wildl. Manage. 77: 1192–1201.

[119] Hovick, T.J., D.K. Dahlgren, M. Papeş, R.D. Elmore and J. Pitman. 2015. Predicting greater prairie-chicken lek site suitability to inform conservation actions. PLoS One. 10(8): e0137021.

[120] Winder, V.L, A.J. Gregory L.B. McNew and B.K. Sandercock. 2015. Responses of male greater prairie-chickens to wind energy development. Condor. 117: 24–296.

[121] Kokko, H. and D.J. Rankin. 2006. Lonely hearts or sex in the city? Density-dependent effects in mating systems. Philos. Trans. R. Soc. Lond., B. 361: 319–334.

[122] Kokko, H., W.J. Sutherland, J. Lindström, J.D. Reynolds and A. MacKenzie. 1998. Individual mating success, lek stability, and the neglected limitations of statistical power. Anim.Behav. 56: 755–762.

[123] Petrie, M. 1994. Improved growth and survival of offspring of peacocks with more elaborate trains. Nature. 371: 598–599.

[124] Petrie, M., H. Tim and S. Carolyn. 1991. Peahens prefer peacocks with elaborate trains. Anim. Behav. 41: 323–331.

[125] Hutchinson, J.M. 2005. Is more choice always desirable? Evidence and arguments from leks, food selection, and environmental enrichment. Biol. Rev. 80: 73–92.

[126] Rintamaki, P.T., R.V. Alatalo, J. Höglund and A. Lundberg. 1995. Male territoriality and female choice on black grouse leks. Anim. Behav. 49: 759–767.

[127] Johnstone, R.A. and D.J. Earn. 1999. Imperfect female choice and male mating skew on leks of different sizes. Behav. Ecol. Sociobiol. 45: 277–281.

[128] Westneat, D.F. and P.W. Sherman. 1997. Density and extra-pair fertilizations in birds: a comparative analysis. Behav. Ecol. Sociobiol. 41: 205–215.

[129] Petrie, M. and B. Kempenaers. 1998. Extra-pair paternity in birds: explaining variation between species and populations. Trends Ecol. Evol. 13: 52–58.

[130] Ryder, T.B., R.C. Fleischer, W.G. Shriver and P.P. Marra. 2012. The ecological–evolutionary interplay: density-dependent sexual selection in a migratory songbird. Ecol. Evol. 2: 976–987.

[131] Stewart, S.L.M., D.F. Westneat and G. Ritchison. 2009. Extra-pair paternity in eastern bluebirds: effects of manipulated density and natural patterns of breeding synchrony. Behav. Ecol. Sociobiol. 64: 463–473.

[132] Charmantier, A. and P. Perrett. 2004. Manipulation of nest-box density affects extra-pair paternity in a population of blue tits (*Parus caeruleus*). Behav. Ecol. Sociobiol. 56: 380–365.

[133] McKellar, A.E., P.P. Marra, P.T. Boag and L.M. Ratcliffe. 2014. Form, function and consequences of density dependence in a long-distance migratory bird. Oikos. 123: 356–364.

[134] Mayer, C. and G. Pasinelli. 2013. New support for an old hypothesis: density affects extra-pair paternity. Ecol. Evol. 3: 694–705.

[135] Van Wijk, S., A. Bourret, M. Bélisle, D. Garant and F. Pelletier. 2016. The influence of iridescent coloration directionality on male tree swallows' reproductive success at different breeding densities. Behav. Ecol. Sociobiol. 70: 1557–1569.

[136] Fort, K. and K.A. Otter. 2004. Territorial breakdown of black-capped chickadees, *Poecile atricapillus*, in disturbed habitats? Anim. Behav. 68: 407–415.

[137] Rosenfeld, R.N., S.A. Sonsthagen, W.E. Stout and S.L. Talbot. 2015. High frequency of extra-pair paternity in an urban population of Cooper's hawks. J. Field Ornith. 86(2): 144–152.

[138] Smith, S.B., J.E. McKay, M.T. Murphy and D.A. Duffield. 2016. Spatial patterns of extra-pair paternity for spotted towhees *Pipilo maculatus* in urban parks. J. Avian Biol. 47: 815–823.

[139] Bonderud, E.S., K.A. Otter, T.M. Burg, K.L.D. Marini and M.W. Reudink. 2018. Patterns of extra-pair paternity in the mountain chickadee. Ethology. 124: 378–386.

[140] Rodriguez-Martínez, S., M. Carrete, S. Roques, N. Rebolo-Ifrán and J.L. Tella. 2014. High urban breeding densities do not disrupt genetic monogamy in a bird species. PLoS One. 9(3): e91314.

[141] Yeh, P.J. 2004. Rapid evolution of a sexually selected trait following population establishment in a novel habitat. Evolution. 58: 166–174.

[142] Price, T.D., P.J. Yeh and B. Harr. 2008. Phenotypic plasticity and the evolution of a socially selected trait following colonization of a novel environment. Am. Nat. 172: S49–S62.

[143] Thompson, K.A., M. Renaudin and M.T.J. Johnson. 2016. Urbanization drives the evolution of parallel clines in plant populations. Proc. R. Soc. B: Biol. Sci. 283: 20162180. https://doi. org/10.1098/rspb.2016.2180

[144] Zanette, L., M. Clinchy and Sung Ha-Cheol. 2009. Food-supplementing parents reduces their sons' song repertoire size. Pro. R. Soc. B. 276: 2855–2860.

[145] Schlaepfer, M.A., M.C. Runge and P.W. Sherman. 2002. Ecological and evolutionary traps. Trends Ecol. Evol. 17: 474–480.

[146] Rodewald, A.D., D.P. Shustack and T.M. Jones. 2011. Dynamic selective environments and evolutionary traps in human-dominated landscapes. Ecology. 92: 1781–1788.

[147] Derryberry, E.P., J.N. Phillips, G.E. Derryberry, M.J. Blum and D. Luther. 2020. Singing in a silent spring: Birds respond to a half-century soundscape reversion during the COVID-19 shutdown. Science. 10.1126/science.abd5777.

Chapter 6

Sociality and
Antipredator Behavior

Amanda K. Beckman[1], Faith O. Hardin[2],
Allison M. Kohler[2] and Michael L. Morrison[2]

Introduction

Sociality, or when two or more individuals closely associate, evolves when the benefits of that close association outweighs the costs. Varying evolutionary pressures have resulted in the diversity and complexity of songbird sociality that can be observed today (e.g., altruism, mutualism, and various types of mating systems). One benefit of sociality is improved predator detection, but other benefits can include improved foraging and thus increased fitness. However, increasing anthropogenic influence on the natural

[1]Ecology and Evolutionary Biology Program, Texas A&M University, Texas, College Station, USA.
[2]Department of Rangeland, Wildlife, and Fisheries Management Texas A&M University, Texas, College Station, USA.

world may cause the cues and signals that songbirds use to communicate to become unreliable. Understanding how con- and heterospecifics interact, how these behaviors evolved, and the communication modalities used to execute these behaviors, is essential to interpret the current and future social world of songbirds. This chapter discusses the fundamentals of understanding songbird social and antipredator behaviors and how these concepts can be used in songbird conservation efforts. Concepts covered will include the impacts of introduced human factors such as the effect of highway noise, artificial lighting, and habitat modification on antipredator and social behaviors. We close by discussing reasons why a species' social system should be considered in management and restoration plans, and describing specific management actions that can be used to mitigate changes to behavior.

Social Behavior

Cooperation

Cooperation is one extreme of sociality where animals are highly gregarious and interactions are often complex; it is defined as a behavior that is selected for because of the benefit to the recipient [1]. Therefore, cooperation encompasses the terms *mutualism*, *altruism*, and *reciprocity*, which are further explained in Table 1. In Richard Dawkins' "The Selfish Gene," he described organisms as "vessels for the gene's survival and perpetuation [2]." This idea pairs well with Darwin's theory of the survival of the fittest, and since an individual's genetic makeup (along with its environment) influences all behaviors, one cannot have a discussion about the evolution of cooperation without first understanding how behaviors are inherited. Alleles coding for behaviors that help further the lineage of said allele are selected for [2, 3]. Understanding this, how then can traits such as altruism and cooperation be selected for under Darwinian selection, which revolves around the survival and reproduction of the fittest individual? In terms of cooperation it is sometimes advantageous to work with others for a common resource; a shared effort equals a shared reward. In some cases individual variation in helping is directly heritable, and fluctuates in relation to the percentage of helping males within a system [4]. Understanding the drivers facilitating the evolution of sociality and cooperation provides a deeper understanding of how these behaviors are passed across generations, and provides the baseline knowledge necessary to understand how these behaviors play a role in conservation and management of songbirds.

Table 1 A summary of the categories of social behaviors discussed in this chapter. For each type of interaction, donors and recipients of that behavior can benefit (+), be negatively impacted (–), or have no effect from (0) the interaction.

Type of Interaction	Donor	Recipient
Mutualism	+	+
Reciprocity	+ (delayed)	+
Altruism	–	+
Selfish	+	–
Spiteful	0 or –	–

Whereas most examples of cooperation found in nature involve interactions between related individuals, there are instances where cooperative behavior did not evolve between kin. There are three possible paths to the evolution of non-kin social groups. The first way is through high adult mortality and turnover and/or high levels of promiscuous mating. In these situations, the high rates of adult mortality, combined with many breeding interactions with individuals from outside of the social group, leads to low group relatedness resulting in few opportunities to benefit from indirect fitness and thus little selection against non-kin individuals within the social group. This can be seen in superb fairy-wrens, where non-related helpers care for nestlings; a result of high adult mortality with family groups that contain step-parents and extra-pair young [5]. The second path to the evolution of non-kin social groups is when individuals (usually males) disperse to join an unrelated breeding pair. Often the male will immigrate to a new area and invest in a female that already has a paired male partner. The new male may or may not successfully breed with the female, but he will continue to care for the female and the brood (to which he is not related) until the following breeding season when he will have the opportunity to inherit a territory or mate. The third and final path for non-kin social group evolution is when individuals disperse from their original natal group to form a coalition; typically young males band together to form said coalition. The benefit of this behavior is that it minimizes within-group conflict and maximizes shared reproductive opportunities, compared to the male acting alone.

Mutualism

Mutualism is when both the donor and the recipient of a behavior benefit from the action and can be observed within species (conspecifics) and between species (heterospecifics). When individuals of some species discover new food patches, they will call conspecifics to the food patch because larger flocks offer protection against predators (discussed in more detail in the Integrating Social and Antipredator Behavior section). House sparrows (*Passer domesticus*) make a "chirrup" call to recruit conspecifics to a food source, but only when the food item is divisible [6]. When many small pieces of food (birdseed and bread crumbs) were presented, the "pioneer" sparrow benefited from the presence of other individuals. When large food items (large pieces of bread) were presented, the costs of competition outweighed the potential benefit of the presence of other individuals, so no chirrup calls were emitted. Large groups of animals can also benefit from mutualistic interactions. For example, groups of colonially nesting or roosting birds can act as information centers to which unsuccessful foragers can return in order to observe successful foragers, resulting in an information transfer [7].

Songbird literature commonly documents heterospecific mutualistic interactions. The two main factors necessary for heterospecific mutualisms to form are (1) that the species' ranges overlap and (2) that the two species historically encountered or currently encounter each other within their range [8]. An example of mutualism is seen between corvids and raptors at carcasses [9]. Black-billed magpies (*Pica hudsonia*) and common ravens (*Corvus corvax*) benefited from the presence of eagles or hawks because they were able to rip open the carcasses otherwise unattainable to the weak-billed corvids. The raptors also benefited from the presence of the corvids; as the corvids collected around a promising carcass they improved visibility of said carcass, reducing the search time that the raptors required. Additionally, research has suggested that heterospecific coordination of birdsong has several mutual benefits: (1) as a signal of the present individual's health or ability to defend resources, (2) indicating that a "neighborhood" is not vulnerable to intruders, and (3) coordinating social bonding [10]. For the purposes of this chapter, an assemblage (or community) will be defined as a group of heterospecifics that occur together and interact with each other in space and time. To determine potential benefits of heterospecific bird song, dusk songs of a songbird assemblage were recorded to measure the spectral overlap between species' songs [10]. The authors hypothesized that if heterospecifics were coordinating song then there would be very little temporal overlap, as opposed to if they were competitively overlapping their neighbor's refractory periods. They observed many species with frequency range overlap, but less temporal overlap than expected by chance. According to this study, a male could receive several possible benefits from establishing a territory in an acoustically-coordinated community (increased song transmissibility could attract more females), and heterospecific individuals in the community would benefit from the addition of a male capable of coordinating with the interspecific chorus.

Reciprocity

Another cooperative behavior is *reciprocity* (or reciprocal altruism). Reciprocity is a system where a donor helps a recipient in anticipation that the recipient will assist in the future, or because the recipient helped them in the past [11]. The exact definition of reciprocity, the criteria used, and the duration of studies used to assess potential cases of reciprocity have been debated (reviewed in [12]). One factor contributing to the controversy and confusion is that reciprocal interactions can be complex; factors like partner choice and prior experience can influence long-term networks between individuals [12].

One well-documented example of reciprocity is found in pied flycatchers (*Ficedula hypoleuca*), which are known to assist each other by mobbing potential threats (Fig. 1). An experiment was conducted with three nest boxes (henceforth referred to as A, B, and C), each of which contained one pair of flycatchers [13]. The authors placed a model owl outside nest box A, experimentally contained the occupants of nest box B in their box, and allowed occupants of C to assist in mobbing (individuals collectively acting to attempt to drive away a potential predator). The researchers performed subsequent experiments with model owls at B and C, and A's reciprocity was favored towards C compared to B (Fig. 1A). They conducted a follow-up study with the nest boxes closer or further apart than the original experiment [14]. When nest boxes were closer, the researchers did not replicate the results of the first experiment (Fig. 1B). Flycatchers no longer favored reciprocity because the threat of the predator was too immediate when the nest boxes were close together. However, they replicated the results when the nest boxes were moved further apart than the original study (Fig. 1C). These studies highlighted that long-term interactions between individuals can shape the behavioral patterns observed, but environmental context (like predation risk) can alter short-term interactions.

Reciprocity is observed in the wild in sedentary and migratory flocks of chaffinches (*Fringilla coelebs*) [15]. The first individual to mob a predator or give an alarm call is the most likely to get injured or eaten (high cost), so without incentive to initiate mobbing the behavior would be eliminated because of cheaters (individuals that benefit from mobbing but never mob themselves). Chaffinches that joined sedentary heterospecific flocks were more likely to initiate mobbing since individuals possessed long-term relationships and would be able to recognize or punish individuals that are cheaters. Migratory flocks contained individuals that were more loosely associated; individuals were less likely to initiate mobbing because they were less likely to receive reciprocal benefits. Over the course of the summer, the frequency of mobbing did not increase in the sedentary flock, but did in the migratory flock as individuals became more familiar with one another.

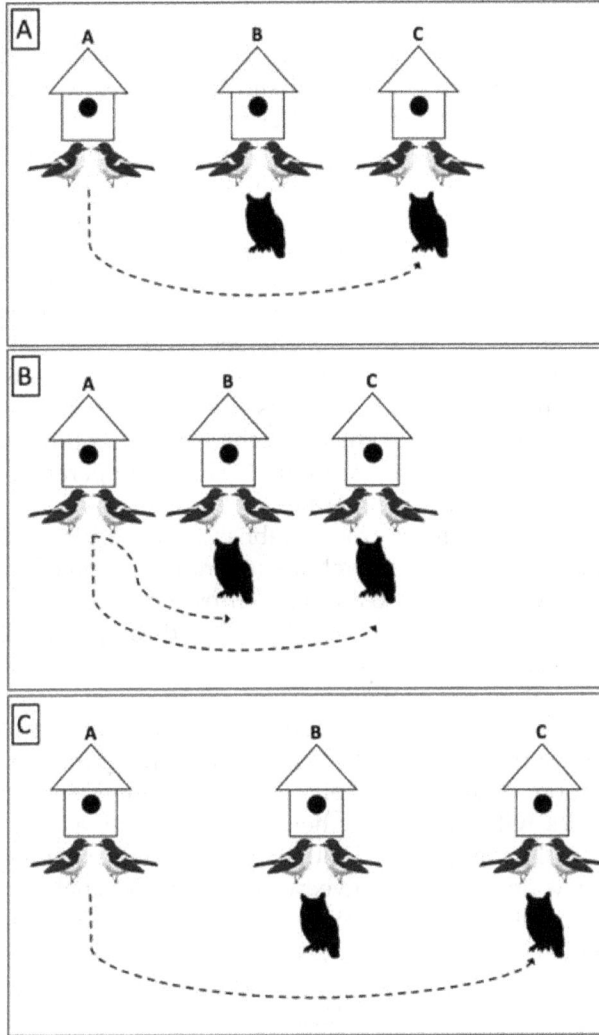

Figure 1 Prior to panels A, B, and C the model owl was placed outside nest box A, and the pair of pied flycatchers in nest box C assisted A by mobbing the owl. (1A) The flycatchers in nest box A reciprocally assisted C by mobbing the owl, but not B. (1B) When the nest boxes were closer together, A assisted B and C since the threat to their own nest box was imminent. (1C) When the nest boxes were further apart, A only helped C due to C's past assistance [13, 14].

Altruism

The last cooperative behavior is *altruism*, which is when an individual decreases its own fitness while increasing another individual's fitness [16]. Reciprocity can be confused for altruism at the early stages of investigation or if interactions are long-term and complex [17]. Often altruism evolves to be directed towards kin due to a high amount of shared genetic material between individuals. If the benefit of helping and the respective relatedness of both individuals are greater than the costs of cooperation, then altruism can evolve. This is referred to as *kin selection*, one mechanism that results in the evolution of altruism within a system. Indirect fitness, reproductive benefits a helper indirectly gains by assisting relatives, is considered altruistic since individuals are forgoing breeding even though there are benefits indirectly.

Cooperative breeding, when helpers (usually last year's young) assist in the construction of nests and/or raising of offspring is a type of altruistic behavior usually directed towards kin. Across species of cooperatively breeding birds, helper investment increased with their relationship to the brood [18]. In Florida scrub-jays (*Aphelocoma coerulescens*), helpers that assisted their parents were physiologically capable of reproducing, but limited opportunities to establish new territories resulted in the evolution of cooperative breeding to indirectly pass on genes [19]. Brown-headed nuthatch (*Sitta pusilla*) helpers were almost always males that assisted their parents with the next year's brood [20]. In this system, sex ratio influenced the prevalence of cooperation present in the population [21]. Wild nuthatch chicks were genotypically sexed and moved to female or male-biased plots in a cross-fostering experiment for two years. Treatments for each plot were then switched to the opposite sex for two years after a one year break. The percentage of individuals cooperating was significantly higher when the plots were male-biased. Males are the dispersing sex in this species and limited habitat prevents many males from establishing territories when the sex ratio is skewed towards males, so the higher level of cooperation in male-biased plots was due to males not having the opportunity to attract their own mate. In addition to the benefits of indirect fitness, helpers can also gain direct benefits through territory acquisition and experience caring for offspring by co-rearing a nest with a dominant individual.

Finally, the cooperative Seychelles warbler (*Acrocephalus sechellensis*) shows an interesting intermediate in cooperation benefits [22]. Helper warblers gain indirect benefits by helping raise their siblings. However, due to rampant extra-paternal mating, the most likely evolutionary driver for their cooperation comes from direct joint nesting benefits; young females will breed with their mother's mate, adding their own egg to their mother's nest. This cooperative relationship is likely the result of the isolated, resource limited islands on which they live, where young Seychelles warblers have little chance of establishing their own territory by their first year. Breeders can also benefit from cooperation; female Seychelles warblers (*Acrocephalus sechellensis*) with nest helpers live longer because of reduced oxidative stress (associated with aging) and thus produce more young than those without helpers [23].

Although it is rarer, altruism can still exist between non-related individuals, such as between alpha and beta male lance-tailed manakins (*Chiroxiphia lanceolata*). Males of this species form long-term partnerships between a dominant (alpha) male that mates with females, and a subordinate (beta) that cooperates to perform complicated displays for females on an exploded lek [24]. Using microsatellites (repetitive and highly variable sections of DNA), alpha and beta males were found to be less likely to be related than by chance, so indirect fitness benefits could not explain the evolution of cooperation in this species. Betas often did not receive any direct benefits (siring offspring) either. It was hypothesized that delayed direct fitness benefits explained cooperation in this system; limited dispersal opportunities led to the evolution of cooperative betas. Long-term behavioral observations found that beta males were more likely than random males to inherit sites from their alpha partner, but other males frequently inherited sites as well. This indicates that a strict queuing system does not explain alpha status acquisition, but betas may receive other benefits from cooperating (territory assessment or future partner affiliations). This also means that this relationship cannot be classified as reciprocity since all betas do not gain a delayed benefit from cooperating.

Box 1 Summary of cooperation

- Cooperative behaviors are selected for because of the benefit to the recipient, and these behaviors can evolve between kin and non-kin
- The three common paths for cooperation to evolve between non-kin are:
  1. High adult mortality and promiscuity
  2. An individual (usually male) disperses and joins a breeding pair
  3. Individuals disperse and form a same-sex coalition
- Mutualisms can evolve between conspecifics, and between heterospecifics whose ranges overlap and frequently encounter each other often
- Reciprocity often involves long-term interactions between individuals where individuals can alternate being the donor and recipient
- Kin selection is a common explanation for the evolution of altruism between related individuals, but altruism between non-kin can evolve because of other selection pressures

---

Selfish Behavior

In some cases it is more beneficial to act *selfishly*, defined as when one individual benefits at the expense of another, such as selfish food hoarding where one individual may steal food from another regardless of the victim's need for the food. Selfishness is the basis of all strategic choices, especially when viewed from the game theory perspective. Game theory, modified from its original economic purpose by Maynard-Smith [25], provides a mathematical model that can determine optimal strategies for competition between adversaries. In behavior, strategies are genetically inherited traits (cooperation, selfishness) that control an individual's actions. These strategies are then balanced and turned against each other in the "game" of competing for resources. Take a common sight on the beach: a gull (Family: Laridae) stealing a piece of bread from another gull. At first glance this may seem like the preferred strategy, as the thief would have obtained a valuable reward with little effort while the victim (the bread gatherer) would have put in effort in vain. However, if there are too many thieves the system would break down since there would be no more gulls to steal from. If the stealing behavior is an inherited trait, natural selection will limit the number of thieves (selfish individuals) relative to the number of gatherers. Likewise, thieves are favored once they become uncommon. This trade-off is best explained by frequency-dependent selection, with the uncommon trait, or strategy, being favored [26].

This conflict and balance is also found between the sexes, in intersexual competition. New Zealand robin (*Petroica australis*) males act selfishly by excluding all females from feeding stations year round, then during the breeding season preferentially feed only the females they have mated with [27, 28]. In doing this, a male excludes all females that may have carried the genetic material of another male, thus ensuring the survival and viability of his own offspring. In another example, evolution has enabled females to avoid egg fertilization with closely related males by rejecting sperm with high homozygosity, thus avoiding inbreeding [29]. This process results from a combination of direct female choice based on phenotypic cues such as sight and smell (resulting in the female forcibly ejecting or accepting the sperm after copulation), and internal physicochemical processes on the molecular level, that result in some sperm being more successful at fertilizing the egg than others [30].

Selfish behavior is also seen in some interspecific interactions. One example of selfishness can be found in obligate brood parasites [31]. Parasitic cuckoos (Family: Cuculidae) will eavesdrop on their host's sexually selected traits (song rate, visual displays, nest size and nest building activity) to choose the best host for their eggs. Instead of stealing physical resources from another individual, as in the example given before with gulls stealing bread, here the cuckoo is stealing valuable information (i.e., fitness information) from the host. The host is attempting to share information with conspecifics and the brood parasite is stealing this information by eavesdropping. The ability to distinguish low and high quality individual hosts benefits the brood parasite by increasing the likelihood their offspring will be raised by competent hosts, but this selfish behavior can have immense consequences for the host species. This is particularly important for species of conservation concern. If cuckoos are targeting the individuals with the highest quality signals in a population, then the individuals that likely have the best genes are less likely to be able to reproduce. In addition to cuckoos, the parasitic North American cowbirds (Family: Icteridae) have similar traits in that they often eavesdrop on their host's behavior for individual benefits [32].

Spiteful Behavior

Spiteful behaviors are behaviors that have no effect or slightly reduce the fitness of the aggressor, but more drastically reduce the fitness of the aggressor's victim [33]. Researchers have contested the existence of spiteful behavior since examples can be difficult to find in nature. However, the evolutionary backstory to the life history traits of some hosts of parasitic cuckoo eggs sheds light on this topic. Hosts of avian parasites are faced with the decision to either reject a foreign egg or accept it. While it may seem obvious to always reject a foreign egg, there are hidden costs to rejection such as accidentally rejecting one's own egg, damaging one's own eggs while attempting to eject the foreign one, or attracting predators while attempting to drive away the nest parasite [34, 35]. Some cuckoo hosts will not reject cuckoo eggs even if they can identify them as being foreign [36]. This behavior may have evolved because some species of nest parasites will come back to the parasitized nest to check on the status of their egg and destroy the nest and eggs of the host if they find their egg rejected, a behavior that has been directly observed and termed the 'mafia hypothesis' [37]. These negative consequences of egg rejection are suspected to have kept the rejection strategy of hosts from reaching fixation in their populations. In extreme cases, the host European magpie (*Pica pica*) selected for the acceptance and rearing of cuckoo chicks, like the great spotted cuckoo (*Clamator glandarius*), even though the parasitic chick would eventually push out all of the host's eggs. The key here is that in later broods, the host pair of magpies would benefit from lower rates of parasitism, while brood parasites spitefully attack pairs that rejected eggs.

Sociality and Mating Systems

Birds exhibit variation in the types and durations of associations with mates or family members [38]. For example, cooperative breeding Florida-scrub jay families live together year-round. Families coordinate vigilance behaviors with one individual acting as a sentinel watching for predators while the rest of the family forages [39]. Alternatively, other species are solitary throughout the year and only associate with conspecifics to mate and raise young. Many of these pair monogamously, and two types of monogamy are found in birds, *social monogamy* and *true* or *genetic monogamy* [40]. *Social monogamy* is often found in passerines and consists of a bonded pair sharing a territory long-term or only for a season, but the pairs are not always exclusive. Over 150 species of birds have extra-pair copulations, in which females or males may mate with individuals outside the pair bond [41]. Out of 100 species studied genetically, 75% used extra-pair fertilizations in which the brood had mixed parentage. This is advantageous for the female as she is able to sample multiple males. Males may also benefit from extra-pair fertilizations. For example, within the same breeding population of chestnut-sided warblers (*Setophaga pensylvanica*), males had higher reproductive success than females because of their extra-pair fertilizations [42].

Much less common in passerines is *true* or *genetic monogamy*, in which there is a genetically-confirmed exclusive mating relationship. This behavior is most commonly found in non-passerine birds that have long reproductive lifespans and in species in which both parents' full effort is required to successfully raise the chicks [43]. Paternal care, in particular, plays a crucial role in the evolution and maintenance of genetic monogamy, as in any other situation there would be less incentive for the male to continue to assist with the rearing of young. A paternity study on burrowing parrots (*Cyanoliseus patagonus*) showed that out of 49 nests, none were of mixed paternity [43]. These long-lived parrots (~20 year) excavate nest burrows into sandstone or limestone cliff faces and the same breeding pair revisit their burrows every year. The female incubates a clutch of two to five eggs, while the male feeds her, and the chicks require extensive care from both male and female after hatching. The chicks are fed for 60 days while in the nest and parental care is continued after fledging for up to four months, which is typical for genetically monogamous species.

Finally, a *polygamous* relationship is characterized by one individual having multiple mates. There are three different forms of polygamy: *polygyny* (male forms pair bonds with multiple females), *polyandry* (female forms pair bonds with multiple males), and *polygyandry* (both males and females form multiple bonds with the opposite sex). Polygyny is most common in birds, whereas polyandry is only found in 1% of avian species. In this 1% of species, including the brown kiwi (*Apteryx australis*), males provide parental care and females take on characteristics typically associated with males in that they are more colorful, tend to abandon their offspring leaving the males to provide care, and will have multiple nests with different mates [44]. Multiple mechanisms drive the form of polygamy observed, the most obvious being the ability to assess the current situation and abandon young for additional reproductive opportunities or simply to reserve energy for future broods [45]. In accordance with Bateman's Principle that eggs are more costly to produce than sperm, most avian species females invest more in their young, which leaves the males free to desert her in favor of having additional young with other females [46]. Sex ratio also influences polygynous systems. If there are more females than males in a population then it makes sense for females to invest in existing offspring since additional mating opportunities may be scarce. Females in male-biased species tend to be more promiscuous, which results in nests with multiple paternity, and since male mating opportunities may be limited there is selection for males to still invest in their social partner.

Lekking is an extremely social form of polygamy in which males collectively display to attract females; females are free to mate with as many males as they desire, but males provide no parental care to their offspring [47]. Leks are typically found in systems typified by biased sex ratios and where males are unable to monopolize the necessary resources to attract females. In several species of manakins (Family: Pipridae), tropical lekking songbirds, female home ranges were 3–7 times larger than adult males', and males settled leks at sites with high levels of female traffic [48]. Other species of lekking birds include the Jackson's widowbird (*Euplectes jacksoni*), the kakapo (*Strigops habroptilus*), and the long-tailed hermit hummingbird (*Phaethornis superciliosus*) [49–51].

Antipredator Behavior

Passerine birds are small and tend to be common prey items. As such, these birds have evolved a variety of sensory mechanisms to detect and respond to predators, including sight, sound, olfaction, and behavioral modifications. Ground foraging passerines, such as the house sparrow, have wide fields of vision that allow them to see predators while foraging [52]. Many songbirds use auditory cues to directly detect predators and alert surrounding individuals to the predator's presence. Research has shown that unwanted eavesdropping by predators on songbird vocalizations has led to passerine alarm calls sounding "seeet-like" or lacking abrupt starts and stops and having no frequency modulation [53]. Predatory birds and mammals have difficulty localizing these types of sounds, which makes them beneficial for alerting surrounding flock members while maintaining secrecy from predators. Recent research has indicated that many birds also have a fairly well-developed sense of smell, suggesting the ability to detect predators by odor is innate in birds. Great tits (*Parus major*) were able to use olfactory cues to detect mustelid predators when inspecting potential roosting cavities [54]. In addition to detecting predators, the Skutch hypothesis predicts that songbirds may also modify their nest sizes and chick feeding behaviors to avoid being detected by predators [55]. Finally, a risky technique used by many species of birds is to attempt a distraction display. The most common technique is for an individual to behave as if injured while drawing the attention of the predator away from its nest, which often puts the adult bird in danger of predation [56].

Non-lethal interactions with predators can still impact songbird social behavior by altering songbird density and distribution. Ovenbirds (*Seiurus aurocapilla*) typically live in the forest interior, but in the presence of high numbers of nest predators, such as the eastern chipmunk (*Tamias striatus*), they preferred to nest along edges contiguous to their preferred mature forests, an area generally considered to be sub-optimal [57]. While it is impossible for songbirds to completely avoid areas with predators, they preferentially choose areas with predators of least concern [58].

Integrating Sociality and Antipredator Behavior

Social groups can also increase predator detection, and as a result mixed flocks tend to be composed of species that forage slightly differently from each other to avoid competition. Large mixed flocks of songbirds have two potential benefits: (1) improved feeding efficiency (discussed earlier with respect to information sharing) and (2) reduced risk of predation. Foraging in mixed flocks not only increases efficiency through the sharing of information about food location; flock members also tend to spend less time scanning for predators and more time foraging when in larger flocks [59]. A comparative analysis found that the leaders (individuals at the front of the flock) of overwintering mixed-flocks tended to be cooperative breeders, with followers more likely to be non-cooperative breeders [60]. One explanation proposed for the presence of followers was that kin-selected behavior like alarm-calling was used more often by cooperative breeding species and could be exploited by non-cooperative species that join the flock.

It was previously discussed how lekking males can choose sites that maximize female visitation, but males in some species will choose locations that enhances their ability to detect predators [61]. Raptor attacks on Guianan cock-of-the-rocks (*Rupicola rupicola*), which congregate in leks of about 50 individuals, were studied to understand the effects of predation on lekking behavior [62]. Raptor attack rate was inversely correlated with group size; the more birds in a lek, the less likely the raptor would be to attack. Cock-of-the-rock alarm calls were predator specific, with lekking birds responding to cues accordingly. In some instances, terrestrial predators were followed from above by individuals that constantly updated the lek with the predator's location. Individuals within the lek were very vocal and startled easily, but lek-wide flushes only occurred during a direct predation event; in some cases, males were seen boldly displaying in front of perched raptors. These behaviors demonstrate an individual's reliance on the communication of the all of the birds on the lek as a whole. Some males may even join mixed-species leks for added protection even though it does not increase their mating opportunities. For example, greater prairie chickens (*Tympanuchus cupido*) do not gain additional mating opportunities by joining sharp-tailed grouse (*T. phasianellus*) leks [63]. Rather, the prairie chicken males benefit through reduced predation pressure from predators they both share, due to combined vigilance.

Individuals can alter antipredator behavior based on their physical position relative to conspecifics. This idea of specialized behavior is also shown in the brain physiology and social hierarchy of birds. When wary of predators, mallards preferentially use unihemispheric slow-wave sleep (the ability to rest half of the brain while the other half remains alert) to remain vigilant to predators while exposed [64]. Electroencephalogram (EEG) recordings of the brain activity of mallards arranged in a row showed that birds exposed at the ends of the row showed a 150% increase in brain function and preferred to sleep with their outside eye open compared to birds in the center. The same effect was found in European starlings (*Sturnus vulgaris*) where individuals on the outside of the foraging flock spent more time scanning the horizon than did those at the center [65]. This effect was compounded by the finding that social hierarchy dictates where flock members spent their time. Dominant individuals spent more time at the centers of flocks where they could maximize foraging with subdominants at the periphery where dangers were higher and were forced to spend less time foraging.

Songbirds have various ways to communicate with each other both visually and acoustically to convey information about predators, including use of conspicuous colors, flash patterns, and sharp calls to facilitate recognition. Red-winged blackbird (*Agelaius phoeniceus*) females listen to distinct changes in male alarm calls for cues on when to leave their nests to feed [58]. The females were more vigilant themselves when they could not hear their mates' calls. To avoid predation, however, birds need to balance communication with maintaining secrecy from potential predators. In flocks composed of multiple species of antwrens (Family: Thamnophilidae), flocking individuals used soft, quiet calls and spread out over large open areas when undisturbed, but began calling loudly and banded together when a predator was detected [66]. Within the flock of antwrens, those with the loudest calls had more muted plumage and vice versa. Finally, in tropical areas mixed species flocks are common during the non-breeding season because of the benefits of mutual vigilance. However, even in temperate areas, loose flocks form during the nonbreeding season, and these flocks are thought to increase the frequency of early-warning calls and deterrence of predators [58].

Songbirds use alarm calls to signal to other songbirds as well as the predator. Alarm calls directed at the predator function as a "pursuit deterrent" in that the prey species is communicating to the predator that they have detected the threat, and a surprise attack would therefore not be successful [67]. Alarm calls also function to mobilize other species that share the same predators. In some regions of Australia, superb fairy-wrens (*Malurus cyaneus*) have learned to eavesdrop on alarm calls of a species of honey-eater, the noisy miner (*Manorina melanocephala*), to increase the number of individuals watching a predator, even though the two birds' songs are quite different acoustically [68]. Superb fairy-wrens in areas where the noisy miner's range did not overlap did not respond to miner alarm calls; this indicates that learning due to range overlap, not acoustical similarities, determines behavioral response. Additionally, different types of calls signal different dangers, for example great tit parents gave distinctly different calls to alert nestlings to jungle crows (*Corvus macrorhynchos*) and Japanese rat snakes (*Elaphe climacophora*) [69]. In response to the "crow" alarm call from parents, chicks crouched down inside their nest boxes, thus avoiding the attack of an invading beak, but jumped out of the nest boxes entirely when the "snake" call was given, to avoid the complete invasion of the nest by a snake.

Sometimes, when a predator is detected in the vicinity but it does not pose an immediate threat, individuals that are potential prey will band together to collectively drive off the predator with the act of mobbing. Alarm calls typical of mobbing behavior are repeated, sharp, loud sounds that are easily localizable and allow other prey species to find and join the mob. Mobbing typically involves the prey species repeatedly diving at, and sometimes making contact with, the predator. This conspicuous antipredator strategy recruits both conspecifics and heterospecific prey species to the area, which is advantageous because the more individuals mobbing the greater chance of driving the predator away [70]. In North America, chickadees are the principal members of mixed flocks that can be made up of nearly 50 different species responding to their iconic "chick-a-dee" alarm call, encoded with information about the size and threat of the predator [71, 72]. Black-capped chickadees (*Poecile atricapillus*) have a mobbing specific alarm call that differs from the "seeet" calls discussed earlier [73]. Seeet calls are given when there is an aerial predator and the appropriate response is to run away from the call, while mobbing alarm calls are given when there is a perched raptor that does not pose an immediate threat. These calls cause other conspecifics to fly towards the sound and recruit an active mob to harass the raptor with the intent of driving it away. A common flock mate of the chickadee, the red-breasted nuthatch (*Sitta canadensis*), routinely eavesdropped on different variations of the chickadee's alarm calls and responded accordingly to cues for either small or large terrestrial predators [74]. The nuthatches came to the aid of the chickadee only if the predator cue was indicative of a small terrestrial predator, as these are the most common cause of nest mortality in nuthatches.

> **Box 2** Summary of integrating social and antipredator behavior
> - Benefits of being in a group
>   - Improved foraging efficiency because of increased information available
>   - Individuals can decrease vigilance and increase foraging due to less risk of predation or because more individuals are present to detect threats
> - Acoustic communication about predators
>   - Alarm calls can warn predators that they have been spotted
>   - Alarm calls can warn nearby individuals that a threat is present, and sometimes heterospecifics will eavesdrop on these calls
>   - Specific alarm calls act to mobilize conspecifics and heterospecifics to drive off predators (mobbing)

Summary of Social and Antipredator Behavior

Songbirds interact with conspecifics, other heterospecific songbirds, and predators in a wide array of social contexts from cooperation to spite. These behaviors help them navigate their environment in search of resources, opportunities for reproduction, as well as helping them avoid predators. These behaviors are heritable, wholly or partially coded for by different alleles arising from mutations that are selected for or against. Many species will take part in mutualistic interactions while searching for food, as seen in the relationship between corvids and raptors at carcasses. Additionally, acting cooperatively under pressure from kin selection provides indirect benefits to individuals even if they do not receive the direct benefit of mating themselves. Alternatively, selfish behaviors and spiteful behaviors have evolved in response to conflict and competition. In attempts to avoid predation, songbirds use these behaviors to signal to other prey and predators alike. Grouping together in flocks can protect individuals while confusing predators and complex vocalizations are used to pass information within flocks. Understanding this remarkable suite of behaviors and interactions is crucial to knowing how to best manage songbirds in a world constantly changing as a result of anthropogenic influences.

Conservation

Many bird species are known for their ability to adjust quickly to changing environments but anthropogenic changes are testing the limits of their ability to adjust [75, 76]. Bird behavior and sociality can be altered by many anthropogenic factors, including noise and light pollution, and the introduction of invasive plants and animals. Although human activities (e.g., fire, deforestation, and hunting) have influenced songbirds for thousands of years, the quantity and intensity of modern anthropogenic factors may pose a substantial threat to the ability to communicate and maintain their social structures. It is important to note, however, that some songbird species benefit from these changes and may prosper in altered areas that are not preferred by the majority of songbirds (i.e., urban exploiters such as northern mockingbirds and house sparrows). In an effort to better understand how anthropogenic factors influence songbirds, researchers have used a wide range of technologies to document the importance of dynamic interactions between con- and heterospecifics at different life stages. For most species, in-depth knowledge of their social world is incomplete, but there is increasing evidence that social interactions are important for survival, reproductive success, and development of adult traits. Social interactions are important to consider in management plans in an effort to fully manage and conserve all behaviors. This section will outline anthropogenic factors that influence songbird sociality and antipredator behavior, explain how a consideration of sociality should be incorporated into current and future management plans, and describe technology that can be used in management plans to measure social and antipredator behaviors of songbirds.

Habitat Alteration

Songbirds rely heavily on specific habitats for their survival. If habitats change, stable interactions between species, like mixed-flock foraging, can be altered. Some Neotropical migrants commonly forage in large mixed-species flocks, and these heterospecific interactions may play a role in winter survival [60]. In two common agroforestry systems in the Andes Mountains, flock foraging activity was positively correlated with canopy cover [77]. Abundances of several migratory bird species within flocks were also positively correlated with tree basal area (cross-sectional area of tree at breast height) and vegetation complexity. More complex agroforest systems may benefit some species, specifically over-wintering Neotropical migrants in mixed flocks, because they provide a variety of species-specific habitats that can accommodate more diverse flocks. This research shows that agroforestry plots are important sites for migratory songbirds, and management plans maintaining a minimum canopy cover and vegetation complexity could be vital for supporting heterospecific overwintering flocks in human-altered landscapes.

Through human-induced landscape modifications, species that have not coevolved with the local natural areas may invade. For example, brown-headed cowbirds (*Molothrus ater*) historically occupied short-grass prairies, but they expanded their range into forest clearings and parasitized a wider variety of host species [78, 79]. Larger landscape patches that were near areas of chronic disturbance were more likely to contain brown-headed cowbirds than smaller patches that were further from areas of chronic disturbance. Introduction of new species or the breakdown of previous natural barriers can also lead to increased rates of hybridization between species; this can result in increased genetic homogenization due to high gene flow or local extinction of native species, but also can create novel combinations of genes not seen in either parent species [80–83]. For example, some species of male ducks force copulate with females, even females of other species. Feral mallards forcibly crossbreed with native species, and due to assortative mating, the resulting young are unwilling to breed with anything other than hybrids. This behavior has caused extinction in local populations of endemic pacific black ducks (*Anas superciliosa* [84]). Additionally, the wild mallards' range has also spread in conjunction with increased agricultural expansion and land clearing, causing them to hybridize with morphologically similar, but less common, American black ducks [85]. Culling of wildlife populations by lethal methods (e.g., shooting, trapping, poisoning) is often proposed as a potential solution to hybridization [86, 87]. This strategy has been implemented for yellow-crowned (*Cyanoramphus auriceps*) and red-fronted parakeets (*Cyanoramphus novaezelandiae*) in the Chatham Islands [88], but it can prove difficult because morphology is not always a reliable indicator of which individuals are hybrids. Furthermore, if the population at risk is small many of the unique genes may be carried by the hybrids, and therefore culling the hybrids may exacerbate the situation [86].

Habitat changes can also alter territory size, which can then change social interactions among con- and heterospecifics. For some boreal songbirds, it has been shown that habitat changes from forest harvesting at the landscape level, in addition to local level, influence species incidence [89]. Areas outside an individual's territory could be important for securing extra-pair copulations or acquiring additional resources for young. Social alterations due to habitat changes can also affect cooperatively-breeding species. When habitat quality of sociable weavers (*Philetairus socius*) was experimentally-manipulated, breeding decisions and helping behaviors of cooperative breeders were influenced [90]. The proportion of birds that acted as nest helpers decreased when resources were abundant. Thus, it is possible that monitoring the number of helpers in a cooperatively-breeding species may be an indicator of habitat changes.

Invasive Species

By studying songbird sociality, it is possible to develop informed management plans to potentially stop and control new invasive species. For example, social species are more likely to be successful invaders, so understanding the sociality of bird species may provide insights for identifying species that have the potential to be invasive [91]. Birds that are invasive to Southeast Asia, such as the Eurasian tree sparrow (*Passer montanus*) and feral pigeon (*Columba livia*), display social behaviors including communal roosts and congregation at food sources [92]. Invasive species can cause serious impacts on native biodiversity and songbird communities. The introduction of several species of invasive rats in the Dominican Republic has reduced the overwintering survival rate and altered antipredator behavior in Bicknell's thrush (*Catharus bicknelli*) [93]. On the island, broad-leaf forests, which are the preferred roosting habitat for the thrushes, had significantly more invasive rats compared to other biomes. The increased predation pressure forced the thrushes to alternatively roost in pine forests in an attempt to avoid predation, which exposed the birds to higher wind speeds, lower temperatures at night, and thus reduced winter survival rates.

Invasive plants also have the potential to alter songbird assemblages in a variety of ways. An invasive plant in India (*Lantana camara*) impacted local bird abundance and distribution [94]. Areas with the invasive plant displayed lower avian species richness, diversity, and abundance, specifically in canopy-foraging and insectivorous birds. Songbird community structure was modified by this invasive plant, and further studies are needed to understand how these altered heterospecific interactions change individual foraging and antipredator behaviors. Additionally, American robins nesting in exotic plants experienced higher predation rates compared to conspecifics nesting in comparable native shrubs [95]. This was due to several differences between the native and exotic plants, including that the exotic plants lacked thorns, resulted in lower nest heights, and perhaps the branch architecture enabled predation. Since native animals did not coevolve with the exotic plants, the robins might have not been able to alter their antipredator behavior to protect their nests or the nests were more conspicuous to predators.

Anthropogenic Noise

Being able to communicate with and hear conspecifics is vital for coordinating social behaviors, attracting a mate, and communicating about potential threats. When surrounded by excess noise, birds are less likely to hear these biologically-important calls. In a Canadian boreal forest region experiencing rapid industrial development, chronic noise produced by pipeline compressors impacted ovenbird mate attraction, antipredator communication, and the age structure of the population [96]. Male ovenbirds were 15% less likely to attract a mate if they chose a territory near a compressor station, likely because their calls were not heard by females. The researchers also hypothesized that alarm call transmissibility was lower near these compressors, which could especially impact females since they nested on the ground and performed all of the egg incubation.

A study of great tits found that the species altered their warning calls in noise-polluted areas in an attempt to compensate for the unnatural noise disturbance [97]. Despite vocal adjustments, the tits were unable to communicate their warning calls effectively, even if the environment was only moderately noise-polluted. In addition to warning calls, noise pollution can disrupt or completely alter a male's mating song. Female great tits were more attracted to males that produced low-frequency calls [98]. Since female tits make sound-based reproductive decisions, there may be crucial fitness consequences as the males must call with low frequencies to attract females or high frequencies to avoid noise masking. The base template for a male's song is inherited in songbirds, and hearing male tutors (usually the father) during a critical period of development solidifies the details of male song. Studies have shown that anthropogenic noise and frequency of the male tutor's song can impact song learning. Juvenile male white-crowned sparrows (*Zonotrichia leucophyrs*) were presented with a low frequency or high frequency tutor song at the same time as playback of anthropogenic noise or no background noise as a control [99]. Chicks exposed to anthropogenic noise playback tended to learn higher-pitched songs than those in a control group because the higher-frequency songs were less masked by anthropogenic noise interference. Furthermore, males exposed to anthropogenic noise produced songs that were of even higher frequency than the original recordings, which indicated that they learned how to increase signal transmission in noise-polluted environments.

If individual species alter their behavior because of noise pollution, this then changes the overall assemblage [100–102]. This could be detrimental to assemblages where species rely on auditory signals to coordinate heterospecific migratory flocks, feeding groups, or antipredator responses. After accounting for detectability and variation of vegetation structure, songbird species richness declined in louder areas, as did the abundances of three out of seven common songbird species of North America [103]. This led to the conclusion that in many cases, a songbird's minimum vocal frequency can be used to predict whether its abundance will decrease with increasing anthropogenic noise disturbance.

Anthropogenic noise pollution, which now permeates natural environments worldwide, may be a strong selecting force that shapes songbird ecology. Noise can be seen as a biological filter that in some areas allows certain species to flourish while it causes others to become extinct [104]. Management plans should consider the surrounding anthropogenic noise of a study area and also consider species-specific vocal frequencies in an effort to better understand and preserve the communication mechanisms of the assemblages. Some techniques used to reduce anthropogenic noise for natural areas include: constructing noise barriers, reducing speed limits, and increasing restrictions on noise emissions [105].

Light Pollution

Similar to noise disturbance, human-produced light (i.e., light pollution) is another factor that can affect songbird sociality and antipredator behavior. Light pollution can impact nighttime migration behavior, and it may increase the risk of predation since predators have more time with enough visibility for hunting and furthermore are supplied with light-producing novel structures to perch on (i.e., lamp posts). In addition, bright light can have a variety of effects on avian flight behavior. Behavioral changes, which are often more energetically expensive, could disrupt the birds' ability to communicate and make them more susceptible to other hazards which can reduce survival [106]. Artificial illumination changed several migration behaviors, causing birds to disperse differently, increase flight speeds, decrease call activity, and rapidly move away from the light source. Peregrine falcons (*Falco peregrinus*) are known predators of songbirds, but direct observation of them hunting migratory birds at night did not occur until observations were made at the Empire State Building [107]. Peregrine falcons were significantly more likely to be present on nights with more than 50 migrant individuals. Peregrine falcons had a 33% prey capture success rate at the Empire State Building, which is not significantly more successful compared to the large range of success rates observed in this species (7%–83%) [108, 109]. Researchers observed that New York City lights disoriented migrating birds, resulting in them circling tall buildings; the peregrines then preyed upon the confused birds. It is possible that the lights disorient the falcons themselves, preventing them from having a higher prey capture success rate.

When taking into consideration the substantial declines in abundance observed in some migratory bird species around the world it is important to understand how artificial light pollution may be impacting these assemblages [110, 111], and studies have addressed the issue of light pollution and have provided solutions to aid in songbird conservation [112–114]. During a 20 day songbird migration study, there were approximately four fatalities at communication towers with flashing lights, which was a significantly smaller number compared to 13 fatalities at towers with steady-burning red lights combined with flashing lights [113]. They predicted that these avian fatalities could be reduced, perhaps up to 70%, by removing steady-burning red lights. The study's authors noted that these changes would not only be inexpensive to obtain, but they would also lower the operation costs of communication towers. The Federal Aviation Administration (FAA) has collected vast amounts of data in an effort to decrease bird strikes, mitigate financial loss, and increase overall safety. These data are intended to track trends and provide a scientific foundation to inform policy, management, and education [115]. For example, the advisory circular entitled "Airport Avian Radar Systems" guides the use of avian radar systems and associated lighting to inform wildlife hazard management plans [116].

Excess light pollution can significantly affect nocturnally active species, including many songbirds. Failed songbird migrations, for example, can have detrimental effects at both individual and population scales [106]. Management for anthropogenic light pollution should focus on reducing excess light throughout the night, especially at sensitive times of the year such as during migration periods. Natural lighting is important for migratory songbirds because they use various cues, including polarized light, to navigate during migration and artificial light can interfere with this process [117]. Several states and cities have implemented laws to reduce light pollution, such as Arizona's statute requiring light shields [118] and Colorado requiring strict considerations before light installation such as energy requirements and potential for light pollution [119]. In addition, the International Dark-Sky Association's dark sky certification assists in avian conservation by mitigating unnecessary light pollution [120]. Additional research exploring how anthropogenic light may affect bird assemblages is warranted and could shed light on how songbirds are affected by this type of pollution. For example, it is unknown whether group versus solo migrants are impacted differently by anthropogenic light.

Integrating Social and Antipredator Behavior in Management Plans

Songbirds exhibit incredible variation in behavior within and between individuals. Several species were discussed in this chapter that possess long-term interactions, multi-species communication, and complex social roles, but this information may not currently be known for species that have not been the subjects of long-term behavioral experiments or observational studies. Even if detailed information about a species' social behavior is not known, behavior can be used to inform and monitor management plans.

Flocking species are typically those that are more vulnerable to predation, suggesting an antipredator role for flocking behavior [121]. In addition, species that lead mixed-flocks tend to act as a source of information about predators for more vulnerable species. Since large flocks may be able to forage in environments that would be risky for smaller groups of individuals, and because certain species often lead these flocks, monitoring flock size and species composition is a valuable management tool. Managers should be aware that social species have specific challenges compared to non-social species. Behaviors may spread through social learning or they may change due to social conformism [122]. This can subdivide what would be considered one population into smaller groups with distinct behavioral differences. Managers can overcome the challenges posed by social species by conducting pilot studies to learn about their social and antipredator behaviors in relation to the management plan, or consider using adaptive management plans [122, 123]. Adaptive management is a way to manage a species acknowledging current uncertainties, and information that is gained while conducting management objectives is integrated in future decisions [123].

When individuals share a limited resource, cooperative breeding may evolve, resulting in young remaining in natal territories near said resource [124]. This can lead to restricted dispersal and habitat specialization, which further limits a species' ability to adapt to a changing environment. The cooperatively breeding, non-migratory brown-headed nuthatch has limited dispersal, with males usually dispersing within 300 m of their natal area; this had led to reduced genetic variation in this species [125]. This decrease in genetic diversity, which is common in cooperative species, resulted in genetic differentiation in geographically close populations, something that could reduce the nuthatches' ability to adapt to changes such as fragmentation, fire suppression, and urban development. Therefore, dispersal distance, natal habitat variables, and number of helpers may be important to monitor when managing cooperative breeding species.

The *dear enemy effect,* or the reduced aggression to an individual on a neighboring territory, is applicable while translocating individuals [126]. After translocations of social groups of the cooperative breeding brown treecreeper (*Climacteris picumnus*), locations with low vegetation cover were preferred by groups at settlement, but survival over 16 months was not influenced by vegetation cover or sex [127]. However, there were significant differences in survival due to social groups. Although the authors did not know the exact social factors influencing survival, this study indicates that social affiliations can be more important than the release site habitat or individual movement after translocation. The success of translocated pied flycatchers was dependent on individual history [128]. Adult male pied flycatchers showed high site-fidelity to previous breeding locations. However, first-year males were more likely to stay in the translocation area compared to males that had previous breeding success at the original site. Though translocation failures may put the blame on lack of habitat or predators, it is important to note that social learning also plays a major role in whether or not individuals of a species will adjust and thrive in a new environment. Asocial individuals may have to learn survival tactics individually where social species could spread this learned knowledge faster, resulting in the social species being better able to adjust to a translocated area [122]. Therefore, previous interactions and social learning may be linked with translocation success in some cases and should be considered in management plans.

Population Viability Analysis (PVA) is a useful management tool to understand the sustainability of a population after conservation efforts (ideally this population has been confirmed via genetic analysis that it is isolated). PVAs often consider abiotic and biotic factors to calculate effective population size, the size of a theoretically ideal population with the same magnitude of variance as the population being considered [129]. Common abiotic factors considered are weather and elevation, while common biotic factors are food availability and vegetation structure. While not usually considered as a factor in this analysis, the way behavior alters effective population size could be one of the most important considerations to make for more behaviorally-minded conservation plans [130].

Nine behavioral mechanisms have been identified that can influence population size, reproductive skew, or population growth rate, and these mechanisms in turn influence the effective population size [130]. Four of these mechanisms relate to social and antipredator behaviors: (1) mating system, (2) social plasticity, (3) mechanism of mate choice, and (4) conspecific attraction. (1) The specific mating system can influence reproductive skew between individuals. Additionally, if there is intense competition among individuals for access to mates then mortality from intraspecific competition may impact population growth rate or population size. (2) Social plasticity is when some species alter their social systems in the presence or absence of certain resources or only associate with con- or heterospecifics in certain contexts (i.e., when predation is high) or at certain times of year [131]. This then influences effective population size estimates compared to when these relationships are not considered. For example, dunnocks (*Prunella modularis*) have an incredibly variable social system, and arguably have one of the most variable breeding systems: individuals can be monogamous, polyandrous, and polygynous [132]. (3) The mechanism of mate choice can also be a factor that impacts population growth rate and reproductive skew. Females usually do not mate randomly, so if males possessing traits that meet a needed threshold are not present in the population, then females may not find a mate or may choose to not reproduce that year [133]. Additionally, some females exhibit mate-choice copying (where one female, often a first year breeder, copies the mate choice of an older or higher-ranked female), so if experienced females are absent, young females may not learn what traits to assess in males and may make maladaptive mate choice decisions. (4) Conspecific attraction, the tendency of individuals of a species to settle near other individuals, may be important for territory establishment and range expansion in some species, which directly affects the number of individuals in the population [134]. The effects of conspecific attraction may be especially important in small, founding populations as individuals search for new food resources and suitable nesting territories [135].

A framework useful for integrating animal behavior and conservation includes: behavioral indicators, anthropogenic impacts on animal behavior, and behavior-based management [136]. Multiple behavioral indicators (evidence of management success or failure) should be used for robust and effective management plans, and incorporating species-specific behavioral indicators could be useful for observing a species' response to management actions. Behavioral indicators can also provide early warning signs of population decline, inbreeding depression, or environmental instability before more serious ramifications can be quantified. The authors of this chapter envision that social behaviors, like the number of helpers present in cooperative species or differential habitat-use based on social context, and aspects of antipredator behavior, like the number of individuals or species composition that respond to alarm calls, could be informative for some management plans. Next, it is important to understand and quantify how anthropogenic changes affect behavior. Novel sounds, lights, or structures can alter an individual's behavior, which in turn may result in changes to a population's social structure or individual mating success. Intra- and interspecific changes in interactions can then impact the local assemblage. For example, precise calculations of Flight Initiation Distance (FID), the distance at which individuals fly away when approached by a predator, can be used to understand how wary songbirds are about humans approaching them [137]. Behavior-based management involves integration of both behavioral indicators and anthropogenic impacts on behavior. For wild birds, conservation goals would include considering the species' natural behaviors in protocols and decisions, and continuing to monitor changes to these behaviors while completing management goals such as increasing or stabilizing population numbers or invasive species removal.

Technology to Understand Social and Antipredator Behaviors

If nothing is known about the social or antipredator behaviors of species of management concern or if there is reason to believe local adaptation or response to anthropogenic influences is altering a typical behavior, how should these behaviors be measured? The first decision that needs to be made is whether the stress of capturing, banding, or tagging of individuals so they are identifiable in the field is necessary for the management outcomes. Banding prior to in-person behavioral observations may be worthwhile if there is reason to believe that variation in individual social interactions, variation in parental care in the presence or absence of extra-pair copulations, or long-term social bonds may impact management outcomes. If management is focused on nesting habitat preferences, response to predators, or how assemblage composition varies over time then identification of individuals may not be necessary.

Technology has made it possible to monitor songbird social behavior even when humans are not physically there to observe them. Popular technologies used to understand the spatial locations of birds include tracking and/or identification devices, such as those that transmit to satellites or on-ground base stations. Proximity loggers, mini GPS tags, and RFID (radio frequency identification) tags can be used to compare the locations of tagged birds in the same time period to infer social interactions [138–140]. The MOTUS wildlife tracking network is a global collaborative network of receiver stations used to monitor movement of tagged animals [141]. The authors of this chapter believe the MOTUS network could be extremely valuable for understanding how the species composition of flocks impacts movement and survival. Acoustic devices attached to birds also allow individual vocalizations to be monitored along with predator interactions [142, 143]. Microphone arrays provide a way to passively monitor the acoustics of multiple individuals in a given area [144]. If individual signatures in calls can be identified with these devices then the exact identity of interacting individuals can be determined. Additionally, the calls and movements immediately before, during, and after any attempted predation event can be recorded and can be used to understand antipredator responses. Some species have localized areas where most of their activities occur each day (like the dance perches of lance-tailed manakins), therefore cameras at set locations can be used to monitor social interactions [145]. Cameras can also be used to monitor the antipredator responses of parents and nestlings at a nest [146]. As technology advances (and becomes smaller) more opportunities to remotely monitor songbirds will become available.

Summary

Songbirds must be able to adjust to persist in a dynamic, human-dominated world with new challenges such as artificial light, anthropogenic noise, altered habitats. Songbird sociality and antipredator behaviors are important to consider when developing conservation and management plans. When designing plans, managers should carefully consider species that exhibit flocking, cooperative breeding, lekking, and/or neighbor conflict behaviors due to the fact that they may be more susceptible to negative consequences than others. Integrating social and antipredator behaviors into conservation efforts will continue as new techniques and technologies are developed to accurately measure and understand these behaviors. The authors believe that future research is warranted on how anthropogenic light and manmade structures affect migration patterns of solo versus group songbird migrants. It is possible that artificial light, for example, may differentially impact solo migrants compared to songbirds that migrate in groups. These potential studies, along with others, are essential in the future understanding of how sociality and antipredator behaviors influence songbird assemblages and may be vital in the conservation of several songbird species.

Acknowledgements

We would like to thank Carla C. Vanderbilt, Terri L. Pope, Darren Proppe, and an anonymous reviewer for their comments on this chapter. We also thank Carmen Rosenthal Struminger for the pied flycatcher images for Fig. 1.

LITERATURE CITED

[1] West, S.A., A.S. Griffin, and A. Gardner. 2007. Evolutionary Explanations for Cooperation. Curr. Biol. 17: R661–R672.

[2] Dawkins, R. 1976. The Selfish Gene. Oxford University Press.

[3] Drickamer, L.C., S.H. Vessey and D. Meikle. 1996. Animal behavior: Mechanisms, Ecology, and Evolution. Wm C Brown Publishers.

[4] Wang, C. and X. Lu. 2018. Hamilton's inclusive fitness maintains heritable altruism polymorphism through rb = c. Proc. Natl. Acad. Sci. USA. 115: 1860–1864.

[5] Dunn, P.O., A. Cockburn and R.A. Mulder. 1995. Fairy-wren helpers often care for young to which they are unrelated. Proc. R. Soc. Lon. B. Biol. Sci. 259: 339–343.

[6] Elgar, M.A. 1986. House sparrows establish foraging flocks by giving chirrup calls if the resources are divisible. Anim. Behav. 34: 169–174.

[7] Brown, C.R. 1988. Enhanced foraging efficiency through information centers: a benefit of coloniality in cliff swallows. Ecology. 69: 602–613.

[8] Axelrod, R. and W.D. Hamilton. 1981. The evolution of cooperation. Science. 211: 1390–1396.

[9] Orr, M.R., J.D. Nelson and J.W. Watson. 2019. Heterospecific information supports a foraging mutualism between corvids and raptors. Anim. Behav. 153: 105–113.

[10] Malavasi, R. and A. Farina. 2013. Neighbours' talk: interspecific choruses among songbirds. Bioacoustics. 22: 33–48.

[11] Alexander, R.D. 1974. The Evolution of Social Behavior. Annu. Rev. Eco. Syst. 5: 525–583.

[12] Carter, G. 2014. The reciprocity controversy. Animal Behaviour and Cognition. 1: 368–386.

[13] Krams, I., T. Krama, K. Igaune and R. Mänd. 2008. Experimental evidence of reciprocal altruism in the pied flycatcher. Behav. Ecol. Sociobiol. 62: 599–605.

[14] Krama, T., J. Vrublevska, T.M. Freeberg, C. Kullberg, M.J. Rantala and I. Krams. 2012. You mob my owl, I'll mob yours: birds play tit-for-tat game. Sci. Rep. 2: 800.

[15] Krams, I. and T. Krama. 2002. Interspecific reciprocity explains mobbing behaviour of the breeding chaffinches, Fringilla coelebs. Proc. R. Soc. Lond. B. Biol. Sci. 269: 2345–2350.

[16] Foster, K.R., T. Wenseleers and F.L.W. Ratnieks. 2006. Kin selection is the key to altruism. Trends Ecol. Evol. 21: 57–60.

[17] Trivers, R.L. 1971. The evolution of reciprocal altruism. Q. Rev. Biol. 46: 35–57.

[18] Green, J.P., R.P. Freckleton and B.J. Hatchwell. 2016. Variation in helper effort among cooperatively breeding bird species is consistent with Hamilton's Rule. Nat. Commun. 7: 12663.

[19] Woolfenden, G.E. and J.W. Fitzpatrick. 1984. The Florida Scrub Jay: Demography of A Cooperative-Breeding Bird. Princeton University Press.

[20] Cox, J.A. and G.L. Slater. 2007. Cooperative breeding in the brown-headed nuthatch. Wilson J. Ornithol. 119: 1–8.

[21] Cox, J.A., J.A. Cusick and E.H. DuVal. 2019. Manipulated sex ratios alter group structure and cooperation in the brown-headed nuthatch. Behav. Ecol. 30: 883–893.

[22] Richardson, D.S., T. Burke and J. Komdeur. 2002. Direct benefits and the evolution of female-biased cooperative breeding in Seychelles warblers. Evolution. 56: 2313–2321.

[23] Hammers, M., S.A. Kingma, L.G. Spurgin, K. Bebbington, H.L. Dugdale, T. Burke, J. Komdeur and D.S. Richardson. 2019. Breeders that receive help age more slowly in a cooperatively breeding bird. Nat. Commun. 10: 1301.

[24] DuVal, E.H. 2007. Adaptive advantages of cooperative courtship for subordinate male lance-tailed manakins. Am. Nat. 169: 423–432.

[25] Maynard Smith, J. 1974. The theory of games and the evolution of animal conflicts. J. Theor. Biol. 47: 209–221.

[26] Ayala, F.J. and C.A. Campbell. 1974. Frequency-dependent selection. Annu. Rev. Ecol. Syst. 5: 115–138.

[27] Menzies, I.J. and K.C. Burns. 2008. Food hoarding in the New Zealand robin: a review and synthesis. pp. 163–183. *In*: E.A. Weber and L.H. Krause (eds). Animal Behaviour: New Research. New York: Nova Science Publishers.

[28] van Horik, J. and K.C. Burns. 2007. Cache spacing patterns and reciprocal cache theft in New Zealand robins. Anim. Behav. 73: 1043–1049.

[29] Denk, A.G., A. Holzmann, A. Peters, E.L.M. Vermeirssen and B. Kempenaers. 2005. Paternity in mallards: effects of sperm quality and female sperm selection for inbreeding avoidance. Behav. Ecol. 16: 825–833.

[30] Løvlie, H., M.A.F. Gillingham, K. Worley, T. Pizzari and D.S. Richardson. 2013. Cryptic female choice favours sperm from major histocompatibility complex-dissimilar males. Proc. Biol. Sci. 280: 20131296.

[31] Parejo, D. and J.M. Avilés. 2007. Do avian brood parasites eavesdrop on heterospecific sexual signals revealing host quality? A review of the evidence. Anim. Cogn. 10: 81–88.

[32] Scardamaglia, R.C., V.D. Fiorini, A. Kacelnik and J.C. Reboreda. 2016. Planning host exploitation through prospecting visits by parasitic cowbirds. Behav. Ecol. Sociobiol. 71: 23.

[33] Pierotti, R. 1980. Spite and altruism in gulls. Am. Nat. 115: 290–300.

[34] Moksnes, A., E. Røskaft and Anders T. Braa. 1991. Rejection behavior by common cuckoo hosts towards artificial brood parasite eggs. Auk. 108: 348–354.

[35] Campobello, D. and S.G. Sealy. 2018. Evolutionary significance of antiparasite, antipredator and learning phenotypes of avian nest defence. Sci. Rep. 8: 10569.

[36] Robert, M., G. Sorci, A.P. Moller, M.E. Hochberg, A. Pomiankowski and M. Pagel. 1999. Retaliatory cuckoos and the evolution of host resistance to brood parasites. Anim. Behav. 58: 817–824.

[37] Soler, M., J.J. Soler, J.G. Martinez and A.P.M Ller. 1995. Magpie host manipulation by great spotted cuckoos: evidence for an avian mafia? Evolution. 49: 770–775.

[38] Cockburn, A. 2006. Prevalence of different modes of parental care in birds. Proc. Biol. Sci. 273: 1375–1383.

[39] McGowan, K.J. and G.E. Woolfenden. 1989. A sentinel system in the Florida scrub jay. Anim. Behav. 37: 1000–1006.

[40] Gowaty, P.A. 1996. Battles of the sexes and origins of monogamy. Partnerships in Birds. 21–52.

[41] Hasselquist, D. 2001. Social mating systems and extrapair fertilizations in passerine birds. Behav. Ecol. 12: 457–466.

[42] Byers, B.E., H.L. Mays, I.R.K. Stewart and D.F. Westneat. 2004. Extrapair Paternity Increases Variability in Male Reproductive Success in the Chestnut-Sided Warbler (*Dendroica Pensylvanica*), A Socially Monogamous Songbird. Auk. 121: 788–795.

[43] Masello, J.F., A. Sramkova, P. Quillfeldt, J.T. Epplen and T. Lubjuhn. 2002. Genetic monogamy in burrowing parrots Cyanoliseus patagonus? J. Avian Biol. 33: 99–103.

[44] Owens, I.P. 2002. Male-only care and classical polyandry in birds: phylogeny, ecology and sex differences in remating opportunities. Philos. Trans. R. Soc. Lond. B. Biol. Sci. 357: 283–293.

[45] Liker, A., R.P. Freckleton and T. Székely. 2014. Divorce and infidelity are associated with skewed adult sex ratios in birds. Curr. Biol. 24: 880–884.

[46] Bateman, A.J. 1948. Intra-sexual selection in *Drosophila*. Heredity 2: 349–368.

[47] Wiley, R.H. 1991. Lekking in birds and mammals: behavioral and evolutionary issues. Adv. Stud. Behav. 20: 201–291.

[48] Théry, M. 1992. The evolution of leks through female choice: differential clustering and space utilization in six sympatric manakins. Behav. Ecol. Sociobiol. 30: 227–237.

[49]   Andersson, S. 1989. Sexual selection and cues for female choice in leks of Jackson's widowbird Euplectes jacksoni. Behav. Ecol. Sociobiol. 25: 403–410.

[50]   Sutherland, W.J. 2002. Science, sex and the kakapo. Nature. 419: 265.

[51]   Stiles, F.G. and L.L. Wolf. 1979. Ecology and evolution of lek mating behavior in the long-tailed hermit hummingbird. Ornithol. Monogr. 27: 1–78.

[52]   Fernández-Juricic, E., M.D. Gall, T. Dolan, V. Tisdale and G.R. Martin. 2008. The visual fields of two ground-foraging birds, House Finches and House Sparrows, allow for simultaneous foraging and anti-predator vigilance. Ibis 150: 779–787.

[53]   Searcy, W.A. and K. Yasukawa. 2017. Eavesdropping and cue denial in avian acoustic signals. Anim. Behav. 124: 273–282.

[54]   Amo, L., M.E. Visser and K. van Oers. 2011. Smelling Out Predators is Innate in Birds. Ardea. 99: 177–184.

[55]   Skutch, A.F. 1985. Clutch Size, Nesting Success and Predation on Nests of Neotropical Birds, Reviewed. Ornithol. Monogr. 575–594.

[56]   Humphreys, R.K. and G.D. Ruxton. 2020. Avian distraction displays: a review. Ibis 5: 83.

[57]   Morton, E.S. 2005. Predation and variation in breeding habitat use in the Ovenbird, with special reference to breeding habitat selection in northwestern Pennsylvania. Wilson J. Ornithol. 117: 327–336.

[58]   Lima, S.L. 2009. Predators and the breeding bird: behavioral and reproductive flexibility under the risk of predation. Biol. Rev. Camb. Philos. Soc. 84: 485–513.

[59]   Thiollay, J.-M. and M. Jullien. 2008. Flocking behaviour of foraging birds in a neotropical rain forest and the antipredator defence hypothesis. Ibis 140: 382–394.

[60]   Sridhar, H., G. Beauchamp and K. Shanker. 2009. Why do birds participate in mixed-species foraging flocks? A large-scale synthesis. Anim. Behav. 78: 337–347.

[61]   Alonso, J.C., J.M. Álvarez-Martínez and C. Palacín. 2012. Leks in ground-displaying birds: hotspots or safe places? Behav. Ecol. 23: 491–501.

[62]   Trail, P.W. 1987. Predation and antipredator behavior at Guianan Cock-of-the-Rock Leks. Auk. 104: 496–507.

[63]   Gibson, R.M., A.S. Aspbury and L.L. McDaniel. 2002. Active formation of mixed-species grouse leks: a role for predation in lek evolution? Proc. Biol. Sci. 269: 2503–2507.

[64]   Rattenborg, N.C., S.L. Lima and C.J. Amlaner. 1999. Facultative control of avian unihemispheric sleep under the risk of predation. Behav. Brain Res. 105: 163–172.

[65]   Jennings, T. and S.M. Evans. 1980. Influence of position in the flock and flock size on vigilance in the starling, Sturnus vulgaris. Anim. Behav. 28: 634–635.

[66]   Wiley, R.H. 1971. Cooperative roles in mixed flocks of antwrens (*Formicariidae*). Auk. 88: 881–892.

[67]   Woodland, D.J., Z. Jaafar and M.L. Knight. 1980. The "pursuit deterrent" function of alarm signals. Am. Nat. 115: 748–753.

[68]   Magrath, R.D. and T.H. Bennett. 2012. A micro-geography of fear: learning to eavesdrop on alarm calls of neighbouring heterospecifics. Proc. Biol. Sci. 279: 902–909.

[69]   Suzuki, T.N. 2011. Parental alarm calls warn nestlings about different predatory threats. Curr. Biol. 21: R15–6.

[70]   Picman, J., M. Leonard and A. Horn. 1988. Antipredation role of clumped nesting by marsh-nesting red-winged blackbirds. Behav. Ecol. Sociobiol. 22: 9–15.

[71]   Gunn, J.S., A. Desrochers, M.-A. Villard, J. Bourque and Jacques Ibarzabal. 2000. Playbacks of mobbing calls of black-capped chickadees as a method to estimate reproductive activity of forest birds. J. Field Ornithol. 71: 472–483.

[72]   Hurd, C.R. 1996. Interspecific attraction to the mobbing calls of black-capped chickadees (*Parus atricapillus*). Behav. Ecol. Sociobiol. 38: 287–292.

[73] Templeton, C.N., E. Greene and K. Davis. 2005. Allometry of alarm calls: black-capped chickadees encode information about predator size. Science. 308: 1934–1937.

[74] Templeton, C.N. and E. Greene. 2007. Nuthatches eavesdrop on variations in heterospecific chickadee mobbing alarm calls. Proc. Natl. Acad. Sci. USA. 104: 5479–5482.

[75] Pennisi, E. 2002. Finches adapt rapidly to new homes. Science. 295: 249–250.

[76] With, K.A. 2015. How fast do migratory songbirds have to adapt to keep pace with rapidly changing landscapes? Landsc. Ecol. 30: 1351–1361.

[77] McDermott, M.E., A.D. Rodewald and S.N. Matthews. 2015. Managing tropical agroforestry for conservation of flocking migratory birds. Agrofor. Syst. 89: 383–396.

[78] Coker, D.R. and D.E. Capen. 1995. Landscape-level habitat use by Brown-Headed Cowbirds in Vermont. J. Wildl. Manage. 59: 631–637.

[79] Mayfield, H. 1965. The Brown-headed Cowbird, with old and new hosts. Living Bird. 4: 28.

[80] Carroll, S.P. and C.W. Fox. 2008. Conservation Biology: Evolution in Action. Oxford University Press.

[81] Grant, P.R. and B.R. Grant. 1994. Phenotypic and genetic effects of hybridization in Darwin's finches. Evolution. 48: 297–316.

[82] Ottenburghs, J., R.C. Ydenberg, P. Van Hooft, S.E. Van Wieren and H.H.T. Prins. 2015. The avian hybrids project: gathering the scientific literature on avian hybridization. Ibis. 157: 892–894.

[83] Rhymer, J.M. 1996. Extinction by hybridization and introgression. Annu. Rev. Ecol. Syst. 27: 83–109.

[84] Guay, P.-J. and J.P. Tracey. 2009. Feral mallards: a risk for hybridisation with wild Pacific Black Ducks in Australia? The Victorian Naturalist. 126: 87–91.

[85] Heusmann, H.W. 1974. Mallard-black duck relationships in the Northeast. Wildl. Soc. Bull. 2: 171–177.

[86] Simberloff, D. 1996. Hybridization between native and introduced wildlife species: importance for conservation. Wildl. Biol. 2: 143–150.

[87] Oogjes, G. 1997. Ethical aspects and dilemmas of fertility control of unwanted wildlife: an animal welfarist's perspective. Reprod. Fertil. Dev. 9: 163–168.

[88] Cade, T.J. 1983. Hybridization and gene exchange among birds in relation to conservation. pp. 288–309. *In*: C.M. Schonewald-Cox, S.M. Chambers, B. MacBryde and W.L. Thomas (eds). Genetics and Conservation. Benjamin/Cummins Publishing Company Inc., Menlo Park, California, US, and London, England, UK.

[89] Taylor, P.D. and M.A. Krawchuk. 2005. Scale and sensitivity of songbird occurrence to landscape structure in a harvested boreal forest. Avian Conserv. Ecol. 1: 5.

[90] Covas, R., C. Doutrelant and M.A. du Plessis. 2004. Experimental evidence of a link between breeding conditions and the decision to breed or to help in a colonial cooperative bird. Proc. Biol. Sci. 271: 827–832.

[91] Chapple, D.G., S.M. Simmonds and B.B.M. Wong. 2012. Can behavioral and personality traits influence the success of unintentional species introductions? Trends Ecol. Evol. 27: 57–64.

[92] Yap, C.A.M. and N.S. Sodhi. 2004. Southeast Asian invasive birds: ecology, impact and management. Ornithol. Sci. 3: 57–67.

[93] Townsend, J.M., C.C. Rimmer and K.P. McFarland. 2009. Investigating the limiting factors of a rare, vulnerable species: Bicknell's Thrush. *In*: Tundra to Tropics: Connecting Birds, Habitats and People. Proceedings of the 4th International Partners in Flight Conference. 91–95.

[94] Aravind, N.A., D. Rao, K.N. Ganeshaiah, R. Uma Shaanker and J.G. Poulsen. 2010. Impact of the invasive plant, *Lantana camara*, on bird assemblages at Malé Mahadeshwara Reserve Forest, South India. Trop. Ecol. 51(2S): 325–338.

[95] Schmidt, K.A. and C.J. Whelan. 1999. Effects of exotic *Lonicera* and *Rhamnus* on songbird nest predation. Cons. Biol. 13: 1502–1506.

[96] Habib, L., E.M. Bayne and S. Boutin. 2007. Chronic industrial noise affects pairing success and age structure of ovenbirds *Seiurus aurocapilla*. J. Appl. Ecol. 46: 176–184.

[97] Templeton, C.N., S.A. Zollinger and H. Brumm. 2016. Traffic noise drowns out great tit alarm calls. Curr. Biol. 26: R1173–R1174.

[98] Halfwerk, W., S. Bot, J. Buikx, M. van der Velde, J. Komdeur, C. ten Cate and H. Slabbekoorn. 2011. Low-frequency songs lose their potency in noisy urban conditions. Proc. Natl. Acad. Sci. USA. 108: 14549–14554.

[99] Moseley, D.L., G.E. Derryberry, J.N. Phillips, J.E. Danner, R.M. Danner, D.A. Luther and E.P. Derryberry. 2018. Acoustic adaptation to city noise through vocal learning by a songbird. Proc. Biol. Sci. 285.

[100] Francis, C.D., C.P. Ortega and A. Cruz. 2009. Noise pollution changes avian communities and species interactions. Curr. Biol. 19: 1415–1419.

[101] Proppe, D.S., M.T. Avey, M. Hoeschele, M.K. Moscicki, T. Ra, C.C. St Clair and C.B. Sturdy. 2012. Black-capped chickadees Poecile atricapillus sing at higher pitches with elevated anthropogenic noise, but not with decreasing canopy cover. J. Avian Biol. 43: 325–332.

[102] Slabbekoorn, H. and W. Halfwerk. 2009. Behavioural ecology: noise annoys at community level. Curr. Biol. 19: R693–R695.

[103] Proppe, D.S., C.B. Sturdy and C.C. St Clair. 2013. Anthropogenic noise decreases urban songbird diversity and may contribute to homogenization. Glob. Chang. Biol. 19: 1075–1084.

[104] Francis, C.D., N.J. Kleist, B.J. Davidson, C.P. Ortega and A. Cruz. 2012. Chapter 4: Behavioral responses by two songbirds to natural-gas-well compressor noise. Ornithol. Monogr. 74: 36–46.

[105] Slabbekoorn, H. and E.A.P. Ripmeester. 2008. Birdsong and anthropogenic noise: implications and applications for conservation. Mol. Ecol. 17: 72–83.

[106] van Doren, B.M. and K.G Horton, A.M. Dokter, H. Klinck, S.B. Elbin and A. Farnsworth. 2017. High-intensity urban light installation dramatically alters nocturnal bird migration. Proceedings of the National Academy of Sciences. 114: 11175–11180.

[107] DeCandido, R. and D. Allen. 2006. Nocturnal hunting by peregrine falcons at the empire state building, New York City. Wilson J. Ornithol. 118: 53–58.

[108] Buchanan, J.B., S.G. Herman and T.M. Johnson. 1986. Success rates of the peregrine falcon (*Falco peregrinus*) hunting dunlin (*Calidris alpina*) during winter. J. Raptor Res. 20: 130–131.

[109] Buchanan, J.B. 1996. A comparison of behavior and success rates of Merlins and Peregrine Falcons when hunting Dunlins in two coastal habitats. J. Raptor Res. 30: 93–98.

[110] Robbins, C.S. J.R. Sauer, R.S. Greenberg and S. Droege. 1989. Population declines in North American birds that migrate to the neotropics. Proceedings of the National Academy of Sciences. 88: 7658–7662.

[111] North American Bird Conservation Initiative. 2016. The State of North America's Birds 2016. Environment and Climate Change Canada, Ottawa.

[112] Adams, C.A., A. Blumenthal, E. Fernández-Juricic, E. Bayne and C.C. St. Clair. 2019. Effect of anthropogenic light on bird movement, habitat selection, and distribution: a systematic map protocol. Environmental Evidence. 8: 1–16.

[113] Gehring, J., P. Kerlinger and A.M. Manville Ii. 2009. Communication towers, lights, and birds: successful methods of reducing the frequency of avian collisions. Ecological Applications. 19: 505–514.

[114] Goller, B., B.F. Blackwell, T.L. DeVault, P.E. Baumhardt and E. Fernández-Juricic. 2018. Assessing bird avoidance of high-contrast lights using a choice test approach: implications for reducing human-induced avian mortality. PeerJ. 6: e5404.

[115] Federal Aviation Administration. 2019. Fact Sheet – The Federal Aviation Administration's (FAA) Wildlife Hazard Mitigation Program.

[116] Federal Aviation Administration. 2010. Advisory Circular 150/5220–25.

[117] Muheim, R., J.B. Phillips and S. Akesson. 2006. Polarized light cues underlie compass calibration in migratory songbirds. Science. 313: 837–839.

[118] 2019. Arizona Revised Statutes. Title 49- The Environment. Chapter 7- Light pollution.

[119] 2016. Colorado Revised Statutes. Title 24 - Government - State State Property Article 82 - State Property Part 9 - Outdoor Lighting Fixtures § 24-82-901. Definitions. Colorado Revised Statute. Title 24–82–901.

[120] Meier, J. 2014. Designating dark sky areas: actors and interests. pp. 177–196. *In*: J. Meier, U. Hasenöhrl, K. Krause and M. Pottharst (eds). Urban Lighting, Light Pollution and Society. Routledge, New York.

[121] Goodale, E., P. Ding, X. Liu, A. Martínez, X. Si, M. Walters and S.K. Robinson. 2015. The structure of mixed-species bird flocks, and their response to anthropogenic disturbance, with special reference to East Asia. Avian Research. 6.

[122] Whitehead, H. 2010. Conserving and managing animals that learn socially and share cultures. Learn. Behav. 38: 329–336.

[123] Williams, B.K. and E.D. Brown. 2014. Adaptive management: from more talk to real action. Environ. Manage. 53: 465–479.

[124] P.B. Stacey and J.D. Ligon. 1987. Territory quality and dispersal options in the acorn woodpecker, and a challenge to the habitat-saturation model of cooperative breeding. Am. Nat. 130: 654–676.

[125] Haas, S.E., J.A. Cox, J.V. Smith and R.T. Kimball. 2010. Cooperatively breeding brown-headed nuthatch (*Sitta pusilla*). Southeast. Nat. 9: 743–756.

[126] Tumulty, J.P. 2018. Dear enemy effect. pp. 1–4. *In*: J. Vonk and T. Shackelford (eds). Encyclopedia of Animal Cognition and Behavior. Springer International Publishing.

[127] Bennett, V.A., V.A.J. Doerr, E.D. Doerr, A.D. Manning, D.B. Lindenmayer and H.-J. Yoon. 2012. Habitat selection and post-release movement of reintroduced brown treecreeper individuals in restored temperate woodland. PLoS One 7: e50612.

[128] Burger, C. and C. Both. 2011. Translocation as a novel approach to study effects of a new breeding habitat on reproductive output in wild birds. PLoS One. 6: e18143.

[129] Hahn, M.W. 2018. Molecular Population Genetics. Oxford University Press.

[130] Anthony, L.L. and D.T. Blumstein. 2000. Integrating behaviour into wildlife conservation: the multiple ways that behaviour can reduce Ne. Biol. Conserv. 95: 303–315.

[131] Lott, D.F. 1991. Intraspecific Variation in the Social Systems of Wild Vertebrates. Cambridge University Press.

[132] Davies, N.B. 1992. Dunnock Behaviour and Social Evolution. Oxford University Press.

[133] Blumstein, D.T. 1998. Female preferences and effective population size. Anim. Conserv. 1: 173–177.

[134] Schlossberg, S.R. and M.P. Ward. 2004. Using conspecific attraction to conserve endangered birds. Endangered Species Update. 21: 132–138.

[135] Reed, J.M. and A.P. Dobson. 1993. Behavioural constraints and conservation biology: conspecific attraction and recruitment. Trends Ecol. Evol. 8: 253–256.

[136] Berger-Tal, O., T. Polak, A. Oron, Y. Lubin, B.P. Kotler and D. Saltz. 2011. Integrating animal behavior and conservation biology: a conceptual framework. Behav. Ecol. 22: 236–239.

[137] Blumstein, D.T. 2006. Developing an evolutionary ecology of fear: how life history and natural history traits affect disturbance tolerance in birds. Anim. Behav. 71: 389–399.

[138] Snijders, L., E.P. van Rooij, J.M. Burt, C.A. Hinde, K. van Oers and M. Naguib. 2014. Social networking in territorial great tits: slow explorers have the least central social network positions. Anim. Behav. 98: 95–102.

[139] Hallworth, M.T., T.S. Sillett, S.L. Van Wilgenburg, K.A. Hobson and P.P. Marra. 2015. Migratory connectivity of a Neotropical migratory songbird revealed by archival light-level geolocators. Ecol. Appl. 25: 336–347.

[140] Bonter, D.N. and E.S. Bridge. 2011. Applications of radio frequency identification (RFID) in ornithological research: a review: RFID Applications in Ornithology. J. Field Ornithol. 82: 1–10.

[141] Taylor, P.D., T.L. Crewe, S.A. Mackenzie, D. Lepage, Y. Aubry, Z. Crysler, et al. 2017. The Motus Wildlife Tracking System: a collaborative research network to enhance the understanding of wildlife movement. ACE. 12.

[142] Gill, L.F., P.B. D'Amelio, N.M. Adreani, H. Sagunsky, M.C. Gahr and A. ter Maat. 2016. A minimum-impact, flexible tool to study vocal communication of small animals with precise individual-level resolution. Methods Ecol. Evol. 7: 1349–1358.

[143] Stowell, D., E. Benetos and L.F. Gill. 2017. On-bird sound recordings: automatic acoustic recognition of activities and contexts. IEEE/ACM Transactions on Audio, Speech, and Language Processing. 25: 1193–1206.

[144] Mennill, D.J., M. Battiston, D.R. Wilson, J.R. Foote and S.M. Doucet. 2012. Field test of an affordable, portable, wireless microphone array for spatial monitoring of animal ecology and behaviour: Field test of a portable wireless microphone array. Methods Ecol. Evol. 3: 704–712.

[145] Vanderbilt, C.C., J.P. Kelley and E.H. DuVal. 2015. Variation in the performance of cross-contextual displays suggests selection on dual-male phenotypes in a lekking bird. Anim. Behav. 107: 213–219.

[146] Pope, T.L., T.J. Conkling, K.N. Smith, M.R. Colón, M.L. Morrison and R.N. Wilkins. 2013. Effects of adult behavior and nest-site characteristics on black-capped vireo nest survival. Condor. 115: 155–162.

Chapter **7**

Optimizing and Competing for Resources

Faith O. Hardin[1], Amanda K. Beckman[2], Alexis D. Earl[2] and Jacquelyn K. Grace[2]

Introduction

Like most animals, songbirds must detect and obtain food, turn it into energy, and balance energetic requirements with predation risk, hetero- and conspecific competition, social dynamics, and anthropogenic pressures. This integrative chapter begins with a discussion of resource acquisition by examining the physiological and behavioral processes that birds employ to obtain food and the evolutionary theories that have driven these tactics. It then moves into conservation and provides an overview of how human-induced changes (e.g., urbanization, agriculture, introduced species, and climate

[1]Department of Rangeland, Wildlife, and Fisheries Management Texas A&M University, Texas, College Station, USA
[2]Ecology and Evolutionary Biology Program Texas A&M University, Texas, College Station, USA

change) are altering these processes. This chapter concludes with specific examples of emerging technologies that have the potential to enhance our understanding of foraging behavior and through this, improve management decisions.

The foraging behavior of songbirds can be summarized by collecting data from five categories: (1) search, (2) attack, (3) foraging site, (4) food taken, and (5) food handling [1]. By using these categories, foraging strategies can be compared within a site and species.

(1) Searching behaviors are the movements and various postures used to find food or identify substrates that hide food. These movements include walking, flying, and climbing, and are usually quantified by the rate or lengths of movements between perches, and the lengths of time between different movements. Searching behaviors can also be classified along a gradient of active to passive searching. For example, wood-warblers (Family: Parulidae) are highly active, quickly jumping from perch to perch, while tyrant flycatchers will passively wait to eat any insects that happen to come near them. Species like the black-capped chickadee (*Poecile atricapillus*) fall somewhere in the middle because their searching rate varies by substrate.

(2) Attack behaviors begin the moment the prey is spotted and end when the capture attempt is made. One category of low-investment attack behaviors are near-perch movements; for example, Bahama orioles (*Icterus northropi*) commonly use the perch-gleaning method (grabbing a food item from a nearby substrate while perched) to obtain insects and berries [2].

(3) The foraging site where the search and attack behaviors take place includes the general habitat, vertical position, horizontal position, foliage density, and precise substrate where the food was taken.

(4) Data on the exact food item taken and whether the attack on the prey was successful can be difficult to collect in the field, especially with small prey. The most basic method to record food obtained while foraging is through direct observation of prey type (crude categories if necessary) and prey size, but eDNA (Environmental DNA) can also be non-invasively obtained to identify prey to the genus-level [3]. eDNA is collected from the environment (like water or soil) as opposed to collected directly from the organism; sources of eDNA include fecal samples, shed skin, and mucous.

(5) After food is obtained, it is handled and eaten, delivered to a mate or offspring, stored for later, or rejected. Measuring food handling time is important because the longer a bird spends handling a food item the costlier it is.

This chapter will discuss each of these stages of foraging, in detail, by providing an overview of resource detection and acquisition (sensory systems, environmental influences, tool use, and group foraging), foraging decision making (optimal foraging theory, foraging in patches, and diet flexibility), and food storage (both external and internal).

Detecting and Obtaining Food

Foraging Sensory Systems

To forage successfully, birds must perceive and utilize information from their environment that will help them to detect and acquire food resources. To procure this information, they rely on their sensory systems. It has long been assumed that birds primarily rely on visual and auditory information and other forms of sensory input (e.g., olfactory, tactile, gustatory) are largely overlooked in the scientific literature. However, recent studies suggest that these sensory modalities are important for resource acquisition in some songbird species and merit additional research.

Songbirds have highly developed tetrachromatic visual systems, with four single-cone retinal photoreceptors for color vision, one double-cone photoreceptor for achromatic motion perception, and a rod for dim light vision [4]. Each of the four single-cone retinal photoreceptors is sensitive to a different wavelength range of light, which allows songbirds to see ultraviolet wavelengths in addition to the visual spectrum seen by humans. As light travels through the avian eye, it passes through oil droplets that enhance color discrimination by filtering the light. Proteins called visual opsins within the cones then convert photons of light into electrochemical signals that are sent to the brain for processing as visual information. In this way, songbirds use both human-visible wavelengths and ultraviolet wavelengths, coupled with excellent color discrimination to see berries, insects, seeds and other food sources [5].

While birds as a group have excellent visual acuity, several factors can affect visual discrimination in songbirds, including environmental factors, the visual field, and specialized adaptations. In terms of the environment, the light (color/brightness) reflected by the food item, the lighting environment, foliage, and other habitat factors can affect contrast and light transmission, thus impacting visual detection and discrimination of food items. For example, in house finches (*Carpodacus mexicanus*) evidence suggests that retinal carotenoid levels, which are visual pigments derived from food that affect color discrimination, are optimized to meet visual needs under specific lighting conditions and can affect foraging performance [6]. Physiologically, bird species differ in the size, shape, and position of the eyes. These differences can affect a bird's visual field (i.e., the area visible when the eye is fixed on a location). Songbirds typically have wide visual fields, although this varies by species [7]. Additionally, some groups of songbirds have developed special visual adaptations. For example, the New World flycatchers (*E. virescens and E. minimus*) have a unique retinal structure that replaces the traditional oil droplet and may allow for more rapid motion tracking in these sit-and-wait insectivores [8].

The complex songs performed by songbirds suggests strong auditory senses, in addition to visual acuity. Songbirds use auditory information in a variety of ways to locate resources. Songbirds hear within a narrower range of frequencies than humans and other mammals. Birds are most sensitive to sound frequencies between 1000 and 4000 Hz (although they can hear outside of that range) [9]. Songbirds, however, compared to humans are less sensitive to frequency within that range [8]. Songbirds, however, have more acute sound recognition than humans. They can resolve sounds received at very short intervals [9]. They are particularly sensitive to variation in pitch, tone, and rhythm, which is useful for recognizing other individual songbirds by their unique songs and calls [10]. Anatomically, birds have ears that are conical to focus sound, located below their eyes and often protected by auricular feathers, while lacking an external ear or pinna. A slightly curved bony tube in the inner ear called the cochlea (or "cochlear duct") contains the sensory hair cells (and supporting cells) for hearing. Each of these cells is sensitive to specific frequencies. The cochlea varies in length by species, with longer cochlea having more auditory receptors and better resolution/sensitivity to a wider range of frequencies. Cochlea length explains variation in the range of frequencies each species can detect [9]. Apart from frequency detection, the sound must also be localized to use it to detect and hunt prey. Localization is accomplished in the horizontal plane via interaural time differences and interaural level differences, and in the vertical plane through sound modifications facilitated by head shape [11].

Songbirds use auditory information to identify the movements and/or vocalizations of prey. For example, American robins (*Turdus migratorius*) use auditory cues to find worms that are underneath the soil and cannot be detected visually [12]. The short, repeated sounds made by worm movement contain a range of frequencies that travel across the bird's tympanic membrane (or "eardrum"). Robins cock their heads from side to side while foraging, which may maximize the distance of sound traveling across the eardrum when acoustically localizing worm prey. Songbirds also use auditory information to discover new food sources through sounds made by conspecifics or heterospecifics hunting or foraging nearby [13]. These sounds can be unintentional (i.e., "eavesdropping") or intentional (i.e., "recruitment") on the part of the actor. For example, common ravens (*Corvus corax*) intentionally recruit conspecifics when feeding through yelling, thus accruing the antipredator benefits of group foraging [14]. Cliff swallows (*Hirundo pyrrhonota*) also call to recruit other foragers to a food source because group foraging increases their foraging efficiency; with more foragers present, it is more likely that insect movements will be detected and tracked because there are more individuals watching for them [15].

Some species of songbirds may also use chemical cues, detected through olfaction, to locate and select food items [16]. This mechanism has gained increased attention in recent years, although it remains understudied. Olfactory cues (specifically herbivore-induced plant volatiles) appear to be used in making foraging decisions by a variety of bird species including the European starling (*Sturnus vulgaris*) and members of the families Corvidae and Turdidae, particularly when variation in visual information is limited in agricultural habitats [16]. Evidence of olfactory-assisted foraging decisions exist for other avian species, as well. For instance, great tits (*Parus major*) also used herbivore-induced plant volatiles to locate trees infested with their prey, herbivorous caterpillars [17]; and black-billed magpies (*Pica hudsonia*) uncovered significantly more caches of food scented with cod liver oil than control, unscented caches [18]. Other studies have shown that birds can use chemical olfactory cues to avoid certain food items [19, 20], including those treated with pesticides [21]. However, some species of songbirds may not use or exclusively rely on olfaction to locate food sources. For example, although zebra finches (*Taeniopygia guttata*) are known to use olfaction in the context of social communication (e.g., use odors to discriminate kin from non-kin), olfactory cues may not be important for them in a foraging context. Given only olfactory cues to find food, zebra finches were not more successful than would be expected by chance, likely because they forage predominantly on seeds which can be easily visually detected [22].

Additional sensory systems that may be used by birds for food detection include tactile (i.e., touch), gustatory (i.e., taste), and magnetic systems. The use of tactile cues in birds appears to be dependent on foraging strategy. For example, tactile cues are heavily used by foraging shorebirds, and perhaps New Zealand robins (*Petroica australis*) and New Zealand fantails (*Rhipidura fuliginosa*). The facial bristles of the latter two species have pressure sensitive mechanoreceptors that may facilitate manipulation of insect prey in the bill or detection of prey in close-quarters [23]. Gustatory cues (taste) are not well studied, but red-winged blackbirds (*Agelaius phoeniceus*) use both visual and gustatory cues in food selection and avoidance of toxins [24], suggesting that more widespread gustatory cue use may occur. Magnetic directional cue use in a foraging context is unstudied in songbirds, however songbirds do use magnetic directional information to orient themselves for migration [25], and may be capable of using these cues for foraging. Magnetic cues are used by other groups, such as homing pigeons (*Columba livia f. domestica*) to locate hidden food sources [26]. Specifically, homing pigeons can be trained to associate food with magnetic anomalies and then use those anomalies to locate hidden food sources (see Fig. 1). More research is needed to fully understand the role that these and other sensory cues play in songbird foraging.

Figure 1 In a simple, but effective, study by Thalau et al. (2007) [26], one group of pigeons was trained to associate food with magnets attached to their bowl (A), while the control group was not. When the food was hidden under sand with a magnetic cue (B), the pigeons previously exposed to the magnets were 4.5 times more likely to search in the correct area than the control group.

Environmental Influences on Food Acquisition

Sensory cues used to detect and acquire food can be masked or amplified by habitat features. For example, vegetation density influences lighting environment and sound propagation, which affects reception of visual and auditory information, respectively. These changes in sensory reception can motivate different foraging strategies based on habitat type, such as foliage gleaning, aerial foraging, bark foraging and ground foraging [27]. Weather and seasonal changes also have strong effects on food detection and acquisition, largely by affecting the type and amount of foraging substrate available. Thus, migratory songbirds in temperate regions exploit spring and summer food abundance in the north, and then escape harsh northern winter conditions and food shortages by migrating south. During migration, birds experience heightened physiological demands due to the unique energy requirements of flight, decreased feeding opportunities (more so as habitat loss reduces the availability of high quality stopover sites), and variable environmental conditions [28] (see "Energy Partitioning and Internal Storage" for further discussion).

Food acquisition is also influenced by other aspects of the environment, such as the presence of heterospecific competitors. These competitors can drive species to utilize different foraging strategies, food types, or microhabitats. This concept of "niche partitioning" was made famous by Robert MacArthur in his classic 1958 paper on competing warbler species within a single habitat type. He found that the warbler species hunted insects at different arboreal latitudes, utilized different hunting behaviors, and staggered their nesting times to mitigate competition and avoid competitive exclusion [29]. In other words, each species was partitioning its specific niche (i.e., ecological role) to avoid overlap, and thus competition, with the other species. More recent studies have lent further support to the concept of niche partitioning. For example, black-capped chickadees and tufted titmice (*Baeolophus bicolor*) partition food sources, substrate, and foraging strategy in an area of overlapping range [30]. Similarly, eastern Australian songbirds appear to partition resources by occupying foraging guilds with differentiated foraging substrates and methods [31], and different species of sympatric Malaysian babblers niche partition based on foraging height and substrate [32].

Tool Use for Food Acquisition

Some songbird species take a highly proactive approach to obtaining food by forcing prey to expose themselves, sometimes through tool use. For example, woodpecker finches (*Cactospiza pallida*) use twigs or cactus spines to pry insects out of holes in trees [33]. In this species, tool use appears to be learned during a critical period of juvenile development, and is not entirely dependent on genetics or social learning [33]. Development of this complex foraging strategy occurs only during the juvenile stage (adults who did not use tools did not learn to do so after observing conspecific tool-use), and juveniles were able to develop tool use without a model individual and regardless of whether their parents used tools [33]. New Caledonian crows (*Corvus moneduloides*) also use a variety of tools for foraging, and are the only non-human animal species known to create hooked tools in the wild [34]. Hooked tools are hypothesized to have arisen in this species due to cumulative tool innovation, and provide benefits like faster foraging time in different tasks. In contrast to tool use development in woodpecker finches, tool use in New Caledonian crows involves a combination of inherited action patterns, cultural/social transmission, and individual problem solving [35, 36]. While tool-use is rare in songbirds, many species have evolved other proactive behavioral mechanisms for finding prey items, including flushing prey from concealed locations. This is a common tactic employed by American redstarts (*Setophaga ruticilla*), that flick their wings and fan their tails along branches and twigs to flush out prey [37].

Foraging in Groups

Group foraging can be beneficial for songbirds by increasing group vigilance from predators, the probability of detecting food resources, and the diversity of available foraging skills, competence, and knowledge compared to solitary foraging [38]. These groups can be formed entirely of conspecifics, or a mix of conspecifics and heterospecifics (i.e., a "mixed-species flock"). Within mixed-species foraging flocks, there are several types of interspecific relationships. Some mixed-species foraging flocks are mutualistic, in which all species benefit from the interaction. For example, problem solving during foraging increased in efficiency and frequency with flock size in mixed-species flocks of blue tits (*Cyanistes caeruleus*) and great tits (*Parus major*) under natural conditions. Group members benefited from higher seed intake, because larger, more diverse flocks were more likely to include birds with high levels of foraging experience [39]. Other flocks can be commensal, in which one species benefits and the other is not helped or harmed. Commensal relationships can also occur between species in different taxonomic classes, for example birds and mammals. For instance, eastern phoebes (*Sayornis phoebe*) consume deer ectoparasites and insects that are flushed from vegetation that the deer walks through, although whether this is commensalistic or mutualistic requires further study [40]. Other interactions can be parasitic, which includes kleptoparasitism (i.e., stealing food from other individuals; [41]. For example, fork-tailed drongos (*Dicrurus adsimilis*) follow flocks of heterospecifics, and use physical attacks or deceptive false alarm calls to steal prey, to their own benefit and the detriment of the heterospecifics [42].

Ecology of Foraging

Optimal Foraging Theory

How do animals decide which food items to eat, and which to ignore? How do they decide where to look for food? Game Theory (discussed more in depth in Chapter 6, but is derived from a mathematical model to determine optimal strategies in competition between adversaries) and specifically *Optimal Foraging Theory* (OFT) attempts to answer these questions. Generally, OFT assumes that organisms will make foraging decisions that maximize energy intake while minimizing costs, thus increasing inclusive fitness [43]. However, before beginning to apply OFT to any specific organism, it is crucial to be aware of its innate shortcomings and assumptions. OFT also assumes that (1) all animal strategies are rational, (2) animals always seek to maximize energy return, and (3) every individual has the same strategy. Despite these relatively simplistic assumptions, OFT has extensive support in the literature and provides a foundation upon which to build more complex and naturalistic models for specific taxa.

Food items come with inherent benefits, typically energy and nutrients; but they also entail inherent costs, including time and energy to search and handle the item. Some food items are very easy to handle (e.g., tomatoes, worms), while others require large amounts of energy to consume (e.g., seeds with heavy shells, prey that can fight back). Some food items are very abundant and easy to find, while others require extensive searching. OFT provides a framework within which to understand how animals integrate these costs and benefits. Within OFT, the *Optimal Diet model* explicitly defines how energy reward, handling time/energy, and search time/energy should be balanced in an optimal world. This model is frequently used to predict whether an animal should invest in a food item that they have already found, or search for a more profitable food item [44]. The Optimal Diet model is presented in the following equation, where the energy in prey$_1$ = E_1, the energy in prey$_2$ = E_2, handling time of prey$_1$ = h_1, and handling time of prey$_2$ = h_2. The profitability of prey$_1$ is thus the ratio of the energy content to the handling required (E_1/h_1). Prey$_1$ should be consumed whenever encountered, and prey$_2$ ignored when:

$$\frac{E_1}{h_1} > \frac{E_2}{h_2}$$

$$\text{\includegraphics{seed}} = E_1 \qquad\qquad \text{\includegraphics{acorn}} = E_2$$

Figure 2a An individual sunflower seed has less energy than an acorn, however, an acorn is much harder to crack open because of its tough shell. When both items are equally accessible, the optimal diet model predicts that it is more profitable for a blue jay to eat the sunflower seeds than expend energy to open the acorn.

However, the animal should eat prey$_2$ if searching for prey$_1$ is too costly. Thus, prey$_2$ should be consumed when encountered only if the energy contained in prey$_1$ is too low to justify the time and energy to search (S_1) for it:

$$\frac{E_2}{h_2} > \frac{E_1}{(h_1 + S_1)}$$

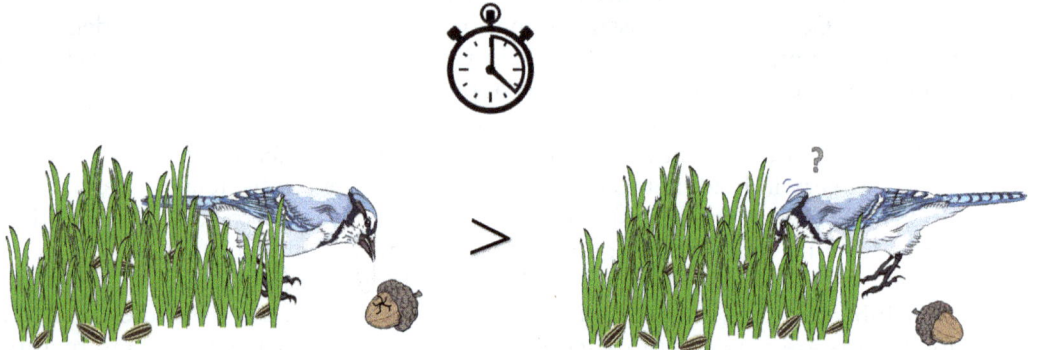

Figure 2b According to the optimal diet model, if the easily opened sunflower seeds are hidden in tall grass, and thus have a longer search time, it may be more profitable to expend the energy to open the acorn.

When the search time for prey items is very long, it can be beneficial to eat a variety of prey items. One can see this by rearranging the equation above, again, to predict how long S_1 must be for an animal to choose to eat both types of prey. When an animal's S_1 reaches this threshold, it is defined as generalists, or chasing/eating many different types of food items they come into contact with. One the other hand, when a species has a long handling time coupled with a short search time the equation becomes more equally balanced, resulting in a specialist, defined as a species that includes only high-profitability items in their diet.

$$S_1 > \left[\frac{E_1 \times h_2}{E_2} \right] - h_1$$

Generalist

$>$

Specialist

$=$

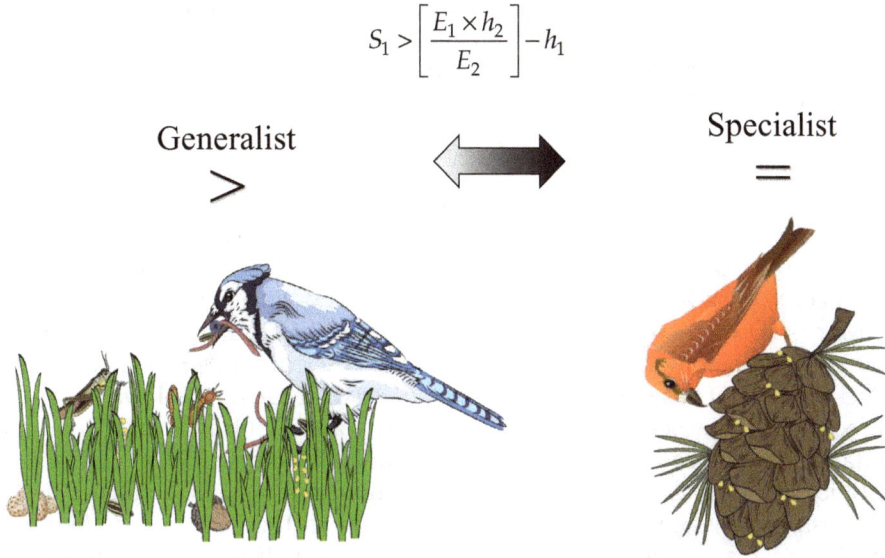

Figure 2c As shown in the equation above, the optimal diet model can work as a gradient. Animals can act as generalists when the search time exceeds the threshold (the equation is unbalanced) and they must look for a variety of foods, like the blue jay, or animals can act as specialists when search time is short, but handling time is long (the equation is balanced). A classic example of a specialist is the red crossbill (*Loxia curvirostra*), whose bill is uniquely shaped to extract conifer seeds.

A classic example of organisms engaging in optimal foraging comes from oystercatchers (*Haematopus ostralegus*) that balance the energy gained from large, nutritionally rich blue mussels (*Mytilus edulis*) with the energetic handling cost required to hammer open the thicker shells created by larger mussels, resulting in medium sized mussels being preferred [45]. Songbird studies also support optimal foraging. For example, when presented with a low energy food item, carrion crows (*Corvus corone*) and common ravens (*Corvus corax*) will continue to search for a higher quality food item, but exhibit a search time threshold after which they will consume either food item [46]. Great tits (*Parus major*) become more selective as the abundance of a preferred prey item increases (decreasing search time), when handling effort is held constant [47].

Resource Patchiness and Optimal Foraging

In nature, food items are often located in patches that provide diminishing returns. How do animals decide where to forage and when to switch foraging patches? The *Marginal Value Theorem* (MVT) attempts to answer this question, while also refining the Optimal Diet model by incorporating predation pressure as a cost of foraging. The MVT is based on the idea that organisms perceive food resources in patches, and after discovering a patch of food they balance the future yield of that patch against the future yield of moving to another, unknown, patch [48]. Thus, an animal should leave behind a certain amount of food within a patch (the "Giving Up Density" discussed, below) once the energy necessary to search out the remaining food is higher than the energy required to move to a fresh patch. The MVT can be generalized to incorporate these costs in addition to predation. An animal should give up on a depletable food patch when the harvest rate (*H*) is no longer greater than the sum of the metabolic cost (*C*), predation cost (*P*), and missed opportunity cost (*MOC;* the energetic value of alternative activities) of foraging [49].

$$H = C + P + MOC$$

The *Giving Up Density* (GUD) refers to the amount of food left in the patch when the animal gives up foraging, and can provide a surrogate estimation of a forager's quitting harvest rate, which is often harder to quantify [49]. The GUD can vary both within and between species, based on these costs and the harvest rate [50]. For example, gerbils have a lower GUD than crested larks (*Galerida cristata*), indicating that gerbils maximize the resource patch before moving on and are more efficient foragers. However, the larks modulated their GUD based on predation risk, thus their GUD was lower in open areas where predation risk was lower, compared to bushy areas [50]. This balance of foraging and predation avoidance was also seen in Carolina chickadees (*Parus carolinensis*) and tufted titmice (*Parus bicolor*). These species preferentially visited food patches with lower predation risk, and when in high-predation risk areas they decreased their foraging efficiency (i.e., increased their GUD), but also increased their harvest rate to compensate for the increased predation cost [51].

The Marginal Value Theorem can also be applied when an animal must travel back to a central location to process food items, for instance when caching or delivering food to young. In these cases, birds can only hold so many food items in their mouth and each additional food item makes it more difficult to pick up the next, thus there is a diminishing return for each additional food item [52]. Imagine trying to balance 15 tennis balls in your arms while bending over to pick up a 16th! This phenomenon is referred to as the *curve of diminishing returns*, and the bird must decide when to give up collecting food so that it can deliver its load and return fresh to the foraging ground. This decision is often based on a trade-off between travel time and load size, such that birds will bring back higher numbers of prey the further the patch is from their central location (e.g., nest) [52].

While it is important to recognize the power of the optimal foraging models discussed here in predicting how songbirds forage in a changing environment, when developing management protocols, one must remember that animal behavior is complex and can be influenced by many (often unknown) variables. For example, the assumption of optimal foraging theory that all individuals within a species have the same strategy is not always true. Individual personality traits can influence feeding strategy and affect conformity to optimal foraging models [53]. For example, great tits consistently vary in exploratory tendencies, with faster exploring individuals more likely to travel farther for food than slower exploring individuals, affecting search time and energy [54]. Differences in great tit personality and associated foraging strategies are also reflected in their problem solving abilities [55], which affects handling time/energy. Similarly, in starlings exploration tendency was positively related to foraging distance, although it was also influenced by home range size and foraging habitat quality [56], again potentially altering search time/energy. It is the opinion of the authors that these models should be used as a guide to study and interpret songbird foraging behavior, not as a rule book. This guide can provide a foundation for understanding foraging strategies, but may need to be modified for specific species and scenarios.

Diet Flexibility

Conformation to optimal foraging theory depends on the level of diet flexibility an organism possesses, or its ability to switch between food items when it is advantageous. Some species have high flexibility, for example, after experimental grazing regimes reduced the abundance of preferred large grass seeds, the common diuca-finch (*Diuca diuca*) and rufous-collared sparrow (*Zonotrichia capensis*) expanded their diet to include small grass and forb seeds, which require more energy to gather and provide less energy than a large seed. However, other species exhibit low diet flexibility, like many-colored chaco finches (*Saltatricula multicolor*) and ringed warbling-finches (*Microspingus torquatus*), and were not able to make this switch [57]. While diet flexibility can enable closer conformation to optimal foraging, it often comes at a cost to search efficiency. To find prey, birds typically use a cognitive representation of prey (i.e., a learned "search image") derived from memory, combined with other information obtained from prior experiences [58]. Tinbergen found that great tit prey preference in the wild reflected prey abundance and that with each interaction with a specific insect species the tits's ability to recognize it became stronger, thus reinforcing and refining the search image. Blue jays (*Cyanocitta cristata*) have also been shown to utilize search images to improve foraging efficiency; when faced with foraging for just moths, just caterpillars, or foraging for both simultaneously on a computer screen, blue jays performed best when searching for one prey type at a time, and were less efficient when splitting their focus [59]. Thus, switching prey items can incur a short-term cost in foraging efficiency until a new search image is produced.

Birds will often switch or shift preferred food items and foraging strategies during reproductive periods. As discussed later in this chapter (see "Energy Partitioning and Storage"), reproduction is one of the most energetically demanding periods of songbird life history, necessitating increased foraging intensity, efficiency, and prey switching to higher protein- and lipid-rich items. Within species, food availability may change within a breeding season and parents will often adjust their feeding strategies according to this shift in resources [60]. Depending on the species, severity of resource depletion, age, and body condition of the bird, parents may decrease foraging intensity (even to the point of abandoning offspring), increase foraging intensity, or preferentially feed offspring by size or condition. Parents and offspring are sometimes at odds regarding feeding strategies (i.e., "parent-offspring conflict"), with parents trying to maximize their fitness by producing many young over their lifetime and at times investing in young differentially, while their young each try to maximize their own individual protection and feeding rates by their parents [61]. The amount of parent-offspring conflict tends to vary based on environmental conditions [62]. For example, smaller and weaker nestlings tend to beg more loudly and frequently for food than larger nestlings, and parents preferentially feed more based on offspring need in good and predictable environmental conditions. However, among bird species that have evolved in environments with unpredictable food resources, parents tend to ignore this increased begging and neglect young in poor condition to invest preferentially in higher quality young who are more likely to survive harsh conditions [62].

Food Storage

Most birds must consume food as soon as it is obtained, but some songbird species are known to *cache* food, or store food for some point in the future, sometimes months later. Caching is common in environments where food sources seasonally fluctuate from abundant to scarce [63]. Caching behavior is present in the families Corvidae, Paridae, and Sittidae, but the locations of caches and types of food stored vary between the different families. For example, scrub jays (Genus: Aphelocoma) commonly cache acorns in the ground or under leaf-litter, while chickadees (Genus: Poecile) will store insects and seeds in trees. However, caching as a storage strategy includes the risk that conspecifics or heterospecifics could come across the cache randomly, or watch where the food was stored and pilfer the cache later [64, 65]. Cache pilfering can lead to the evolution of cache protection behaviors [66]. For example, western scrub-jays (*Aphelocoma californica*) move their cached items if another individual saw where the item was cached [67]. Interestingly, only individuals who have previous experience being cache thieves themselves adopt this strategy; naive birds do not move their caches if an observer is present [67].

It is hypothesized that the physiological and behavioral mechanisms that affect cache recovery should be shaped by natural selection since the ability to cache and locate those caches impacts survival [68]. A review of avian caching behavior supported this hypothesis and found that selection for food hoarding results in the evolution of enhanced memory and changes in some structures in the hippocampus [68]. Similar to cache retrieval abilities, the internal processes promoting cache protection could also be under selection [69].

Energy Partitioning and Internal Storage

After birds have obtained and consumed a food resource, they then have to decide how to partition the energy within that resource into different behavioral activities and physiological processes, or whether to store that energy for later use, typically as fat or muscle. Partitioning decisions will depend on life history stage, food availability, and current energetic requirements. The activities that demand the greatest amounts of energy for migratory songbirds are typically: reproduction, migration, and overwintering survival. Songbirds exhibit a large degree of phenotypic flexibility regarding energy investment in these areas, depending on environmental, social, and internal cues, and will often trade-off between these activities and less critical processes.

Reproduction in songbirds is largely synchronized with food availability [70]; for example, blue tit females decide when to lay eggs based on peak insect abundance from the previous year [71]. Energy partitioning decisions are critical for breeding birds because reproduction is energetically expensive, requiring mating behaviors, nest building, territorial defense, egg formation, incubation and brooding (with limited foraging opportunities), predator defense, and food delivery to nestlings in addition to self-maintenance activities. Moreover, nestlings require foods with high lipid and protein loads, and substitution with low energy foods results in heavier digestive organs and lower lipid reserves in nestlings, and higher provisioning demands on parents [72]. Consequently, many granivorous passerines feed their nestlings weakly chitinized invertebrates (flies and spiders), and then advance to highly chitinized invertebrates, followed by slowly introducing seeds to allow the nestlings' digestive systems time to adjust [73]. These foraging shifts ensure maximum energy availability for critical processes during the breeding season.

Energy partitioning decisions are also of utmost importance for songbirds during migration, because of the high energy investment involved; a songbird flying at minimum energy level is estimated to expend greater than twice the energy of a running mammal [74]. This demand is even heavier during migration when songbirds fly for extended periods of time [74] without the ability to consistently forage. In preparation for this extreme energy expenditure, songbirds exhibit hyperphagia (increased eating) and store up to 50% of their body weight in adipose fat, and excess protein in muscles and digestive organs, as they prepare for fasting periods during flight [74]. For example, migrating semipalmated sandpipers (*Calidris pusilla*) double their size by gorging on burrowing amphipods (*Corophium volutator*), and naturally "dope up" on unsaturated fatty acids that increase lipid transport to enhance muscle performance [75]. This rapid storing and utilization of fats and proteins highlights the importance of stopover points, or areas with historically high food availability used by migrating flocks, which are critical for maintaining and replenishing energetic stores required for migratory flight.

During migration, birds prioritize recovering protein loads before fat and reduce the size and efficiency of their digestive systems, resulting in slow rates of mass gain at stopover sites [76]. This process is developmentally plastic; juvenile songbirds have heavier digestive organs, smaller flight muscles, and faster metabolic rates than adults, which allows them to eat a more diverse diet and compensate for poor foraging and hunting skills [28]. But energy storage is only one part of the energetic puzzle during migration; songbirds must also optimize foraging choices by choosing foods with the most efficient calories. Many songbirds switch from feeding on insects (high protein) to fruits (low protein] during their migration because protein inhibits the fattening processes [77]. Additionally, migratory birds like garden warblers (*Sylvia borin*) may prefer plants with high concentrations of long-chain unsaturated fatty acids which promote body mass recovery [77].

Careful partitioning of resources becomes increasingly important during winter for temperate songbirds, as food becomes less available while energetic costs of thermoregulation rise. Birds maintain relatively high internal temperatures and thermoregulation under cold temperatures is energetically expensive, requiring increased caloric input from foraging [78]. However, as during any season, foraging activity in winter must be balanced against the risk of predation. One model that attempts to describe this risk-reward balance is the Optimal Body Mass model [79]. This model predicts that when resources are predictable it is more beneficial to remain lean in an effort to reduce predation (too fat, you get caught) and when food is unpredictable it is more beneficial to fatten up to avoid starvation. For example, dark-eyed juncos (*Junco hyemalis*) and American tree sparrows (*Spizella arborea*) that forage in winter on the ground where deep snow makes food less predictable, were more likely to have large fat reserves when compared to closed cover foragers like the spotted towhee (*Pipilo maculata*) and the swamp sparrow (*M. georgiana*) [80]. To review topics covered in this section, see Box 1.

Box 1   Foraging ecology and evolution

Detecting Obtaining Food

- Songbird foraging behaviors can be grouped into five categories: (1) search, (2) attack, (3) foraging site, (4) food taken, and (5) food handling.
- Songbirds have highly developed senses that they use to detect and localize prey.
  ○ Vision: color discrimination, wide visual fields, and can see ultraviolet wavelengths.
  ○ Hearing: sensitive to variation in pitch, tone, and rhythm, and use sound to localize prey.
  ○ Other senses such as smell, touch, taste, and magnetic systems have been understudied, but are used to detect and obtain food, while avoiding toxins.

Environmental Influences and Techniques of Food Acquisition

- Seasonal changes and con/hetero-specific competition influence food acquisition.
- Songbirds can use tools and con/hetero-specific foraging tactics to obtain food more efficiently and safely.

Ecology of Foraging

- Optimal Foraging Theory (OFT) predicts that animals will maximize energy intake while minimizing costs.
- The Optimal Diet model defines how food energy, handling time, and search time should be balanced in an optimal world.
- The Marginal Value Theorem predicts where animals decide to forage and when to switch from one patch of food to another.
- Diet flexibility allows songbirds to conform to optimal foraging, by switching food preferences and foraging strategies when faced with varying external pressures.

Energy Partitioning and Storage

- Caching (food storage) is most common in unstable environments, and requires enhanced memory and cognition.
- The activities that demand the greatest amounts of energy for migratory songbirds are typically: reproduction, migration, and overwintering survival.
- When food is scarce, songbirds must weigh foraging against the risk of predation. This is predicted by the Optimal Body Mass model.

Conservation

The modern world is dominated by anthropogenic changes, which alter the diversity and abundance of food sources for songbirds. Some species have thrived in these changing environments, especially those with generalist foraging strategies (e.g., America crows (*Corvus brachyrhynchos*)), highly developed cognitive abilities (e.g., blue jays (*Cyanocitta cristata*)), and/or bold behavioral profiles (e.g., urban mountain chickadees (*Poecile gambeli*)). In contrast, other species have struggled to adapt [81]. This section will discuss anthropogenic factors that affect songbird foraging behavior, such as expanding urbanization, changes to the plant community, and climate change, while outlining ways to integrate foraging behavior into practical applications for management, including the most recent technological advances in the field.

Impacts of Urbanization

As of 2017, urban areas (i.e., the total area within the administrative boundaries of a city, including all the impervious surfaces, vegetated areas, barren land, and water bodies) cover almost 3% of the terrestrial Earth; a percentage that is constantly increasing [82, 83]. Urbanization often leads to declines in biodiversity, as most species populations' decline while a few are able to take advantage of these altered habitats [84]. In birds, urbanization results in changes to the acoustic environment, predator and competitor densities, resource patch distribution, and food item availability/density, all of which can impact foraging decisions and strategies.

As was discussed earlier in this chapter (see "Detecting and Obtaining Food: Foraging Sensory Systems"), songbirds often rely on auditory cues to locate food. Urban and suburban environments are noisy compared to natural areas [85], imposing restraints on auditory detection and localization during foraging. Bird species with low-frequency contact calls may be drowned out by vehicular noise, causing birds to call louder and more frequently, which is energetically expensive and increases risk of predator exposure [85]. Urbanization can also affect predator and competitor densities, including those of non-native species, altering the costs and benefits of foraging decisions. This can result in increased vigilance, reduced feeding rates, and species-specific changes in social foraging, and predator avoidance [86]. Additionally, the urban matrix (i.e., the combination of roads, bridges, parking lots, and buildings) can have interactive effects, altering the way songbirds move through resource patches [87]. For example, the effect of a single road or bridge has little impact on songbird movements, but both the urban exploiter black-capped chickadee and the riparian specialist yellow warbler (*Setophaga petechia*) were unable to permeate dense urban matrixes [88]. This urban matrix also influences the Giving Up Density (GUD) of foraging birds. Increased fragmentation and distance from edge habitat, and decreased connectivity and canopy cover can lead to higher GUDs (i.e., less efficient foraging within a patch), and can restrict movement while increasing energy expenditure as birds must cross undesirable habitat to reach the next isolated resource patch [89].

Urbanization can also directly affect food abundance, with differing effects on insectivores, granivores, and generalists. For insectivores, prey abundances increase or decrease in urban environments, depending on the specific prey item. For example, the abundance of insect orders commonly eaten by birds (e.g., Lepidoptera and Hymenoptera) decreases toward urban centers, corresponding with increases in impervious surfaces (e.g., concrete, asphalt) [90–92]. However, the abundance of some insect orders (e.g., Diptera and Hemiptera) increase in urban areas due to reduced predation and favorable environments [90–92]. Some insectivores can also successfully modify their hunting tactics to exploit anthropogenic structures, such as European magpies (*Pica pica*) that opportunistically eat caddisflies (Order: Trichoptera) disoriented by the polarized light from glass buildings [93]. Granivorous songbird species tend to experience an increase in year-round food availability in urban and suburban areas, because they can rely on anthropogenically supplied food (i.e., bird feeders) during winter when other food items are scarce [83, 94]. Generalists also benefit from anthropogenically supplied food from feeders, as well as gardens and trash bins [95]. Because of this, areas with high densities of bird-feeders exhibit increased bird species richness [94], but urban and suburban habitat in general has repeatedly been shown to exhibit lower species richness compared to undisturbed habitat [96]. Bird abundance, however, is often higher in urbanized areas due to the proliferation of urban exploiters [97, 98].

These shifts in food abundance in anthropogenically affected areas have a direct effect on the behavior of local birds; in blue tits, a shortage of caterpillars in an urban area resulted in increased travel distances for parents, reduced offspring weights, and numbers of successful offspring [99]. Alternatively, great tits in urban areas had shorter travel times due to a local increase in caterpillars, but needed to compensate for lower than average carotenoid levels (a proxy for food quality) in their prey by feeding their chicks more frequently [100]. The ability to gain nutrients in a highly variable environment (such as urban areas) and take advantage of novel foraging opportunities while adjusting for novel risks (traffic, anthropogenically altered light, increased background noise) is a defining characteristic of urban adapted species [101]. On a college campus, red-winged starlings *Onychognathus morio* experience large swings in the types and abundance of food on weekends opposed to weekdays [102]. This bird's ability to thrive in this environment was characterized by its ability to switch from discarded human food to natural food within the span of a week.

Impacts of Altered Plant Communities

Increased global conversion of natural areas to agriculture and timber production has produced large swathes of homogenized plant communities, which then leads to homogenization of local animal communities. This reduces songbird food sources including native seed, fruit and arthropod diversity, and thus limits many songbird species' reproductive success, an issue that is compounded by the poor nesting habitat that plant monocultures provide [103]. In Uganda, songbird communities centered around agriculture and logging were characterized by low biodiversity, containing larger bodied, frugivorous birds, while regenerating forests and mature forests had richer avian communities that included small, insectivorous songbirds [104]. Birds that do persist in agricultural landscapes may have to alter their foraging strategies, in some cases to more energetically expensive tactics. For example, the conversion of oak savannas across the midwestern United States to corn and soybean crop fields has led to fundamental changes in the foraging techniques of endemic flycatchers. Both the eastern wood-pewee (*Contopus virens*) and the great crested flycatcher (*Myiarchus crinitus*) prefer to forage in the upper canopy, but in open crop fields they must forage at the shrub and ground level and use more energetically demanding attack maneuvers to flush and pursue prey [105]. However, some omnivorous and/or generalist species are able to exploit the homogenous food resources provided by agriculture and logging, and thrive in these habitats. For example, blackbirds (Family: Icteridae) are often attracted to rootworm beetles (*Diabrotica virgifera*) and armyworms (*Spodoptera frugiperda*) in cornfields, grain stored in cattle feedlots, and young fish in hatcheries [106].

In addition to agriculture and timber production, humans alter plant communities through the introduction of non-native species, which can also impact seed, fruit, and arthropod diversity and dispersal [103]. Overall, exotic plant communities have less diverse insect communities, given that non-native plants – which did not adapt to their introduced environment – are often poor hosts to native animals [107]. Similar to agricultural impacts, these altered plant communities can reduce food availability for certain songbird species, while increasing it for others [108]. For example, grasshoppers are less abundant within areas infested by the exotic forb spotted knapweed (*Centaurea maculosa*). Grasshoppers are the primary food source for chipping sparrows (*Spizella passerina*) during the breeding season. In areas with high numbers of spotted knapweed, sparrows display delayed nesting initiation, which reduces the probability of laying a second brood and lowers overall fecundity [109]. Chestnut-collared longspurs (*Calcarius ornatus*) also exhibit negative fitness effects of non-native vegetation. Nestlings in exotic crested wheatgrass monocultures gain mass more slowly, fledge at a smaller size, and have lower survival rates than nestlings in native prairie grasses, even though longspurs do not differentiate between these habitats when nesting [110]. These findings suggest that the exotic habitat type may be an ecological trap; appearing to be high quality habitat, but after nesting, birds were unable to find enough food to sustain nestlings or fledglings. Exotic vegetation can appear to provide high quality habitat for songbirds because of its high foliage density, which some bird species have evolved to use as a cue for food density. For example, the red-eyed vireo (*Vireo olivaceus*) uses foliage density as a proxy for caterpillar density and adjusts territory sizes accordingly [111]. However, areas of dense, tangled, non-native monocultures, such as Japanese barberry (*Berberis thunbergii*), may trick birds into nesting in areas that are unlikely to support adequate nesting resources. This can be a widespread issue in areas where humans have suppressed fire [112]. On the other hand, some songbird species benefit from exotic vegetation. For instance, gray catbirds (*Dumetella carolinensis*) experience lower predation rates in introduced honeysuckle thickets (Family: Caprifoliaceae), and are better able to provision nestlings with the shrub's berries than in native vegetation [113]. Thus, the life histories of all relevant species should be considered when developing management strategies for invasive plant removal.

Many songbirds display a high degree of dietary flexibility (see "Diet Flexibility"), which can be beneficial when confronting introduced and novel food sources [114]. However, as birds take advantage of these novel resources they can aid in the dispersal of non-native and occasionally harmful plants [115]. For example, introduced honeysuckle fruits are present throughout winter, thus facilitating range expansion in frugivorous birds and serving as an important stop-over resource for birds returning to their wintering grounds [116]. These birds then deposit honeysuckle seeds later in migration, enabling honeysuckle range expansion through seed dispersal, and thus increasing the suitability of additional habitat for these bird species in a "snowball" effect. Although songbird mediated seed dispersal can benefit invasive plants in some cases, it can also benefit natives [117]. In managed oak woodlands in Spain, Eurasian jays (*Garrulus glandarius*) were the only species capable of crossing large expanses of open areas hostile to the oak, facilitating oak reestablishment through sprouting of hoarded acorns (which had higher chances of survival than those dispersed by wind) [118]. In the United States, Clark's nutcrackers (*Nucifraga columbiana*) are crucial to establishing and stabilizing native pine populations, such as ponderosa pine (*Pinus ponderosa*), in the Rocky Mountains [119]. They can carry up to 150 pine seeds at a time, as far as 32 km, and cache them about 2.5 cm below ground, an optimal depth for pine seed germination [120]. On sites with sun exposure and excessive wind, nutcracker-dispersed pines serve as nurse trees for other shrubs and trees, which in turn provide additional foraging opportunities for songbirds [121].

Impacts of Climate Change

Warming temperatures and other climatic shifts are changing ecological processes and altering songbird behaviors such as egg laying and nestling provisioning, migration, and range expansion [122, 123]. By understanding how songbirds modify their foraging behavior in response to changing environments, managers will be able to predict consequences and respond with applicable solutions.

Elevated near-surface air temperatures alter the composition of plant communities and drive plants to produce buds, leaves, and seeds earlier in the year. Thus, food sources for insectivorous, frugivorous, and granivorous songbirds peak earlier as well [124]. Some animals are unable to shift their breeding timing to match this earlier peak in resources, which can be particularly detrimental for those that depend on currently available resources for breeding (i.e., "income breeders"), such as small songbirds [125]. This is demonstrated clearly in the temporal variation of breeding blue tits in the Mediterranean. Populations inhabiting the evergreen forests of Corsica (a mountainous island off the coast of France) time egg laying with evergreen leaf renewal in early June, which coincides with peak caterpillar abundance [126]. In contrast, populations inhabiting the oak regions of southern France time their breeding with the spring leaf flushes in early May [127]. This difference in breeding timing appears to be genetic with limited plasticity within individuals. When a mismatch occurs between breeding date and peak caterpillar abundance, blue tit parents are forced to forage beyond their sustainable metabolic limit, resulting in decreased parental and nestling survival and decline in nestling mass [125]. Such mismatches between reproductive phenotype and resource phenology are expected to occur more frequently as warmer temperatures push greening dates earlier, beyond the limits of songbird phenotypic plasticity.

Higher temperatures and extended spring and summer seasons can alter annual resource availability and distribution, leading to latitudinal and elevational changes in avian ranges. For example, in the Arctic, Lapland longspurs (*Calcarius lapponicus*) forage primarily on ground-dwelling beetles and large spiders [128], while white-crowned sparrows (*Zonotrichia leucophrys*) glean flies and small spiders from tall shrubs [129]. Warming temperatures in Arctic regions facilitate woody encroachment, which subsequently increases the abundance of flies and small spiders, and decreases that of beetles and large spiders, which prefer open habitat [130]. Thus, as temperatures continue to warm, white-crowned sparrows are expected to exhibit latitudinal range expansion and a northward shift, tracking increased resources in woody encroachment, while Lapland longspurs are predicted to experience equivalent reductions in range size [130]. Birds can also shift their elevational ranges to track changes in climate and resource availability. For example, over the past century 48 out of 53 avian species in the Sierra Nevada mountains of California, USA, exhibited tracking of their temperature and precipitation niches along elevational gradients [131]. These results indicate that climatic niche models can forecast future range shifts for bird species, and may be useful in directing management efforts.

Impacts of Anthropogenic Toxins

The introduction of toxins in songbird habitats can dramatically alter the availability of food, foraging behaviors, and the way songbirds process nutrients. While many insecticides do not directly affect birds (most insecticides kill insects through inhibiting chitin-synthesis), they may suffer indirect effects such as reductions in their preferred foods. The use of diflubenzuron (a widely-used insecticide, targeting Lepidopterans) has not shown to directly impact songbird populations, but in areas where the toxin was used, the gut contents of several warbler species was reduced and/or diets shifted away from Lepidopteran larvae towards Dipterans, and Coleopterans which are more difficult to catch and have lower dietary value [132]. Additionally, the Louisiana waterthrush (*Parkesia motacilla*) expanded both its diet richness and foraging niches to compensate for reduced aquatic insect communities from increased stream acidification [133]. Other anthropogenic contaminants, such as high exposure to heavy metals, can alter foraging behavior by changing an individual's behavioral suite. Great tits at a site polluted by high levels of lead and cadmium were slower to explore new environments, potentially indicating damaged neurological functions [134]. To review topics covered in this section, see Box 2.

Box 2 Anthropogenic effects on foraging

Urbanization

- Urban areas have decreased biodiversity, both in songbirds and their resources.
- Increased noise production impairs songbird's ability to detect and localize food.
- Habitat fragmentation leads to less efficient foraging.
- Generalist species tend to thrive in urban areas as they are able to switch foraging strategies and food types, whereas specialists are often pushed out.

Altered Plant Communities

- Agriculture, timber production, and introduced non-native species, produce homogenized environments, leading to reduced biodiversity and decreased foraging opportunities.
- Non-native plant communities can dupe songbirds into nesting, but then do not provide nutritional/abundant food for their young, resulting in ecological traps.
- Songbirds with high degrees of diet flexibility may facilitate the spread of native and non-native species as they forage.

Climate Change

- Changes in air temperature cause plants to produce buds, leaves, and seeds earlier in the year, altering foraging (and thus nesting time) resources. This is especially detrimental for species unable to shift breeding timing.
- Higher temperatures alter resource distribution and can change avian ranges and foraging tactics.

Anthropogenic Toxins

- Toxins from agriculture and urbanization can impact food availability and behavior of songbirds.

Management Solutions

As urbanization accelerates globally, working within urban systems to make them habitable for wildlife and facilitate coexistence with humans is just as important as preserving what is left of wild spaces. One way of making urban spaces songbird-friendly is to mitigate the loss of foraging habitat due to urban sprawl by developing and maintaining urban green spaces. Urban green spaces are important for residential and migratory species alike, as migratory species rely on them for stopovers [135]. Migration may be the most challenging period of a songbird's annual cycle, thus conservation plans should consider stopover sites along migration routes as high priority. Urban areas can provide high-quality stopover sites. For example, one of the best known stopover locations in North America is Central Park in New York City, USA, which is positioned under the Atlantic Flyway and consists of a diverse community of native woodlands, meadows, ponds, and gardens that provide resources including sap, fruit, seeds, insects and more for over 250 bird species [136]. A growing number of studies show that birds heavily utilize a few strategic areas as migratory stopover locations. For example, in the Americas, prime stopover sites include the Orinoco grasslands (Llanos), the Sierra Nevada de Santa Marta (Colombia), and the Yucatan Peninsula [135, 137]. Efforts should be made to conserve these and other high priority areas. Additionally, management efforts should be focused on creating green spaces in areas such as Central America (particularly SE Mexico, Costa Rica, and Panama) that create a geographic bottleneck for migrating warblers [137].

Urban planners developing green spaces should prioritize connectivity of landscapes by creating interconnected urban greenspace networks (especially along migratory routes) rather than creating and maintaining large, isolated bird-friendly patches [88]. Given that perceived predation risk and habitat connectivity influence foraging behaviors, managers can encourage songbird foraging in urban parks by enhancing habitat connectivity and providing suitable cover for protection from predators [89]. For example, suitable habitat for the common redstart (*Phoenicurus phoenicurus*) is greatly increased in green space networks by augmenting short-cut lawns with large canopy covering hardwoods [138].

Green roofs featuring native plants also mitigate the effects of urbanization and other types of human development (e.g., agriculture) on songbirds by providing foraging habitat for resident birds and stopover habitat for migrating songbirds [139]. Green roofs can be rich in songbird food resources like seeds and fruit, while also providing habitat for songbird prey species (e.g., arthropods) [140]. However, the locations and heights of green roofs should be chosen with consideration for bird collision potential, noise pollution, and height relative to bird foraging flight. For example, attracting birds to green roofs within or near airport environments could increase the risk of bird collisions with aircrafts [141]. Acoustic signals that birds rely on for foraging may also be masked by background noise near airports [142], or other areas of high noise pollution. Birds residing near high noise pollution areas are also at greater risk of damage to auditory receptors due to constant exposure to elevated sound pressure levels. This can result in hearing loss and disadvantage birds that use hearing to forage (e.g., some gleaning insectivores) [142]. Additionally, native bird species use green roofs that are less than 60 m from the ground to a greater extent than higher roofs, likely due to the effort required to reach high roofs and increased wind currents [143]. When designing a roof garden, plant diversity should be determined based on the seasonal dietary requirements and foraging behavior of the target bird species because specific types of plants attract different songbird species. Including trees in green spaces is also recommended because many birds use trees for foraging and resting [143].

Food subsidies (e.g., bird feeders) also affect native bird populations by influencing resource availability, although these effects are nuanced and vary by species [144]. Supplemental feeding can have a positive effect on some native birds by alleviating starvation, improving body condition, reducing stress, promoting increased reproduction, and even (indirectly) reducing disease-related mortality due to improved physiological health [145]. Alternatively, food subsidies can negatively affect native birds by attracting predators, attracting nonnative competitors, or enhancing disease transmission [146]. Managers should consider supplemental feeding to support native songbirds, however, due to the diverse effects of food subsidies on different species, it is crucial to carefully monitor supplemental feeding efforts to determine the effects on the target songbird populations. Additionally, seasonal variation in the dietary requirements of target species should dictate the foods provided.

It is important for managers to understand the dietary needs of each species of songbird that they intend to protect, and how those dietary requirements vary seasonally, so that foraging habitats that support those specific dietary needs can be protected. For example, some migratory birds may rely on fruit resources from coastal shrubland habitats to meet their energetic requirements during migratory stopovers [114], but rely on insect prey during the breeding season. Thus, management decisions regarding foraging habitat should be specific to the season and species or guild.

Native plants provide important foraging habitat for songbirds. Management actions that involve removal of non-native introduced species need to be accompanied by restoration of native plants that provide critical resources for birds. For example, land managers typically eradicate bush honeysuckles (*Lonicera* spp.) because they diminish native plant richness and create monocultures. However, honeysuckles provide foraging and nesting habitat for frugivorous birds, even through the winter when other sources of fruit are limited. To support songbirds while removing honeysuckles, simultaneous reestablishment of the native shrub understory is critical [116]. Native plants including trees, diverse, dense shrubbery, and native grasses should also replace wasted spaces like lawns and open fields of non-native grasses to support songbirds by providing foraging habitat [147].

In lawns, open field areas, and farmland, pesticide use should also be reduced or avoided due to the severe negative impacts on songbird health and survival that result from pesticide poisoning and depleted populations of songbird prey items [148, 149]. Insecticides, including the commonly used neonicotinoid insecticide, imidacloprid, are linked to insectivorous bird population declines [150]. A review of 122 studies found that pesticides were more detrimental to songbird populations in farmland across North America than habitat loss and fragmentation (which came in second), mowing and harvesting, and disturbance related to livestock grazing and reduced food supply [151]. Insectivorous songbirds frequently hunt insects that feed on crops, thus the need for insecticide can be reduced by promoting high densities of these songbirds in agricultural areas [152]. Organic farming rather than conventional farming is recommended and can be especially effective at supporting songbird populations in homogeneous landscapes [153].

Songbirds can benefit farmers by hunting crop-eating insects, pollinating fruiting plants, and dispersing seeds. In return, farmers can support songbird communities by reducing pesticide inputs through integrated pest management (i.e., ecologically based pest control strategies; [154] which includes cultivating higher quality habitat with native plants [151]. Farmers can also plant cover crops like cereal rye between primary crop seasons to provide habitat for songbirds [155, 156]. Reduced tillage or no-till farming is another important mechanism to support songbird diversity in farmland by maintaining vegetation density and structure for grassland bird foraging habitat [157]. Grassland bird nests in no-till soybean farms occurred at a higher density and exhibited greater survival rates than those in tilled soybean farms in east-central Illinois [157]. Farmland songbirds also benefit from access to field edges, shelter belts, and fence rows [158], and minimized weed control in grass fields [159]. Combining all or many of these practices will provide the best support for resident and migratory songbirds in agricultural areas.

A challenging and critical aspect of songbird conservation and management is identifying the most suitable and productive areas for restoration [160]. Typically, high priority areas include agricultural lands, forest cover gaps and edges, as well as land in close proximity to rivers. These areas are of particular importance for neotropical, migrant songbirds that are often of high conservation concern. Migratory songbirds like the yellow warbler rely on riparian zones for nesting and the unique foraging opportunities that these zones provide [161]. Songbird species in restored riparian areas appear to have similar nest success to species in natural reference areas, suggesting that restoration efforts are effective for short-term reproductive success. However, overall species density was lower in restored areas for three songbird species in California, indicating a lower carrying capacity [161]. Restored sites had lower foliage volumes, resulting in lower densities of insect prey, which corresponded with larger songbird territory sizes and lower density [161]. These results suggest that while restoration is important, prevention of habitat degradation may be higher priority. However, longer term monitoring of restored areas is necessary to assess the full impact of restoration.

Restoration can benefit some species while harming others [150], thus restoration efforts should focus on the community or guild level. This may be a more efficient and effective management strategy than species-specific management [151]. Anthropogenic disturbance can alter interaction networks, which subsequently alters community composition. A meta-analysis of 170 studies on terrestrial mixed-species foraging flocks demonstrated the importance of multi-species interaction networks and the presence of "nuclear species" (i.e., leaders, ranked based on consistency to lead and number of followers) for songbirds globally (on every continent except Antarctica) and suggested that conservation target interactions, and particularly species that are "strong interactors" [151]. Species interactions are important to consider when determining conservation priorities because co-extinction is a possibility if mutualistic interactions are interrupted.

Finally, land managers and conservationists should engage in public outreach whenever possible. Conservation requires public support and there is evidence that exposure to nature determines sensitivity to conservation issues [162]. Environmental education and outreach in communities that are less otherwise exposed to nature is essential. Community-based land stewardship programs that emphasize the importance of native plants and working with schools or in urban and suburban areas to create greenspaces for native songbirds to forage can provide the opportunity for children and adults who are not regularly exposed to nature to form environmental awareness and appreciation. One simple option can be to teach outdoors in green spaces or in botanical gardens to provide hands-on training in maintaining plant diversity for songbird foraging habitat. In rural communities near managed land, managers should recruit local volunteers, involve community members, and give public talks and tours to foster support and engagement of local conservation efforts.

Emerging Methods for Studying and Managing Foraging Behaviors

This chapter has discussed research that demonstrates the wide breadth of methods used to study the foraging behaviors of songbirds. The techniques used to measure foraging behavior can broadly be placed into two categories: direct and indirect measures of foraging. Direct measures involve researchers or managers using traditional field methods, advanced technology, or both, to observe and track temporal patterns of food choice and foraging behavior, including interactions with conspecifics and heterospecifics. Indirect measures of foraging consist of inferring diet or foraging behavior from lab-based analyses of physical samples that were collected in the field (e.g., fecal samples, gut analysis, eDNA). The research question or management goal should determine whether to use a direct versus indirect method. Direct methods will more clearly shed light on the ways in which songbirds interact with their environment and other organisms while foraging, whereas indirect methods will provide information on specific diet composition. Discussed below are several new technologies used to directly monitor songbird foraging behavior and several indirect methods to quantify diet.

Direct Measures of Foraging Behaviors

Foraging behavior can be measured directly by traditional human field observation; however, it can be particularly challenging outside of the context of offspring and mate provisioning to visually follow fast-paced foraging songbirds from a distance great enough to avoid disrupting the songbird's behavior. Additionally, human observation imposes logistical constraints and observer error. Humans cannot be physically present for observation at all times, may miss less obvious foraging behaviors (e.g., in dense brush or at night), and error varies by observer so dividing observational duties between field researchers or managers can add confounding variance to the data. To bypass some of these limitations, remote-sensing technology like video cameras are commonly used to record songbird foraging. Commercially-available camcorders can be used to continuously record foraging or food delivery at a specific location, often a nest. However, camcorders are typically not weatherproof (although weatherproofing can be accomplished), can only be used for short periods of time due to battery life and storage capacity (but setups specific for long-term video monitoring of nests do exist), and continuous recording requires labor-intensive analysis [163, 164]. Camera traps and infrared, motion-activated cameras have also been used to monitor feeding at regularly used food sources (e.g., bird feeders), and to study interspecific visitation patterns, food item selectivity, seed dispersal, and foraging associations between different species [165, 166]. These cameras can typically be deployed for longer periods of time than traditional video recorders because they record only when activated, like when prey is delivered to nestlings [167]. However, motion-activated cameras can be sensitive to environmental influences (e.g., sunbeams, moving vegetation), potentially causing incomplete recordings. Passive acoustic monitoring (stationary recording devices) provides a broad-scale picture of songbird and prey presence or absence, allowing for evaluation of correlations in occurrence [168, 169]. Passive acoustic detectability is comparable to point counts, and provides a standardized way to record data at an even larger scale.

When individual visitation rates at food sources or nests are desired, but specific foraging behavior observation is not required, radio-frequency identification (RFID) can be a useful tool. RFID consists of a tag placed on or inside the animal being monitored, and a tag reader placed at the nest or food source to record the identity of individuals present. RFID readers at nests and foraging sites can allow researchers to quantify the costs of different foraging behaviors linked to offspring provisioning [170], determine the impact of different environments on foraging routines [171], and investigate individual differences in problem-solving ability [39]. RFID readers, however, will only record an ID when an individual comes close to the reader, and thus can only be used when individuals frequent a specific location (e.g., a bird feeder or a nest).

When songbird movement patterns are desired, radio frequency transmitters are the most traditional and lightweight option, but typically require researchers to spend many hours searching for bird signals with a portable antenna [172]. However, recent technology is drastically increasing the amount of spatial and temporal information gained from radio telemetry, while minimizing human labor. Stationary radio towers like the MOTUS network (an international network of radio towers) are used to understand migratory routes and timing of songbirds [173]. MOTUS technology has also been used to measure migration speed and understand the contribution of energy obtained at wintering sites to birds' overall migratory journeys [174]. Researchers can also deploy their own grids of small radio receivers when birds are expected to remain in one general area (e.g., foraging or breeding territories) to obtain continuous movement data within the area to quantify foraging behaviors [172, 175].

The power demanded by GPS tags for small scale foraging studies often requires batteries that are too large and heavy for most songbirds to carry, and smaller geolocator tags (which rely on light levels to infer location) require the bird to be recaptured as they cannot transmit the data remotely [176, 177]. However, recent innovations in GPS technology may be bypassing this issue. For example, wireless biologging networks to record spatial location and interactions between individuals have been tested on bats [178]. Their small size (as small as 1 gram), long lasting battery life, and frequent recording (up to every 30 seconds) make these devices useful for future songbird studies. Another recent GPS innovation that could be used in future studies to monitor songbird behavior are ICARUS tags, which can record 3D movements and rely on a dedicated satellite for date recording and transmission [179]. GPS devices also equipped with accelerometers (devices that measure acceleration in 2–3 directions) can be used to quantify changes to speed and direction [180]. These devices have also been too large for songbirds until recently, but have been used to study migratory stopovers and estimating energy demands. As technology develops further and is used in more songbird species, researchers and managers will gain wider knowledge of songbird foraging behavior and foraging associations.

Indirect Measures of Foraging Behaviors

Researchers and managers can also quantify diet and foraging behavior using indirect methods, without direct observation; physiological samples can be collected from wild songbirds, allowing diet and foraging patterns to be determined in the lab. The most basic of these analyses is gut dissection, in which food fragments are identified to determine the most recently eaten items, however this method requires lethal take of birds. Much diet information can also be obtained from non-lethal sampling. Stable isotopes (i.e., different forms of the same element that vary in atomic mass because of varying neutron numbers) of elements present in feathers, blood, and feces can be compared to isotopically distinct food sources to determine diet. Isotopes of carbon and nitrogen, for instance, are used to determine the timing of diet changes in migratory songbirds [181]. In addition, mercury concentration in blood and feathers provides information about foraging choices and associated avian health. For example, one study found that ground-foraging species accumulated the highest levels of mercury compared to other foraging guilds [182]. Gut microbiota (i.e., bacterial communities found in digestive structures, which can be measured from fecal samples), can also be affected by many factors including diet and social interactions [183]. Further research on how diet, health, foraging sociality, and gut microbiota are intertwined is warranted.

DNA can also be used to indirectly detect and identify songbird prey items through a variety of different sampling methods. DNA can be amplified from regurgitated gut contents [184], feces [133], and nestling fecal sacs [185]. For example, DNA from regurgitation was used to identify seed species consumed by Baird's sparrows (*Ammodramus bairdii*) and grasshopper sparrows (*A. savannarum*) and compare that to the soil seed bank [184]. eDNA is a useful tool for identifying species presence or absence in foraging areas. Sources of eDNA include feces, carcasses, and shed skin that are deposited in the environment. For example, analysis of eDNA extracted from water samples collected across Europe detected species commonly associated with water like wood pigeons (*Columba palumbus*) and marsh warblers (*Acrocephalus palustris*) [186].

Future Research

Science-based conservation and management actions informed by songbird behavioral research can result in more targeted and effective solutions. Foraging behavior of songbirds has been studied and managed for decades, but there are still many unknowns for some species, and older studies need to be revisited and retested as food sources and local ecosystems are constantly changing. New food sources and vegetation have also been introduced as a component of rapid urban change, so diet and behavior in some areas of a species' range may be different than populations in more natural areas. The authors of this chapter believe that valuable future studies could range from basic ecological knowledge of foraging behavior of individual species, to complex foraging temporal patterns or associations. For some species (e.g., those with a limited distribution, or that nest/forage high in the canopy), basic information about foraging including secondary food sources or offspring provisioning is not known. Recent technologies should be applied to these species, because effective conservation and management cannot occur without an understanding of diet and foraging; only then can critical prey, plants, and foraging associations be intentionally managed.

Understanding breeding, non-breeding, and stopover site foraging ecology of migratory songbirds can require studies spanning multiple continents. These types of studies are critical for species undergoing rapid habitat loss, because flexibility in diet and foraging behavior can be used to predict responses to alteration of wintering and migratory stopover sites [187]. Ideally, studies should use combinations of direct and indirect methods to understand foraging behavior of migratory songbirds, and the importance of different food sources at migratory stopover sites. For example, spatial data from MOTUS radio towers, in combination with indirect sampling of gut contents by collecting fecal samples at various points along migratory routes, could be used to monitor how diet changes along routes or based on time of arrival at the site.

Community science (i.e., scientific data collected primarily by non-professional scientists) can also be a valuable resource for researchers and managers. Worldwide data regarding bird spatial and temporal distributions are publicly available through projects such as eBird and the North American Breeding Bird Survey [136, 188]. Managers can use such data to determine where a species of management interest is regularly sighted, or how migratory stopover site-use has changed over time. Since backyard bird watching has become a worldwide hobby, community scientists can also assist in research and management projects by maintaining cameras, video recorders, acoustic recorders, and feeding stations to understand foraging behavior and associations [165, 189]. For example, the Cornell Lab of Ornithology's and Birds Canada's joint-effort, project feederwatch, organized over 20,000 volunteers that maintained feeders, bird baths, or bird-friendly plants and recorded species presence and behavior to understand species' abundances, distributions, and wintering ranges [190].

The effects of rapid environmental change can be buffered by behavioral flexibility and diversity within songbird populations [191]. When this diversity is lost it can inhibit the adaptation potential of a population to disturbances such as urbanization and climate change [192]. Consequently, monitoring and conserving behavioral diversity should be a management goal, alongside conservation of genetic diversity [193]. However, little is currently known regarding temporal trends in foraging behavioral flexibility and diversity of most songbird species. In particular, more research is needed on songbird foraging behavioral variation within and across populations, as well as methods for conservation and measurement of behavioral diversity. To review topics covered in this section, see Box 3.

Box 3   Management solutions

Increasing Forage in Urban Areas
- If maintaining urban green spaces, urban planners should prioritize landscape connectivity (rather than large isolated patches), and stock them with native plants.
- Migratory pathways and stop-over sites should be conserved to protect forage for migratory birds.
- Managers should assess disease and predation risks, as well as seasonal dietary needs, before implementing supplementary food sources (e.g., bird feeders).

Offsetting Altered Plant Communities
- Removal of non-native introduced species need to be paired with the restoration of native plants that provide critical resources.
- In lawns and farmlands alike, insecticides should be reduced and avoided due to adverse effects on foraging resources. Instead, encouraging native populations of insectivorous birds can offset insect damage, increase pollination, and disperse seeds.
- Farmers can alternate cover crops, such as cereal rye, between rotations and reduce tillage to increase foraging opportunities.

Emerging Technologies for Foraging Studies
- Direct Measures: active tracking and/or observation.
  - Camera traps and acoustic monitoring can be used at common foraging grounds to give information on songbirds and their prey.
  - Radio-frequency identification (FRID) can infer costs of songbird foraging tactics.
  - The MOTUS network (an international network of radio towers) can be used to understand migratory routes and feeding grounds.
  - GPS tags are often too heavy and geolocator tags are too imprecise for foraging studies, but wireless biologging networks and ICARUS tags can be used to record fine scale movements.
- Indirect Measures: inferring diet or foraging behavior from external information.
  - Analyzing gut contents gives information on diet and foraging patterns, but is lethal.
  - Stable isotopes taken from feathers, blood, and feces, can be used to determine diet.
  - eDNA can be taken directly from the environment can give clues to community composition in both birds and their food.

Future Research
- Old studies on diets and foraging behaviors should be revisited with modern methods.
- Many species with limited distribution or difficult to reach nesting locations still lack basic foraging knowledge.
- Studies spanning multiple continents are required to fully understand foraging behaviors during migration.
- Community science offers untapped opportunities for worldwide data collection.
- More information is needed on songbird foraging behavioral variation within and across populations, given its ability to resist environmental change.

Conclusion

Songbird foraging behavior is driven by a complex suite of mechanisms including sensory cues used to locate food, optimization of foraging techniques and diet choices, storage of food, and partitioning of food energy once consumed while adjusting for external environmental influences. By understanding the evolutionary basis of these behaviors, through Optimal Foraging Theory and the Optimal Diet Model, managers will have the tools to develop strategies that reduce the detrimental effects of a changing climate, urban sprawl, and associated alterations to plant communities and other resources. When designing conservation or management plans, adequate consideration should be given to individuals, populations, species, and mixed-species foraging guilds, along with the connectivity of green spaces and the replacement of invasive plants with native food bearing plants to enhance and encourage foraging. By preemptively determining research needs or management goals, researchers can use direct or indirect measures of foraging through traditional or modern technological methods mentioned in this chapter. Future research on this topic should focus on how foraging behaviors are changing due to anthropogenic pressure, and the growing database provided by community-based science projects will allow for more long-term studies on the impacts of a human altered landscape on songbird foraging strategies.

LITERATURE CITED

[1]   Remsen, J.V. and S.K. Robinson. 1990. A classification scheme for foraging behavior of birds in terrestrial habitats. Studies in Avian Biology. 13: 144–160.

[2]   Price, M.R. and W.K. Hayes. 2017. Diverse habitat use during two life stages of the critically endangered Bahama Oriole (*Icterus northropi*): community structure, foraging, and social interactions. PeerJ. 5: e3500.

[3]   Moran, A.J., S.W.J. Prosser and J.A. Moran. 2019. DNA metabarcoding allows non-invasive identification of arthropod prey provisioned to nestling Rufous hummingbirds (*Selasphorus rufus*). PeerJ. 7: e6596.

[4]   Hart, N.S. 2001. The visual ecology of avian photoreceptors. Prog. Retin. Eye Res. 20(5): 675–703.

[5]   Church, S.C., A.S. Merrison and T.M. Chamberlain. 2001. Avian ultraviolet vision and frequency-dependent seed preferences. J. Exp. Biol. 204(Pt 14): 2491–2498.

[6]   Toomey, M.B. and K.J. McGraw. 2011. The effects of dietary carotenoid supplementation and retinal carotenoid accumulation on vision-mediated foraging in the house finch. PLoS One. 6(6): e21653.

[7]   Fernández-Juricic, E., P.E. Baumhardt, L.P. Tyrrell, A. Elmore, S.T. DeLiberto and S.J. Werner. 2019. Vision in an abundant North American bird: The Red-winged Blackbird. Auk. 136(3): 1–27.

[8]   Tyrrell, L.P., L.B. Teixeira, R.R. Dubielzig, D. Pita, P. Baumhardt, B.A. Moore, et al. 2019. A novel cellular structure in the retina of insectivorous birds. Sci. Rep. 9(1): 1–10.

[9]   Köppl, C. 2015. Avian hearing. pp. 71–87. *In*: C.G. Scanes (ed.). Sturkie's Avian Physiology. Academic Press.

[10]   McMillan, N., M.T. Avey, L.L. Bloomfield, L.M. Guillette, A.H. Hahn, M. Hoeschele, et al. 2017. Avian vocal perception: bioacoustics and perceptual mechanisms. pp. 270–295. *In*: C. ten Cate and S.D. Healy (eds). Avian Cognition. Cambridge University Press.

[11] Schnyder, H.A., D. Vanderelst, S. Bartenstein, U. Firzlaff and H. Luksch. 2014. The avian head induces cues for sound localization in elevation. PLoS One. 9(11): e112178.

[12] Montgomerie, R. and P.J. Weatherhead. 1997. How robins find worms. Anim. Behav. 54(1): 143–151.

[13] Pollock, H.S., A.E. Martínez, J.P. Kelley, J.M. Touchton and C.E. Tarwater. 2017. Heterospecific eavesdropping in ant-following birds of the Neotropics is a learned behaviour. Proc. R. Soc. B: Biol. Sci. 284(1865): 20171785.

[14] Heinrich, B. and J.M. Marzluff. 1991. Do common ravens yell because they want to attract others?

[15] Brown, C.R., M.B. Brown and M.L. Shaffer. 1991. Food-sharing signals among socially foraging cliff swallows.

[16] Rubene, D., M. Leidefors, V. Ninkovic, S. Eggers and M. Low. 2019. Disentangling olfactory and visual information used by field foraging birds. Ecol. Evol. 9(1): 545–552.

[17] Amo, L., J.J. Jansen, N.M. van Dam, M. Dicke and M.E. Visser. 2013. Birds exploit herbivore-induced plant volatiles to locate herbivorous prey. Ecol. Lett. 16(11): 1348–1355.

[18] Buitron, D. and G.L. Nuechterlein. 1985. Experiments on olfactory detection of food caches by black-billed magpies. Condor. 87(1): 92–95.

[19] Mason, J.R. and G. Linz. 1997. Repellency of garlic extract to European starlings. Crop Prot. 16(2): 107–108.

[20] Nakamura, K. and H. Yokoyama. 1995. The odor of *p*-dichlorobenzene repels feral pigeons *Columba livia* from their food sites. Japanese J. Ornithol. 44(1): 13–19.

[21] Avery, M.L., P. Nol and J.S. Humphrey. 1994. Responses of three species of captive fruit-eating birds to phosmet-treated food. Pestic. Sci. 41(1): 49–53

[22] Krause, E.T., J. Kabbert and B.A. Caspers. 2016. Exploring the use of olfactory cues in a nonsocial context in zebra finches (*Taeniopygia guttata*). pp. 177–187. *In*: B. Schulte, T. Goodwin and M. Ferkin (eds). Chemical Signals in Vertebrates 13. Springer, Cham.

[23] Cunningham, S.J., M.R. Alley and I. Castro. 2011. Facial bristle feather histology and morphology in New Zealand birds: implications for function. J. Morphol. 272(1): 118–128.

[24] Werner, S.J. and F.D. Provenza. 2011. Reconciling sensory cues and varied consequences of avian repellents. Physiol. Behav. 102(2): 158–163.

[25] Thord, F., S. Jakobsson, P. Johansson, C. Kullberg, J. Lind and A. Vallin. 2001. Magnetic cues trigger extensive refuelling. Nature. 414: 35–36.

[26] Thalau, P., E. Holtkamp-Rötzler, G. Fleissner and W. Wiltschko. 2007. Homing pigeons (*Columba livia* f. *domestica*) can use magnetic cues for locating food. Naturwissenschaften. 94(10): 813–819.

[27] Castaño-Villa, G.J., R. Santisteban-Arenas and H.-A. Jaramillo,. 2019. Foraging behavioural traits of tropical insectivorous birds lead to dissimilar communities in contrasting forest habitats. Wildl. Biol. 1: 1–6.

[28] McCabe, B.J. and C.G. Guglielmo. 2019. Migration takes extra guts for juvenile songbirds: energetics and digestive physiology during the first journey. Front. Ecol. Evol. 7: 381.

[29] MacArthur, R.H. 1958. Population ecology of some warblers of northeastern coniferous forests. Ecology. 39(4): 599–619.

[30] Correia, J. and S.L. Halkin. 2018. Potential winter niche partitioning between Tufted Titmice (*Baeolophus bicolor*) and Black-capped Chickadees (*Poecile atricapillus*). Wilson J. Ornithol. 130(3): 684–695.

[31] Remešová, E., B. Matysioková, L. Turčoková Rubáčová and V. Remeš. 2020. Foraging behaviour of songbirds in woodlands and forests in eastern Australia: resource partitioning and guild structure. Emu - Austral Ornithol. 120(1): 22–32.

[32] Mansor, M.S. and R. Ramli. 2017. Foraging niche segregation in Malaysian babblers (Family: Timaliidae). PLoS One. 12(3): e0172836.

[33]   TTebbich, S., M. Taborsky, B. Fessl and D. Blomqvist. 2001. Do woodpecker finches acquire tool-use by social learning? Proc. R. Soc. B: Biol. Sci. 268(1482): 2189–2193.

[34]   St Clair, J.J., B.C. Klump, S. Sugasawa, C.G. Higgott, N. Colegrave and C. Rutz. 2018. Hook innovation boosts foraging efficiency in tool-using crows. Nat. Ecol. Evol. 2(3): 441–444.

[35]   Knward, B., C. Rutz, A.A. Weir and A. Kacelnik. 2006. Development of tool use in New Caledonian crows: inherited action patterns and social influences. Anim. Behav. 72(6): 1329–1343.

[36]   Kenward, B., A.A. Weir, C. Rutz and A. Kacelnik. 2005. Tool manufacture by naive juvenile crows. Nature. 433(7022): 21–121.

[37]   Scott, K. Robinson and Richard T. Holmes. 1982. Foraging behavior of forest birds: the relationships among search tactics, diet, and habitat structure. Ecology. 63(6): 1918–1931.

[38]   Farine, D.R., L.M. Aplin, B.C. Sheldon and W. Hoppitt. 2015. Interspecific social networks promote information transmission in wild songbirds. Proc. R. Soc. B: Biol. Sci. 282(1803): 20142804.

[39]   Morand-Ferron, J. and J.L. Quinn. 2011. Larger groups of passerines are more efficient problem solvers in the wild. Proc. Natl. Acad. Sci. USA. 108(38): 15898–15903.

[40]   Baruzzi, C., D.P. Chance, J.R. McCollum, G.M. Street, C. Brookshire and M.A. Lashley. 2017. Invited cleaners or unsolicited visitors: Eastern phoebes use white-tailed deer to forage. Food Webs. 13: 38–39.

[41]   Broom, M. and G.D. Ruxton. 1998. Evolutionarily stable stealing: game theory applied to kleptoparasitism. Behav. Ecol. 9(4): 397–403.

[42]   Flower, T.P. and M. Gribble. 2012. Kleptoparasitism by attacks versus false alarm calls in fork-tailed drongos. Anim. Behav. 83(2): 403–410.

[43]   Emlen, J.M. 1966. The role of time and energy in food preference. Am. Nat. 100(916): 611–617.

[44]   Stephens, D.W., J.S. Brown and R.C. Ydenberg. 2007. Foraging: Behavior and Ecology. University of Chicago Press.

[45]   Meire, P.M. and A. Ervynck. 1986. Are oystercatchers (*Haematopus ostralegus*) selecting the most profitable mussels (*Mytilus edulis*)? Anim. Behav. 34(5): 1427–1435

[46]   Hillemann, F., T. Bugnyar, K. Kotrschal and C.A. Wascher. 2014. Waiting for better, not for more: corvids respond to quality in two delay maintenance tasks. Anim. Behav. 90: 1–10.

[47]   Krebs, J.R. 1977. Song and territory in the great tit *Parus major*. pp. 47–62. *In*: B. Stonehouse and C. Perrins (eds). Evolutionary Ecology. Macmillan Education UK, London.

[48]   Charnov, E.L. 1976. Optimal foraging, the marginal value theorem. Theor. Popul. Biol. 9(2): 129–136.

[49]   Brown, J.S. 1988. Patch use as an indicator of habitat preference, predation risk, and competition. Behav. Ecol. Sociobiol. 22: 37–47

[50]   Brown, J.S., B.P. Kotler and W.A. Mitchell. 1997. Competition between birds and mammals: a comparison of giving-up densities between crested larks and gerbils. Evol. Ecol. 11(6): 757–771

[51]   Lee, Y.-F., Y.-M. Kuo and E.K. Bollinger. 2005. Effects of feeding height and distance from protective cover on the foraging behavior of wintering birds. Can. J. Zool. 83(6): 880–890.

[52]   Kacelnik, A. 1984. Central place foraging in starlings (*Sturnus vulgaris*). I. Patch residence time. J. Anim. Ecol. 53: 283–299.

[53]   Gosling, S.D. 2001. From mice to men: what can we learn about personality from animal research? Psycholog. Bull. 127(1): 45–86.

[54]   van Overveld, T. and E. Matthysen. 2010. Personality predicts spatial responses to food manipulations in free-ranging great tits (*Parus major*). Biol. Lett. 6(2): 187–190.

[55]   Zandberg, L., J.L. Quinn, M. Naguib and K. Van Oers. 2017. Personality-dependent differences in problem-solving performance in a social context reflect foraging strategies. Behav. Processes. 134: 95–102.

[56] Minderman, J., J.M. Reid, M. Hughes, M.J. Denny, S. Hogg, P.G. Evans, et al. 2010. Novel environment exploration and home range size in starlings Sturnus vulgaris. Behav. Ecol. 21(6): 1321–1329.

[57] Marone, L., M. Olmedo, D.Y. Valdés, A. Zarco, J.L. de Casenave and R.G. Pol. 2017. Diet switching of seed-eating birds wintering in grazed habitats of the central Monte Desert, Argentina. Condor. 119(4): 673–682.

[58] Tinbergen, L. 1960. The natural control of insects in pinewoods. Archives Néerlandaises de Zoologie. 13(3): 265–343.

[59] Dukas, R. and A.C. Kamil. 2001. Limited attention: the constraint underlying search image. Behav. Ecol. 12(2): 192–199.

[60] Rice, A.M., N. Vallin, K. Kulma, H. Arntsen, A. Husby, M. Tobler, et al. 2013. Optimizing the trade-off between offspring number and quality in unpredictable environments: testing the role of differential androgen transfer to collared flycatcher eggs. Horm. Behav. 63(5): 813–822.

[61] Ekman, J. and B. Rosander. 1992. Survival enhancement through food sharing: a means for parental control of natal dispersal. Theor. Popul. Biol. 42(2): 117–129.

[62] Caro, S.M., A.S. Griffin, C.A. Hinde and S.A. West. 2016. Unpredictable environments lead to the evolution of parental neglect in birds. Nat. Commun. 7: 10985.

[63] Roberts, R.C. 1979. The evolution of avian food-storing behavior. Am. Nat. 114(3): 418–438.

[64] Bednekoff, P.A. and R.P. Balda. 1996. Observational spatial memory in Clark's nutcrackers and Mexican jays. Anim. Behav. 52(4): 833–839.

[65] Burnell, K.L. and D.F. Tomback. 1985. Steller's jays steal gray jay caches: field and laboratory observations. Auk. 102(2): 417–419.

[66] Dally, J.M., N.S. Clayton and N.J. Emery. 2006. The behaviour and evolution of cache protection and pilferage. Anim. Behav. 72(1): 13–23.

[67] Dally, J.M., N.J. Emery and N.S. Clayton. 2010. Avian theory of mind and counter espionage by food-caching western scrub-jays (*Aphelocoma californica*). Eur. J. Dev. Psychol. 7(1): 17–37.

[68] Pravosudov, V.V. and T.C. Roth II. 2013. Cognitive ecology of food hoarding: the evolution of spatial memory and the hippocampus. Annu. Rev. Ecol. Evol. Syst. 44(1): 173–193.

[69] Van Horik, J. and K.C. Burns. 2007. Cache spacing patterns and reciprocal cache theft in New Zealand robins. Anim. Behav. 73(6): 1043–1049.

[70] Daan, S., C. Dijkstra, R. Drent and Meijer. 1989. Food supply and the annual timing of avian reproduction. Proceedings of the International Ornithological Congress, Vol. 19. pp. 392–407. University of Ottawa Press, Ottawa.

[71] Withgott, J. 2002. Last year's food guides this year's brood. Science. 296: 29.

[72] Dahdul, W.M. and M.H. Horn. 2003. Energy allocation and postnatal growth in captive elegant tern (*Sterna Elegans*) chicks: responses to high-versus low-energy diets. Auk. 120(4): 1069–1081.

[73] Orłowski, G., A. Wuczyński, J. Karg and W. Grzesiak. 2016. The significance of seed food in chick development re-evaluated by tracking day-to-day dietary variation in the nestlings of a granivorous passerine. Ibis. 159(1): 124–138

[74] McWilliams, S.R., C. Guglielmo, B. Pierce and M. Klaassen. 2004. Flying, fasting, and feeding in birds during migration: a nutritional and physiological ecology perspective. J. Avian Biol. 35(5): 377–393.

[75] Maillet, D. and J.-M. Weber. 2006. Performance-enhancing role of dietary fatty acids in a long-distance migrant shorebird: the semipalmated sandpiper. J. Exp. Biol. 209(Pt 14): 2686–2695.

[76] Pierce, B.J. and S.R. McWilliams. 2004. Diet quality and food limitation affect the dynamics of body composition and digestive organs in a migratory songbird (*Zonotrichia albicollis*). Physiol. Biochem. Zool. 77(3): 471–483.

[77] Bairlein, F. 1998. The effect of diet composition on migratory fuelling in Garden Warblers *Sylvia borin*. J. Avian Biol. 29(4): 546–551.

[78] Petit, M., S. Clavijo-Baquet and F. Vézina. 2016. Increasing winter maximal metabolic rate improves intrawinter survival in small birds. Physiol. Biochem. Zool. 90(2): 166–177.

[79] Lima, S.L. 1986. Predation risk and unpredictable feeding conditions: determinants of body mass in birds'. Ecology. 67(2): 377–385.

[80] Rogers, C.M. 2015. Testing optimal body mass theory: evidence for cost of fat in wintering birds. Ecosphere. 6(4): 1–12.

[81] Kozlovsky, D.Y., E.A. Weissgerber and V.V. Pravosudov. 2017. What makes specialized food-caching mountain chickadees successful city slickers? Proc. Biol. Sci. 284(1855):

[82] Liu, Z., C. He, Y. Zhou and J. Wu. 2014. How much of the world's land has been urbanized, really? A hierarchical framework for avoiding confusion. Landsc. Ecol. 29(5): 763–771.

[83] Ciach, M. and A. Fröhlich. 2017. Habitat type, food resources, noise and light pollution explain the species composition, abundance and stability of a winter bird assemblage in an urban environment. Urban Ecosyst. 20(3): 547–559.

[84] Marzluff, J.M. 2001. Worldwide urbanization and its effects on birds. pp. 19–47. *In*: J.M. Marzluff, R. Bowman and R. Donnelly. (eds). Avian Ecology and Conservation in an Urbanizing World. Boston, MA, Springer US.

[85] Ortega, C.P. 2012. Effects of noise pollution on birds: a brief review of our knowledge. Ornithol. Monogr. 74(1): 6–22.

[86] Peck, H.L., H.E. Pringle, H.H. Marshall, I.P. Owens and A.M. Lord. 2014. Experimental evidence of impacts of an invasive parakeet on foraging behavior of native birds. Behav. Ecol. 25(3): 582–590.

[87] Ricketts, T.H. 2001. The matrix matters: effective isolation in fragmented landscapes. Am. Nat. 158(1): 87–99.

[88] Tremblay, M.A. and C.C. St. Clair. 2011. Permeability of a heterogeneous urban landscape to the movements of forest songbirds: songbird movements in urban landscapes. J. Appl. Ecol. 48(3): 679–688.

[89] Visscher, D.R., A. Unger, H. Grobbelaar and P.D. DeWitt. 2018. Bird foraging is influenced by both risk and connectivity in urban parks. J. Urban Ecol. 4(1): 1–6.

[90] McIntyre, N.E. 2000. Ecology of urban arthropods: a review and a call to action. Ann. Entomol. Soc. Am. 93(4): 825–835.

[91] Avondet, J.L., R.B. Blair, D.J. Berg and MA. Ebbert. 2003. *Drosophila* (Diptera: *Drosophilidae*) response to changes in ecological parameters across an urban gradient. Environ. Entomol. 32(2): 347–358.

[92] Philpott, S.M., J. Cotton, P. Bichier, R.L. Friedrich, L.C. Moorhead, S. Uno, et al. 2014. Local and landscape drivers of arthropod abundance, richness, and trophic composition in urban habitats. Urban Ecosyst. 17(2): 513–532.

[93] Robertson, B., G. Kriska, V. Horvath and G. Horvath. 2010. Glass buildings as bird feeders: urban birds exploit insects trapped by polarized light pollution. Acta Zool. Academ. Sci. Hung. 56(3): 283–293.

[94] Tryjanowski, P., F. Morelli, P. Skórka, A. Goławski, P. Indykiewicz, A.P. Møller, et al. 2015. Who started first? Bird species visiting novel birdfeeders. Sci. Rep. 5: 11858.

[95] Bellebaum, J. 2005. Between the herring gull *Larus argentatus* and the bulldozer: black-headed gull *Larus ridibundus* feeding sites on a refuse dump. Ornis Fenn. 82: 166–171.

[96] Crooks, K.R., A.V. Suarez and D.T. Bolger. 2004. Avian assemblages along a gradient of urbanization in a highly fragmented landscape. Biol. Conserv. 115(3): 451–462.

[97] Beissinger, S.R. and D.R. Osborne. 1982. Effects of Urbanization on Avian Community Organization. Condor. 84(1): 75–83.

[98] Blair, R.B. 1996. Land use and avian species diversity along an urban gradient. Ecol. Appl. 6(2): 506–519.

[99] Jarrett, C., L.L. Powell, H. McDevitt, et al. 2020. Bitter fruits of hard labour: diet metabarcoding and telemetry reveal that urban songbirds travel further for lower-quality food. Oecologia. 193(2): 377–388.

[100] Isaksson, C. and S. Andersson. 2007. Carotenoid diet and nestling provisioning in urban and rural great tits Parus major. J. Avian Biol. 38(5): 564–572.

[101] Sol, D., O. Lapiedra and C. González-Lagos. 2013. Behavioural adjustments for a life in the city. Anim. Behav. 85(5): 1101–1112.

[102] Stofberg, M., S.J. Cunningham, P. Sumasgutner and A. Amar. 2019. Juggling a "junk-food" diet: responses of an urban bird to fluctuating anthropogenic-food availability. Urban Ecosyst. 22(6): 1019–1026.

[103] Ostfeld, R.S. and K. Logiudice. 2003. Community disassembly, biodiversity loss, and the erosion of an ecosystem service. Ecology. 84(6): 1421–1427.

[104] Naidoo, R. 2004. Species richness and community composition of songbirds in a tropical forest-agricultural landscape. Anim. Conserv. 7(1): 93–105.

[105] Hartung, S.C. and J.D. Brawn. 2005. Effects of savanna restoration on the foraging ecology of insectivorous songbirds. Condor. 107: 879–888.

[106] Bodenchuk, M.J. and D.L. Bergman. 2020. Grackles. Wildlife Damage Management Technical Series. USDA, APHIS, WS National Wildlife Research Center. Fort Collins, Colorado.

[107] Burghardt, K.T., D.W. Tallamy and W. Gregory Shriver. 2009. Impact of native plants on bird and butterfly biodiversity in suburban landscapes. Conserv. Biol. 23(1): 219–224.

[108] Nelson, S.B., J.J. Coon, C.J. Duchardt, J.D. Fischer, S.J. Halsey, A.J. Kranz, et al. 2017. Patterns and mechanisms of invasive plant impacts on North American birds: a systematic review. Biol. Invasions. 19(5): 1547–1563.

[109] Ortega, Y.K., K.S. McKelvey and D.L. Six. 2006. Invasion of an exotic forb impacts reproductive success and site fidelity of a migratory songbird. Oecologia. 149(2): 340–351.

[110] Lloyd, J.D. and T.E. Martin. 2005. Reproductive success of chestnut-collared longspurs in native and exotic grassland. Condor. 107(2): 363–374.

[111] Marshall, M.R. and R.J. Cooper. 2004. Territory size of a migratory songbird in response to caterpillar density and foliage structure. Ecology. 85(2): 432–445.

[112] Brooks, M.L. 2008. Effects of fire suppression and postfire management activities on plant invasions. pp. 269–280. *In*: K. Zouhar, J.K. Smith and S. Sutherland (eds). Wildland Fire in Ecosystems: Fire and Nonnative Invasive Plants. U.S. Department of Agriculture, Forest Service, Rocky Mountain Research Station.

[113] Gleditsch, J.M. and T.A. Carlo. 2014. Living with aliens: effects of invasive shrub honeysuckles on avian nesting. PLoS One. 9(9): e107120.

[114] Parrish, J.D. 2000. Behavioral, energetic and conservation implications of foraging plasticity during migration. Stud. in Avian Biol. 2053: 70.

[115] Bartuszevige, A.M. and D.L. Gorchov. 2006. Avian seed dispersal of an invasive shrub. Biol. Invasions. 8(5): 1013–1022.

[116] McCusker, C.E., M.P. Ward and J.D. Brawn. 2010. Seasonal responses of avian communities to invasive bush honeysuckles (*Lonicera* spp.). Biol. Invasions. 12(8): 2459–2470.

[117] Pesendorfer, M.B., T.S. Sillett, W.D. Koenig and S.A. Morrison. 2016. Scatter-hoarding corvids as seed dispersers for oaks and pines: a review of a widely distributed mutualism and its utility to habitat restoration. The Condor: Ornithol. Appl. 118(2): 215–237.

[118] Purves, D.W., M.A. Zavala, K. Ogle, F. Prieto and J.M.R. Benayas. 2007. Environmental heterogeneity, bird-mediated directed dispersal, and oak woodland dynamics in Mediterranean Spain. Ecological Monographs. 77(1): 77–97.

[119] Lesser, M.R. and S.T. Jackson. 2013. Contributions of long-distance dispersal to population growth in colonising Pinus ponderosa populations. Ecol. Lett. 16(3): 380–389.

[120] Lorenz, T.J., K.A. Sullivan, A.V. Bakian and C.A. Aubry. 2011. Cache-site selection in Clark's Nutcracker (*Nucifraga Columbiana*). Auk. 128(2): 237–247.

[121] Baumeister, D. and R.M. Callaway. 2006. Facilitation by *Pinus flexilis* during succession: a hierarchy of mechanisms benefits other plant species. Ecology. 87(7): 1816–1830.

[122] Thomas, C.D. and J.J. Lennon. 1999. Birds extend their ranges northwards. Nature. 399: 213.

[123] Hüppop, O. and K. Hüppop. 2003. North atlantic oscillation and timing of spring migration in birds. Proc. R. Soc. Lond. B. Biol. Sci. 270(1512): 233–240.

[124] Vsser, M.E., A.V. Noordwijk, J.M. Tinbergen and C.M. Lessells. 1998. Warmer springs lead to mistimed reproduction in great tits (*Parus major*). Proc. R. Soc. Lond. B: Biol. Sci. 265(1408): 1867–1870.

[125] Thomas, D.W., J. Blondel, P. Perret, M.M. Lambrechts and J.R. Speakman. 2001. Energetic and fitness costs of mismatching resource supply and demand in seasonally breeding birds. Science. 291(5513): 2598–2600.

[126] Blondel, J., P.C. Dias, P. Perret, M. Maistre and M.M. Lambrechts. 1999. Selection-based biodiversity at a small spatial scale in a low-dispersing insular bird. Science. 285(5432): 1399–1402.

[127] Dias, P.C. and J. Blondel. 1996. Breeding time, food supply and fitness components of blue tits parus cueruleus in mediterranean habitats. Ibis. 138: 644–649.

[128] Drury, W.H. 1961. Studies of the breeding biology of horned lark, water pipit, lapland longspur, and snow bunting on bylot island, northwest territories, Canada. Bird Banding. 32(1): 1–46.

[129] Hussell, D.J.T. and R.D. Montgomerie. 2002. Lapland longspur (*Calcarius lapponicus*), version 2.0. *In*: A.F. Poole and F.B. Gill (eds). The Birds of North America. Cornell Lab of Ornithology, Ithaca, NY, USA. https://doi.org/10.2173/bna.656

[130] Boelman, N.T., L. Gough, J. Wingfield, S. Goetz, A. Asmus, H.E. Chmura, et al. 2015. Greater shrub dominance alters breeding habitat and food resources for migratory songbirds in alaskan arctic tundra. Glob. Chang. Biol. 21(4): 1508–1520.

[131] Tingley, M.W., W.B. Monahan, S.R. Beissinger and C. Moritz. 2009. Birds track their *Grinnellian niche* through a century of climate change. Proc. Natl. Acad. Sci. USA. 106. Suppl 2: 19637–19643.

[132] Sample BE, R.J. Cooper and R.C. Whitmore. 1993. Dietary shifts among songbirds from a diflubenzuron-treated forest. Condor. 95(3): 616–624.

[133] Trevelline, B.K., T. Nuttle, B.D. Hoenig, N.L. Brouwer, B.A. Porter and S.C. Latta. 2018. DNA metabarcoding of nestling feces reveals provisioning of aquatic prey and resource partitioning among Neotropical migratory songbirds in a riparian habitat. Oecologia. 187(1): 85–98.

[134] Grunst, A.S., M.L. Grunst, B. Thys, T. Raap, N. Daem, R. Pinxten, et al. 2018. Variation in personality traits across a metal pollution gradient in a free-living songbird. Sci. Total Environ. 630: 668–678.

[135] Mehlman, D.W., S.E. Mabey, D.N. Ewert, C. Duncan, B. Abel, D. Cimprich, et al. 2005. Conserving stopover sites for Forest-Dwelling Migratory Landbirds. Auk. 122(4): 1281–1290.

[136] Sullivan, B.L., C.L. Wood, M.J. Iliff, R.E. Bonney, D. Fink and S. Kelling. 2009. eBird: a citizen-based bird observation network in the biological sciences. Biol. Conserv. 142(10): 2282–2292.

[137] Bayly, N.J., K.V. Rosenberg, W.E. Easton, C. Gomez, J.A.Y. Carlisle, D.N. Ewert, et al. 2018. Major stopover regions and migratory bottlenecks for nearctic-neotropical landbirds within the neotropics: a review. Bird Conserv. Int. 28(1): 1–26.

[138] Droz, B., R. Arnoux, T. Bohnenstengel, J. Laesser, R. Spaar, R. Ayé, et al. 2019. Moderately urbanized areas as a conservation opportunity for an endangered songbird. Landsc. Urban Plan. 181: 1–9.

[139] Eakin, C.J., H. Campa III, D.W. Linden, G.J. Roloff, D.B. Rowe and J. Westphal. 2015. Avian response to green roofs in urban landscapes in the midwestern USA: avian conservation potential of green roofs. Wildl. Soc. Bull. 39(3): 574–582.

[140] Partridge, D.R. and J. Alan Clark. 2018. Data from: Urban green roofs provide habitat for migrating and breeding birds and their arthropod prey. Dryad Digital Repository.

[141] Washburn, B.E., R.M. Swearingin, C.K. Pullins and M.E. Rice. 2016. Composition and diversity of avian communities using a new urban habitat: green roofs. Environ. Manage. 57(6): 1230–1239.

[142] Beason, R.C. 2004. What can birds hear? Proceedings of the Vertebrate Pest Conference, 21. Retrieved from https://escholarship.org/uc/item/1kp2r437

[143] Belcher, R.N., K.R. Sadanandan, E.R. Goh, J.Y. Chan, S. Menz and T. Schroepfer. 2019. Vegetation on and around large-scale buildings positively influences native tropical bird abundance and bird species richness. Urban Ecosyst. 22(2): 213–225.

[144] Malpass, J.S., A.D. Rodewald and S.N. Matthews. 2016. Species-dependent effects of bird feeders on nest predators and nest survival of urban American Robins and Northern Cardinals. Efectos especie-dependientes de los comederos de aves sobre los depredadores de nidos y la supervivencia de los nidos en dos aves urbanas. Condor. 119(1): 1–16.

[145] Fischer, J.D. and J.R. Miller. 2015. Direct and indirect effects of anthropogenic bird food on population dynamics of a songbird. Acta Oecol. 69: 46–51.

[146] Moyers, S.C., J.S. Adelman, D.R. Farine, C.A. Thomason and D.M. Hawley. 2018. Feeder density enhances house finch disease transmission in experimental epidemics. Philos. Trans. R. Soc. Lond. B Biol. Sci. 373(1745): 20170090.

[147] Paker, Y., Y. Yom-Tov, T. Alon-Mozes and A. Barnea. 2014. The effect of plant richness and urban garden structure on bird species richness, diversity and community structure. Landsc. Urban Plan. 122: 186–195.

[148] Décarie, R., J.L. DesGranges, C. Lépine and F. Morneau. 1993. Impact of insecticides on the American robin (*Turdus migratorius*) in a suburban environment. Environ. Pollut. 80(3): 231–238.

[149] Mitra, A., Chatterjee, C. and F.B. Mandal. 2011. Synthetic chemical pesticides and their effects on birds. Res. J. Environ. Toxicol. 5(2): 81–96.

[150] Hallmann, C.A., R.P.B. Foppen, van C.A.M. Turnhout,. 2014. Declines in insectivorous birds are associated with high neonicotinoid concentrations. Nature. 511(7509): 341–343.

[151] Stanton, R.L., C.A. Morrissey and R.G. Clark. 2018. Analysis of trends and agricultural drivers of farmland bird declines in North America: a review. Agric. Ecosyst. Environ. 254: 244–254.

[152] Whelan, C.J., Ç.H. Şekercioğlu and D.G. Wenny. 2015. Why birds matter: from economic ornithology to ecosystem services. J. Ornithol. 156(1): 227–238.

[153] Dänhardt, J., M. Green, A. Lindström, M. Rundlöf and H.G. Smith. 2010. Farmland as stopover habitat for migrating birds – effects of organic farming and landscape structure. Oikos. 119(7): 1114–1125.

[154] Flint, M.L. and R. van den Bosch. 1981. Man, pests, and the evolution of IPM: an introduction. pp. 1–8. *In*: M.L. Flint and R. van den Bosch (eds). Introduction to Integrated Pest Management. Springer, Boston, MA.

[155] Parish, D.M.B. and N.W. Sotherton. 2004. Game crops and threatened farmland songbirds in Scotland: a step towards halting population declines? Bird Study. 51(2): 107–112.

[156] Wilcoxen, C.A., J.W. Walk and M.P. Ward. 2018. Use of cover crop fields by migratory and resident birds. Agric. Ecosyst. Environ. 252: 42–50.

[157] VanBeek, K.R., J.D. Brawn and M.P. Ward. 2014. Does no-till soybean farming provide any benefits for birds? Agric. Ecosyst. Environ. 185: 59–64.

[158] LaRose, J. and R. Myers. 2018. Use of cover crops to promote soil health. SARE Grants and Education to Advance Innovations in Sustainable Agriculture. http://dx.doi.org/10.19103/ as.2017.0033.26

[159] Buckingham, D.L., W.J. Peach, D.S. Fox. 2006. Effects of agricultural management on the use of lowland grassland by foraging birds. Agric. Ecosyst. Environ. 112(1): 21–40.

[160] Holzmueller, E.J., M.D. Gaskins and J.C. Mangun. 2011. A GIS approach to prioritizing habitat for restoration using neotropical migrant songbird criteria. Environ. Manage. 48(1): 150–157.

[161] Stephens, J.L. and S.M. Rockwell. 2019. Short-term riparian restoration success measured by territory density and reproductive success of three songbirds along the Trinity River, California. Condor. 121(4):

[162] Savard, J.-P.L., P. Clergeau and G. Mennechez. 2000. Biodiversity concepts and urban ecosystems. Landsc. Urban Plan. 48(3-4): 131–142.

[163] Bowers, E.K., D. Nietz, C.F. Thompson and S.K. Sakaluk. 2014. Parental provisioning in house wrens: effects of varying brood size and consequences for offspring. Behav. Ecol. 25(6): 1485–1493

[164] McQuillen, L.H. and L.W. Brewer. 2000. Methodological considerations for monitoring wild bird nests using video technology. J. Field Ornithol. 71(1): 167–172.

[165] Galbraith, J.A., D.N. Jones, J.R. Beggs, K. Parry and M.C. Stanley. 2017. Urban bird feeders dominated by a few species and individuals. Front. Ecol. Evol. 5: 81.

[166] Hruska, A.M., S. Souther and J.B. Mcgraw. 2014. Songbird dispersal of American ginseng (*Panax quinquefolius*). Écoscience. 21(1): 46–55.

[167] Steen, R. 2009. A portable digital video surveillance system to monitor prey deliveries at raptor nests. J. Raptor Res. 43(1): 69–74.

[168] Darras, K., P. Batáry, B.J. Furnas, I. Grass, Y.A. Mulyani and T. Tscharntke. 2019. Autonomous sound recording outperforms human observation for sampling birds: a systematic map and user guide. Ecol. Appl. 29(6): e01954.

[169] Jeliazkov, A., Y. Bas, C. Kerbiriou, J.F. Julien, C. Penone and I. Le Viol. 2016. Large-scale semi-automated acoustic monitoring allows to detect temporal decline of bush-crickets. Global Ecol. Conserv. 6: 208–218.

[170] Mariette, M.M., E.C. Pariser, A.J. Gilby, M.J. Magrath, S.R. Pryke and S.C. Griffith. 2011. Using an electronic monitoring system to link offspring provisioning and foraging behavior of a wild passerine. Auk. 128(1): 26–35.

[171] Pitera, A.M., C.L. Branch, E.S. Bridge and V.V. Pravosudov. 2018. Daily foraging routines in food-caching mountain chickadees are associated with variation in environmental harshness. Anim. Behav. 143: 93–104.

[172] Whitworth, D., S. Newman, T. Mundkur and P. Harris. 2007. Wild Birds and Avian Influenza: An Introduction to Applied Field Research and Disease Sampling Techniques. Food and Agriculture Organization of the United Nations, Rome.

[173] Taylor, P., T. Crewe, S. Mackenzie, D. Lepage, Y. Aubry, Z. Crysler, et al. 2017. The motus wildlife tracking system: a collaborative research network to enhance the understanding of wildlife movement. Avian Conserv. Ecol. 12(1): 8.

[174] Bayly, N.J., D.R. Norris, P.D. Taylor, K.A. Hobson and A. Morales-Rozo. 2020. There's no place like home: tropical overwintering sites may have a fundamental role in shaping migratory strategies. Anim. Behav. 162: 95–104.

[175] Schofield, L.N., J.L. Deppe, T.J. Zenzal Jr, M.P. Ward, R.H. Diehl, R.T. Bolus, et al. 2018. Using automated radio telemetry to quantify activity patterns of songbirds during stopover. Auk. 135(4): 949–963.

[176] Bridge, E.S., K. Thorup, M.S. Bowlin, P.B. Chilson, R.H. Diehl, R.W. Fléron, et al. 2011. Technology on the move: recent and forthcoming innovations for tracking migratory birds. Bioscience. 61(9): 689–698.

[177] Fudickar, A.M., M. Wikelski and J. Partecke. 2012. Tracking migratory songbirds: accuracy of light-level loggers (geolocators) in forest habitats: accuracy of geolocators. Methods Ecol. Evol. 3(1): 47–52.

[178] Ripperger, S.P., G.G. Carter, R.A. Page, N. Duda, A. Koelpin, R. Weigel, et al. 2020. Thinking small: next-generation sensor networks close the size gap in vertebrate biologging. PLoS Biol. 18(4): e3000655.

[179] Sommer, B., S. Feyer, D. Klinkhammer, K. Klein, J. Wieland, D. Fink, et al. 2019. BinocularsVR A VR experience for the exhibition from lake constance to Africa, a long distance travel with ICARUS. Electronic Imaging. 2019(2): 177(1)–177–8(8).

[180] BBäckman, J., A. Andersson, L. Pedersen, S. Sjöberg, A.P. Tøttrup and T. Alerstam. 2017. Actogram analysis of free-flying migratory birds: new perspectives based on acceleration logging. J. Comp. Physiol. A. 203(6-7): 543–564.

[181] Podlesak, D.W., S.R. McWilliams and K.A. Hatch. 2005. Stable isotopes in breath, blood, feces and feathers can indicate intra-individual changes in the diet of migratory songbirds. Oecologia. 142(4): 501–510.

[182] Knutsen, C.J. and C.W. Varian-Ramos. 2019. Explaining variation in Colorado songbird blood mercury using migratory behavior, foraging guild, and diet. Ecotoxicology. 29: 1268–1280.

[183] Grond, K., B.K. Sandercock, A. Jumpponen and L.H. Zeglin. 2018. The avian gut microbiota: community, physiology and function in wild birds. J. Avian Biol. 49(11): e01788.

[184] Titulaer, M., A. Melgoza-Castillo, A.O. Panjabi, A. Sanchez-Flores, J.H. Martínez-Guerrero, A. Macías-Duarte, et al. 2017. Molecular analysis of stomach contents reveals important grass seeds in the winter diet of Baird's and Grasshopper sparrows, two declining grassland bird species. PLoS One. 12(12): e0189695.

[185] Rytkönen, S., E.J. Vesterinen, C. Westerduin, T. Leviäkangas, E. Vatka, M. Mutanen, et al. 2019. From feces to data: a metabarcoding method for analyzing consumed and available prey in a bird-insect food web. Ecol. Evol. 9(1): 631–39.

[186] Thomsen, P.F., J.O.S. Kielgast, L.L. Iversen, C. Wiuf, M. Rasmussen, M.T.P. Gilbert, et al. 2012. Monitoring endangered freshwater biodiversity using environmental DNA. Mol. Ecol. 21(11): 2565–2573.

[187] Weber, T.P., A.I. Houston and B.J. Ens. 1999. Consequences of habitat loss at migratory stopover sites: a theoretical investigation. J. Avian Biol. 30: 416–426.

[188] Bystrak, D. 1981. The North American breeding bird survey. pp. 34–41. *In*: C.J. Ralph and J.M. Scott (eds). Estimating Numbers of Terrestrial Birds. Cooper Ornithological Society. Allen Press.

[189] Hill, A.P., P. Prince, J.L. Snaddon, C.P. Doncaster and A. Rogers. 2019. AudioMoth: A low-cost acoustic device for monitoring biodiversity and the environment. HardwareX. 6: e00073.

[190] Bonter, D.N. and C.B. Cooper. 2012. Data validation in citizen science: a case study from Project FeederWatch. Front. Ecol. Environ. 10(6): 305–307.

[191] Berger-Tal, O. and D. Saltz. 2016. Introduction: the whys and the hows of conservation behavior. pp. 3–35. *In*: Conservation Behavior: Applying Behavioral Ecology to Wildlife Conservation and Management. Cambridge University Press.

[192] Rubenstein, D.I. 2016. Anthropogenic impacts on behavior: the pros and cons of plasticity. pp. 121–146. *In*: Conservation Behavior: Applying Behavioral Ecology to Wildlife Conservation and Management. Cambridge University Press.

[193] Caro, T. and P.W. Sherman. 2012. Vanishing behaviors. Conserv. Lett. 5(3): 159–66.

Chapter 8

Human Impacts on Avian Communication

Sharon A. Gill[1], Erin E. Grabarczyk[2]
and Dominique A. Potvin[3]

Introduction

With extensive worldwide transformation of wild lands has come equally expansive changes in ambient sound environments. Human-built environments contribute sounds from vehicle traffic, buildings, energy infrastructure and extraction, construction, and military activities, such that noise has increased and few quiet spaces remain. This realization led to researchers around the world to consider the impact of human-generated

[1]Department of Biological Sciences, Western Michigan University.
[2]Southeast Watershed Research Laboratory, USDA, ARS.
[3]School of Science and Engineering, University of the Sunshine Coast.

or anthropogenic noise on animal communication, rapidly building a body of knowledge of noise impacts on signal structure [1, 2]. We know less at this moment about other aspects of communication, but researchers have increasingly turned their attention to noise impacts on receivers. Recent research has examined noise effects on single individuals within dyadic interactions [3–5], as well as multiple individuals within complex signal networks [6]. Despite sustained attention on noise impacts on avian signaling since the first works on this issue [7, 8], questions remain, not the least of which is what we can do about it. How do we manage or mitigate this widespread pollutant?

In this chapter, we describe foundational principles of songbird communication and highlight main findings regarding the impacts of anthropogenic activities on vocal communication. We then focus on management and mitigation strategies in the context of anthropogenic noise pollution. We examine how understanding communication adjustments in noise can guide management, by honing in on what we currently understand about songbird noise exposure. We review important considerations regarding the characterization of noise, spatial and temporal variation in noise, and threshold noise levels for behavioral adjustments. We then discuss species-level and area-based approaches to identify and conserve quiet spaces and to mitigate noise exposure. We end the chapter by identifying gaps in knowledge. Songbirds may not only serve as indicators of decline in habitat quality and suitability due to human-generated noise, but could also be used to inform the effectiveness and future direction of noise management and conservation of quiet areas.

A Brief Introduction to Communication

Communication is a fundamental behavior for all animals, including songbirds. In biology, communication is defined as the transmission of information (the signal) from one individual (sender) to another (receiver), either of the same or another species, prompting a change or response in the receiver. In songbirds, the primary mode of signaling is acoustic—songs and calls—however, visual, olfactory, and tactile signaling also play important roles in conspecific communication [9]. Across modalities, communication is essential for establishing and maintaining social relationships, attracting and retaining mates, and alerting conspecifics to dangers, all of which promote the survival of individuals and populations. Individuals typically rely on acoustic signals when broadcasting over large distances, such as during mate attraction or territory defense, but they may use additional modalities, including visual, olfactory, and tactile signals, for communication over shorter distances, such as between mates or between parents and offspring. Studies of bird communication often focus on the signals themselves and interactions between signalers and receivers, but additional behaviors, including habitat choice (where to live and breed) and selection of a signaling location [10], are involved in communication as well.

Birds use acoustic signals in a variety of social contexts, such that different types of signals have unique functions and thus carry specific information. Songs, for example, are used by males and females to attract mates or to advertise territory boundaries to neighbors. Thus, songs often contain information about the quality (health, condition, genetic background) of the signaler (Otter et al., Chapter 5), and are often directed and are required to be heard over moderate distances [11]. Social or contact calls may be simple signals providing information about the signaler's identity or proximity to others in a flock. Alarm and mobbing calls indicate the presence of danger—usually a predator—not only to conspecifics, but also to other species in the immediate area, and may convey information about predator type or proximity [12, 13]. The repertoire of acoustic signals made by songbirds is extensive, and can include many other forms and functions. Importantly, they all serve to promote an individual's fitness and a population's survival.

Structures of acoustic signals evolve under diverse selective pressures. Signal traits, such as frequency (perceived as pitch) or duration, are influenced by ecological selection on bill size (e.g., [14]), receiver auditory systems [15], and the environment through which the signal travels [16, 17]. As signals travel through the environment their structure degrades and energy contained within the signal decays or attenuates [18]. Habitats affect the rate of signal decay and therefore the distance over which a signal can be heard. In forests, acoustic signals scatter across or are absorbed by vegetation [16, 17], whereas visual signals may be blocked entirely by vegetation. In open habitats, acoustic signals are modified by wind turbulence [18], although visual signals may be detected over larger distances than in forests. Forest-dwelling species are thus more likely to produce low frequency, tonal calls that travel over greater distances than trills, which are broad bandwidth notes given in rapid succession. However, for species in open habitats, trills are advantageous as they may be transmitted over large distances [19]. Birds may further modify the structure of their signals via pitch, amplitude, or composition to transmit short or long distances, depending on the location of the intended receivers (reviewed in [20]).

From the receiver perspective, auditory physiology and receptor sensitivity are critical. Communication systems (signalers, receivers, and their signals) are tuned in ways that increase the probability that receivers differentiate information contained within signals from background noise. For example, the auditory systems of birds are tuned to hear sounds (i.e., pressure waves) at certain frequencies better than others. Auditory sensitivity, or the ability to differentiate the signal from background noise, varies among species and sexes, and can even change seasonally (reviewed in [15, 21, 22]). Passerines, on average, tend to best hear acoustic signals between 1–6 kHz, but there are notable exceptions. For example, the auditory system of tufted titmice (*Baeolophus bicolor*) is more sensitive to signals given at higher frequencies compared to many other birds, possibly because titmice give high frequency alarm call vocalizations in response to predators [23].

Anthropogenic Impacts on Acoustic Communication

Human-directed environmental change affects acoustic communication in diverse ways, with direct effects better understood than indirect ones. Indirect effects refer to cases in which environmental change directly affects one species or system, which then interacts with and affects avian communication. Indirect effects on animal signals could be widespread [24]. For example, the introduction of invasive plant species or wild bird feeding at feeders can shift bird diets, changing the pattern of selection on bill size, which in turn affects song structure. In medium ground finches (*Geospiza fortis*), a bimodal distribution of bill sizes occurs, with larger and smaller billed birds singing different song variants [25]. However, supplemental feeding at bird feeders has favored convergence in bill size, such that bill size in the population has become more similar over time [26]. Because bill size correlates with song features [27], selection on bill size due to diet likely contributes to changes in song structure. The bills of birds also function in temperature regulation, with the possibility that climate change could indirectly affect bird song as well. Under increased temperatures predicted to occur due to climate change, we might expect shifts in song structure associated with selection on bill size in relation to heat exchange [28, 29].

Human transformation of habitat structure can also have an indirect effect on communication, through the link between habitat and signal transmission. Invasive species alter the physical structure of plant communities, impacting the space through which signals must travel. For example, the perennial legume kudzu (*Pueraria montanaui* var. *lobata*), introduced to the United States in the late 1800s, grows rapidly on supporting vegetation altering forest canopy structure, particularly along edges, and leading to dense layers of reflective substrates [30]. As signals reverberate across these dense layers of leaves, signals should rapidly degrade and be less detectable over space. Similarly, the transformation of wildlands to cities results in abundant urban structures, buildings and impervious surfaces, such as roads and parking lots, which reverberate acoustic signals, resulting in an echoing effect and a distortion of sound [31, 32]. Urbanization is a profound habitat change, increasing the area and density of highly reflective surfaces, which fundamentally changes the acoustic environment and may favor signal adjustments that differ from those due to noise [33–35]. Changes in native species may alter habitat as well, such as the intense browsing by white-tailed deer (*Odocoileus virginianus*) in temperate deciduous habitats of North America. Deer browsing leads to habitats with better sound propagation, measured as fidelity of signals to original form, although amplitude is unaffected [36].

Indirect effects, however, are likely to be dwarfed by the direct impacts of environmental noise in highly populated areas [33]. To date, the majority of research on human impacts on bird communication have focused on the effects of anthropogenic noise on acoustic signals [1, 2, 37]. Noise from traffic, infrastructure, and industry is characteristically loud (>50 dB) and low frequency [38], with peaks below 3 kHz ([39]; Fig. 1). High amplitude noise that occurs at the same frequency as signals results in masking, which results in signals being more difficult to differentiate from background noise (Fig. 2). Not only can noise interfere with communication between adult birds, but it can also disrupt signals between adults and their young. Begging signals, for example, may be masked by noise, resulting in young not being fed, as has been observed in tree swallows, *Tachycineta bicolor* [40]. Communication of danger by parents to young (and other adults) may also be masked, increasing the risk of being depredated [41, 42]. Tree swallow young, for example, fail to react appropriately by crouching in the nest in response to adult alarm calls in a noisy environment [41]. Additionally, young birds may not learn their species' songs accurately when subjected to noise during development, with potential consequences for reproductive success later on in life [43]. Zebra finches (*Taenopygia guttata*) in captivity attempting to learn their species' song in moderate to high noise conditions demonstrate an inability to copy proper syntax of songs [43]. The lack of fidelity during song learning has implications for song evolution [44].

Even if the signal itself is not masked by noise, signal content may not be communicated effectively if noise masks information (reviewed in [45]). Information masking occurs when noise that does not overlap the signal diverts the attention of the listener [46, 47], resulting in missed information. Distractions that occur across signaling modalities can lead to cross-modal interference, in which unimportant signals of one type disrupts perception of signals of a second type (reviewed by [45, 48]). For example, dwarf mongoose (*Helogale parvula*) respond with less vigilance to predator fecal presentations (olfactory signal) when given in combination of high ambient noise levels, suggesting that noise distracts mongooses [49]. Noise as a distraction or simply as an extra stimulus may also be considered a source of chronic stress [50], especially for young birds learning to communicate. Stress can affect vocal learning in some species [51], therefore noise may indirectly affect the ability to learn to communicate, although experimental evidence for this is yet to be found (e.g., [43]).

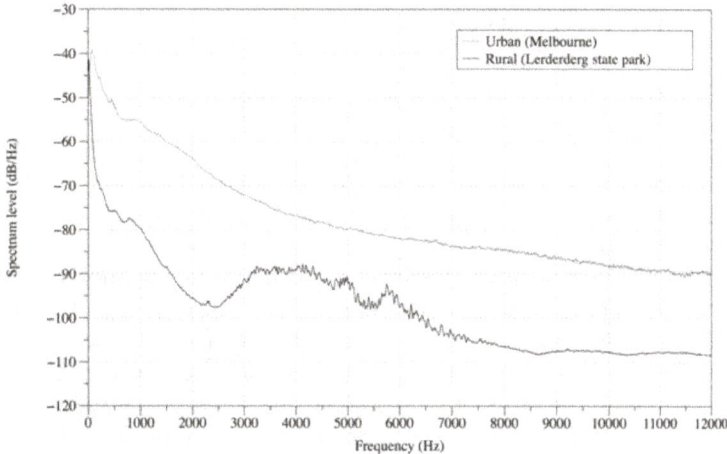

Figure 1 Noise profiles from an urban park (Melbourne, Australia) and a rural park (Lerderderg State Park, Australia) during morning hours. Noise profiles reveal the energy present in background noise across different frequencies. In this profile, the energy from bird song, between 2500–6000 Hz, is visible against background noise in the rural park, but not urban park (from Potvin 2017; reprinted with permission from Springer Link).

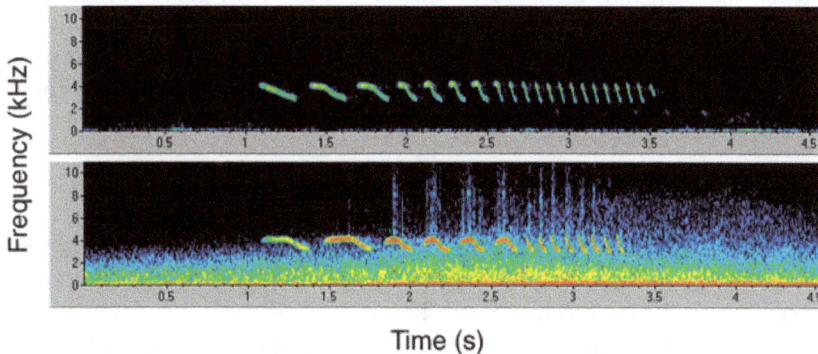

Figure 2 Spectrograms of field sparrow songs, recorded at a rural nature preserve (top) and alongside a road at urban nature preserve (bottom). In the bottom image, sounds from a passing car overlap or mask the field sparrow song.

Noise may also disrupt signals from potential predators or prey [52]. In a playback experiment, northern cardinals (*Cardinalis cardinalis*) in quiet areas responded to tufted titmice alarm calls by ceasing movement, whereas in noisy areas individuals failed to respond with typical anti-predator behavior, even when within 1 m of the playback speaker [47]. Failure to respond could be due to signal masking of alarm calls at some distances, but when signal-to-noise ratios were high, noise may also have distracted cardinals [45, 47]. Birds with songs or calls in the same frequency range as anthropogenic noise may find it difficult to communicate and as a result may be excluded from noisy habitats [53, 54]. Thus, it is clear that environmental noise can have far reaching consequences for birdlife.

As with any potentially disruptive force, animals may adjust their behaviors in response to noise in order to continue to communicate. How do these adjustments happen? Birds respond to environmental noise in one of three main ways: *avoidance*, *adjustment*, or *adaptation* [55]. Next, we discuss each of these responses, with examples that demonstrate the complex ways in which birds (and other animals) deal with anthropogenic noise.

Avoidance

Avoiding anthropogenic noise requires birds to find alternate habitats or to selectively use their existing habitat to avoid noise. It is difficult to discern from observations alone whether a species' absence from a noisy area is due to birds actively avoiding the location or due to extirpation when a population is unable to reproduce in an area and counter natural (or unnatural) death rates. However, experiments have shown that even over a short time frame, individual birds avoid areas of high noise during migratory stopovers (with or without the associated human infrastructure [56]) and during the breeding season [50, 57, 58]. Of course, selecting habitat to avoid noise in the Anthropocene isn't necessarily simple, given the scale of land transformation and fragmentation [59, 60], and inevitable increases in noise exposure. As such, birds may be forced to settle in noisy areas. If noise avoidance proves to be widespread, many species may avoid otherwise good-quality or restored habitat, complicating habitat management (see section 'Noise Management and Mitigation Strategies'). However, as noise levels vary over even small spatial scales, such as territories and home ranges [34, 61], birds could selectively avoid noisy places even once they have settled in a habitat. Individuals move away from noise [62], spend less time in noisier areas of leks [63], and avoid preferred singing locations when noise is introduced near them (Gill and Job, unpublished data). Together, these observations suggest a hierarchy of decisions regarding habitat choice: if possible, birds might avoid noisy habitat altogether, but if not, they may change how they use space to avoid noisy parts of their territories to minimize exposure to noise.

Adjustment

Instead of avoiding noisy habitat, birds may adjust behavior in relation to noise. Behavioral flexibility allows individuals to change their behavior, within their lifetimes, to accommodate a changing environment. To be heard by receivers, for example, signalers may adjust their spatial orientation, singing in a particular direction and closer to a specific receiver [64, 65]. In response to artificial noise added to their nest boxes, female great tits (*Parus major*) respond more slowly to their mates' calls, and as a result, males restore communication by singing from perches closer to nest boxes [65]. Birds may also rapidly change signal structure when environmental conditions change [66–68], while still maintaining information in signals. For example, simply singing louder may be an automatic and very rapid response to allow a signal to be heard above potential masking noise [69–72]. Birds may also sing or call at a pitch that avoids the masking frequencies of background noise, usually singing higher frequencies to avoid low-pitched noise [67]. This phenomenon has been studied and reviewed widely (e.g., [55, 73]), and whether or not it is a conscious effort or able to be performed by all species, is still under study [74]. If it is the case that only some species are able to employ this tactic [75], some species will likely be lost or filtered from noisy habitats [53].

Animals react differently to intermittent bursts of noise than to noise emitted continuously from a stationary source, although both noise types can be considered disruptive and induce changes in signal types, frequency and timing [76]. Spatial proximity to noise source influences individual responses [77]. For example, short, loud sounds with a rapid onset produced near an animal, such as a gunshot, may startle a singing bird, resulting in an immediate change in calling behavior. Indeed, these methods have been employed to deter birds in certain contexts, such as excluding them from dangerous areas like airfields [78]. In contrast, in response to stationary and predictable sounds, such as those produced by a building generator or at a well site, individuals may adjust the structure of their signals to reduce masking, avoid areas in close proximity to the noise source, or alter the timing of vocalizing to avoid noise. Timing adjustments can occur on a short or immediate scale (termed "gap calling", for calling in a gap of noise), a behavior well-established in frogs [79] but still being researched in birds [80]. We do know, however, that entire choruses of birds may shift temporally in response to periods of increased noise. For example, dawn choruses by some species now begin earlier to avoid rush hour noise [81, 82] and other species increase night singing in noisier areas [83]. Other species change their song timing not necessarily as a noise-avoidance strategy, but simply as an effect of artificial light. Light would normally indicate dawn, acting as a trigger for birds to begin their dawn chorus. In noisy areas where artificial light is prevalent, birds may start singing much earlier in the day [84, 85], but see [86].

From the receiver perspective, signal changes in noisy or acoustically challenging areas may improve signal detection. Receivers find it difficult to discriminate between high- and low-quality signals when they are masked by noise, which can have follow-on effects by influencing mate-choice decisions [87], (also see Otter et al. Chapter 5). To increase detectability, individuals may change the location where they receive signals, such as moving to a higher perch, which could improve signal-to-noise ratios [65, 88]. As such, feedback or the lack thereof from intended receivers may trigger additional adjustments by signalers that restore communication [65, 89]. Furthermore, receivers may also experience spatial release from masking (i.e., improved signal-to-noise ratios) by changing the orientation of their ear relative to the signal(er) [90, 91].

Adaptation

If behavioral adjustments are beneficial, such that they positively influence reproductive success and survival, they may lead to long-term changes in signaling traits within populations over generations (i.e., adaptations). We might expect over time, for example, higher frequency songs in noisy habitat as discussed earlier. Research comparing song traits of white-crowned sparrows (*Zonotrichia leucophrys*) over a 30-year period found that increased minimum frequency of songs in parallel with increased noise [92]. The key here is understanding whether higher frequency songs actually convey a fitness benefit; that is, the signal change, whatever it entails, results in higher mating success, higher reproductive success or higher survival, and that benefits of signal adjustments outweigh potential costs associated with altered signals. To date, most studies of noise-induced song changes have not been designed to collect fitness data, pointing to an important gap in our knowledge [93].

As noise exposure varies over time and space, populations may experience different selective pressures from noise. If noise varies across populations and gene flow is low, populations could adapt to local conditions and vocally diverge from one another, as has been observed in silvereyes, *Zosterops lateralis* [35] and little greenbuls, *Andropadus virens* [94]. Noise-induced evolution of signals that lead to population differentiation could result in a mosaic of signal traits over space and time. Selection may also favor signal flexibility, rather than specific signal adjustments. Individuals that are able to flexibly adjust their signals according to noise levels at the time of signaling might experience higher fitness than those that cannot match changing conditions. By being flexible, individuals would adjust signaling only when needed during times of high noise levels, but switch back to typical signaling when noise levels are low [68, 95].

Clearly, human-caused environmental change, and anthropogenic noise in particular, alters communication behavior in birds. Questions remain about fitness impacts of noise-induced vocal adjustments [93], but evidence suggests that noise contributes in whole or in part to reduced pairing success [96], smaller populations near roadsides [97], and filtering of bird communities based on vocal traits [53, 98]. The problems often lie in the ability of species to cope with noise, whether by adjusting signals or habitat choice—when species are unable to do either, they may be extirpated from noisy areas, regardless of habitat quality. We next articulate the problems posed by anthropogenic noise in the context of communication, considering issues facing individual species as well as the habitats in which they breed and survive.

Anthropogenic Impacts on Visual and Other Communication Modalities

Our body of knowledge about how visual (and other) forms of communication by songbirds are impacted by anthropogenic activities is extremely limited when compared to acoustic communication. We know that visual signals are predominantly used for mate attraction in songbirds [99]. Although some parent-offspring communication is visual (large gaping mouths as a begging signal, for example), such close-proximity visual signals are unlikely to be impacted by large-scale environmental change, such as those present in human-dominated areas. Visual sexual signals may include physiology-based characteristics (such as plumage or bill coloration) and behavior-based signals (such as dances). Physiological signals such as bright plumage can indicate body condition or individual health especially of males during the mating season [100]. These signals may be hormonally determined, either through development or seasonally 101, 102], or may be diet-dependent [103]. Thus, if anthropogenic activities disrupt any aspect of these processes, there may be downstream effects on visual communication efficacy. For example, diet changes and heavy metal ingestion in urbanized or polluted habitats may affect body condition and thus the ability to produce carotenoid-based or other plumage colors [104]. A detailed description of such changes and their effects can be found in Otter et al. Chapter 5.

While acoustic and visual signals are the dominant modes of songbird communication, there is increasing evidence that olfaction may also play a part in communication [105]. Decades of studies have demonstrated individuals' abilities to choose genetically compatible mates from the surrounding population; however, how such compatibility is communicated is largely unknown. Recent evidence shows that genetic make-up of an individual may influence its scent through volatile compounds present in preen oil [105, 106]. This suggests that songbirds may have sophisticated olfactory morphological and physiological structures that enable the perception of discrete, possibly subtle odors communicated through pheromones [107, 108]. In non-avian taxa, anthropogenic activities such as chemical use and changing air quality may impact the transmission and perception of chemical signals, again disrupting another avenue of communication [109]. Whether human activities may impact songbird olfactory communication is currently unknown, but a prudent approach would take chemical pollution and air changes into account when investigating the use of olfactory signals by songbirds in changing environments.

The Noise Problem

As explored earlier, signals have evolved under diverse selection pressures, including habitat structure and diet, but also ambient sounds. Ambient sound environments are not necessarily quiet and species that live in naturally high noise environments, such as torrent frogs (*Amolops tormotus*) along rivers [110] or sparrows dealing with the sounds generated by ocean surf [111], produce signals that must be detectable against naturally occurring noise. When exposed to anthropogenic noise, it appears that most songbird species examined do alter signals, by singing more loudly, at higher frequencies or between noise gaps, as discussed earlier. Of concern are those species that do not alter signals at all or enough to avoid masking or that are unable to detect and discriminate signals in noise. In either scenario, individuals may suffer fitness costs, which might lead to the extirpation of species from otherwise good quality habitat (i.e., noise filtering; [53, 54, 98, 112]). Reduced reproductive success in noisy areas has already been identified in some species, contributing to population declines [3, 113] and threatened and endangered species that do not alter signals would be particularly vulnerable. Noise filtering ultimately could lead to more homogenous bird communities, as has been observed with urbanization [114, 115].

On a landscape scale, noise may decrease the quality of areas protected for the maintenance of biodiversity and its associated benefits, including nature's contributions to people [116]. Thus, assessment of the acoustic quality of protected areas would be beneficial. If birds avoid noisy habitat, then the area of protected land suitable for birds may be overestimated (whether due to noise impacts on signaling behavior or for other reasons, e.g., stress), which again might heighten concern for threatened and endangered species. As countries move closer to meeting targets set by the Convention on Biological Diversity that at least 17% of land is protected [117], maintaining ecological integrity and avoiding degradation of these areas due to noise is of prime concern. In many regions of the world, protected areas experience heavy human impacts with little oversight and enforcement [118, 119]. Furthermore, the presence of noise in protected lands alone may degrade otherwise suitable habitat. Without noise management, we may be even further from conservation targets than we realize.

From a management perspective, the nuance regarding noise impacts is important. Noise management and mitigation may be aimed at protecting certain species, particularly listed ones, or may be aimed at the landscape-level. Protected areas designated primarily for the conservation of biodiversity (IUCN categories I and II) should be high priorities for noise mitigation, whereas areas designated for dual purposes of conservation and human activities (categories III and IV) might be reasonably excluded from noise management efforts. Thus, we need to consider how we might mitigate effects of noise on communication, either by taking species-specific or area-based approaches. Before turning our attention to noise management and mitigation, we next consider features of noise itself that will be relevant to the development of conservation plans.

Characterizing Noise

The properties and structure of noise should inform management and conservation decisions. Therefore, in this section, our aim is to identify salient noise features (see Box 1), including ambient or baseline sound levels and threshold noise levels above which communication is affected. Throughout this chapter, we have conveniently discussed noise as if it was a singular phenomenon. But the structure of noise differs in quantifiable ways depending on the source(s) of noise: noise from traffic differs in amplitude, frequency and timing from noise emanating from well compressors or buildings or military noise [120]. In addition to frequency and amplitude, noise may differ in other, less commonly quantified ways. For example, noise may have rapid or gradual onset, may be continuous or not, and may occur with some predictability (such as rush hour) or may fluctuate irregularly (Fig. 3; [95]). Each of these features of noise may have different impacts on communication [95, 121]. Measurements of noise pollution that capture the range of its properties, coincident with the timing and location of signaling birds, would improve our understanding of noise impacts and help to identify the specific features of noise to which species react. Such information is critical for noise mitigation aimed at individual species.

Although amplitude is the most commonly reported feature of noise in communication studies', we could learn more by quantifying noise in nuanced ways [95, 120, 122]. Given species differences in auditory sensitivity, measuring amplitudes at frequencies relevant to the species in question, as is common in studies of marine mammals [123], would allow the use of auditory sensitivities of birds, rather than our own, to identify low-noise environments for conservation. If masking is an issue, noise amplitude in the frequencies that match the signals and auditory sensitivity of a particular species is of concern. Analyzing amplitude at certain frequencies (e.g., octave or 1/3 octave frequency bands) could be useful in landscape-level management, as managers could understand sound energy across bands occupied by species within communities of interest. Another approach involves determination of statistical noise levels (i.e., L_n), which quantify the sound pressure levels (SPLs) that are exceeded (n) proportion of time [120]. Median SPLs (L_{50}) are commonly reported, but ambient levels (L_{90}) in disturbed habitat provides needed information in relation to threshold values. Infrequent (L_{10}) or rare (L_1) SPLs can be useful in teasing out the regularity of uncommon, intermittent noise events, which may be important if noise bursts negatively affect a species of interest. These SPLs can be visualized using decibel duration curves (DDCs), which summarize the patterns of noise levels over time [120]. Measurements of noise variation over space and time in conjunction with DDCs can be useful when applied to recordings taken across sites or at points within sites that are of management concern and used in combination with noise exposure-response curves.

An unresolved question is whether there are consistent noise levels above which effects occur, but below which they do not. Determining such threshold levels of responses, whether from signaler or receiver perspectives, is critical for developing management goals. For many species, only observational studies exist that show changes

Box 1 Glossary of selected terms used in the description and quantification of acoustic signaling and anthropogenic noise

Amplitude: Relative strength of a sound pressure wave, perceived as loudness or volume. Measured in decibels (dB).

Attenuation – loss of energy in acoustic signal over distance, as a result of the spread of energy over space as signal moves from its source (known as spherical spreading), as well as features of the environment, such as temperature, wind speed, and scattering and absorption by vegetation.

Continuous noise – condition in which sound pressure levels (SPLs) are consistently elevated above ambient, baseline levels.

Decibel duration curves (DDCs) – a visualization approach that illustrates the percentage of time that L_n values are equaled or exceeded; does not reveal regularity of exceedance, but is used to illustrate overall pattern of L_n values at a sampling location.

Degradation – loss in signal quality, which affects detection and discrimination of signals.

Intermittent noise – condition in which sound pressure levels (SPLs) are elevated above ambient, baseline levels at regular or irregular intervals. Often refers to impulsive noise or noise bursts.

Frequency: Also referred to as 'pitch', this refers to the number of times that a sound pressure wave repeats or oscillates per second. High frequencies oscillate faster than low frequencies. Measured in Hertz (Hz).

Masking: Auditory masking is the disruption of a focal sound by the presence of other sounds (noise). Masking usually occurs when the focal sound is within the frequency range of simultaneous noise at an amplitude high enough to affect perception of a receiver.

Noise consistency – the percentage of time that SPLs exceed a defined amplitude threshold; reveals patterns of continuous to intermittent noise.

Noise onset – the time for SPLs to reach a defined amplitude threshold, such as maximum noise threshold or a recommended noise level.

Noise threshold: The power level of a signal below which noise is likely to obscure the signal and above which the signal is discernible.

Regularity of sound pressure levels (SPLs) – describes the pattern over time with which SPLs exceed a defined amplitude threshold.

Signal: In the context of communication, an act or structure that conveys information from a sender to a receiver, that can alter the behavior of the receiver.

Signal-to-noise ratio: The ratio of the strength (relative amplitude) of a signal carrying information to that of unwanted background noise. Higher signal-to-noise ratios may result in greater signal clarity for receivers.

Sound pressure levels (SPLs) – sound pressures converted to decibels (dB) relative to a reference pressure value; in air, the reference pressure value is 20 μPa, although the reference value may not always be reported.

Statistical noise levels, or L_n values – sound pressure levels (SPLs) that are equaled or exceeded n percentage of the time

- L_{90} – a SPL that is equaled or exceeded 90% of the time; reflects continuous ambient SPL at sampling location
- L_{50} – a SPL that is equaled or exceeded 50% of the time; reflects median SPL at sampling location
- L_{10} – a SPL that is equaled or exceeded 10% of the time; reflects intermittent peak SPL at sampling location

Sources: Gill et al. 2015, 2017, McKenna et al. 2016, Wiley 2015.

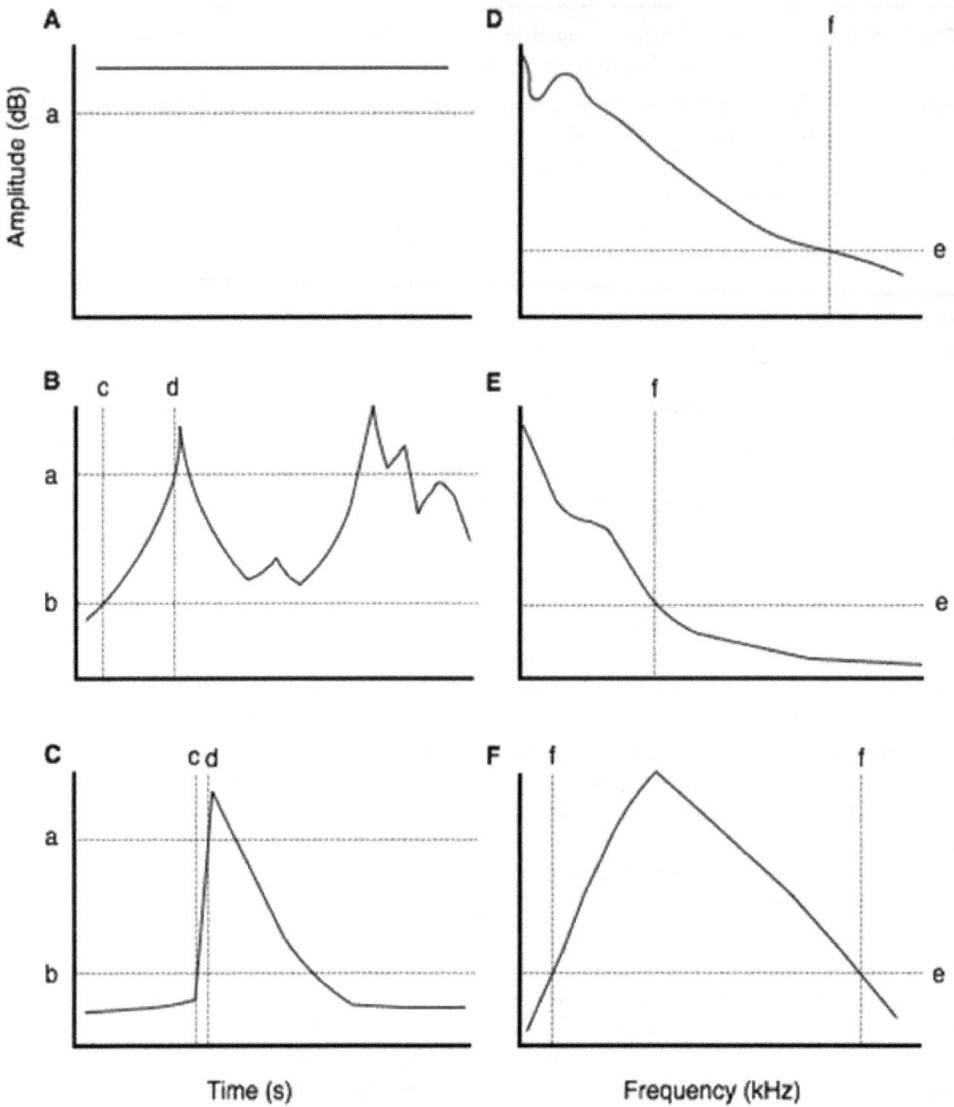

Figure 3 Different sources generate different characteristics of noise, as illustrated in hypothetical amplitude envelopes (A-C) and power spectra (D-F). Images A and D reflect continuous, high-amplitude, broadband noise, which could be generated by a well-compressor or building ventilation systems. B and E illustrate a variable noise environment, such as generated by vehicular traffic. C and F reflect a short-term, high-amplitude noise event, such as generated by a short-gun blast. Lower case letters represent noise features: a: maximum threshold energy value; b: minimum threshold energy value, c: beginning of noise onset, d: end of noise event; e: baseline energy level below which anthropogenic and ambient noise cannot be distinguished; f: frequency range of noise above ambient (from Gill et al. 2015 [95]; reprinted with permission from Oxford University Press).

in signal structure in relation to ambient sound levels (see 'Adjustment'). In most cases, signals or responses change in a linear fashion with increasing noise, but curvilinear responses are possible that would point to threshold levels, as has been found for noise-induced flushing distances in some species [124, 125]. Playback studies aimed at determining threshold noise levels associated with altered signaling or reception would be invaluable to help us to understand how animals respond to increasing levels of constant noise, as well as how such responses compare to those given in response to episodic or intermittent noise events. Together such information would provide critical information regarding whether amplitude threshold levels exists that trigger changes in behavior across species [126]. At the time of writing, studies controlling and monitoring threshold levels of noise amplitude and associated responses in songbirds in real time have yet to be published. Moreover, much less is known regarding noise thresholds for detection, discrimination, and perception by signal receivers and how those in turn may affect interactions between individuals.

Noise levels may vary temporally and spatially with important implications for both communication and management. Noise may vary seasonally, over the course of a day, or very rapidly over minutes or seconds [95]. Of principal concern are noise peaks that coincide in time with behaviors that require sending or receiving of signals, such as attracting a breeding partner or avoiding a predator, respectively. Therefore, when considering the impacts of noise on birds, we should take a matched approach, considering the timing of behaviors that if affected would be detrimental, as well as the timing of noise that is most disruptive. Periods of time considered disruptive for communication may vary for different species, but generally should include times of year associated with mating, nesting and brood rearing, as well as foraging [127]. In addition to temporal fluctuations in noise, noise varies over space. Noise may vary within and between natural areas on large spatial scales [128, 129], but noise levels can vary on small spatial scales as well, such as within an individual bird's territory [34]. In part, spatial variation in noise depends on whether the source is stationary or in motion. Features of the landscape, such as habitat structure and topography [129], may also impact spatial variation in noise.

Noise Management and Mitigation Strategies that Facilitate Communication

As concern grows regarding the effects of anthropogenic noise for humans and wildlife alike, so do calls for noise management and mitigation. We envision two broad categories of approaches to noise management and mitigation in the context of avian communication (summarized in Box 2). Managers could adopt *species-specific* approaches, with monitoring and management aimed at mitigating noise effects on particular species [125, 130] or *area-based* approaches, with monitoring and management aimed at mitigating noise in the total environment, particularly in protected areas [128]. In both approaches, carefully considered and articulated management goals are needed. Is the target to decrease noise masking for threatened species? Species-based approaches will be needed because masking varies across species depending on vocal frequencies and auditory sensitivities. Or is the aim to decrease noise overall, to benefit multiple species? If so, then area-based approaches would be appropriate. The approaches we discuss would likely benefit other aspects of avian behavior [131], ecology (e.g., fitness), and physiology (e.g., stress). Noise management and mitigation plans require additional policy pertaining to regulation and enforcement, but these topics are beyond the scope of this review.

Box 2   Steps for species-based and area-based noise mitigation and management approaches

Species-based approaches	Area-based approaches

• Identify time of day and year, and location of behaviors that rely on communication and are linked to fitness:
  o reproduction: mating, nesting and brood rearing
  o survival: foraging and predator detection
• Estimate current ambient noise levels for the area inhabited by species of interest
	• Compare undisturbed baseline and current noise levels
	  o baseline may be derived from historical records, estimates of noise levels when little or no human activity occurs in area, or nearby sites with little or no human activity
	• Use difference between baseline and current noise levels to guide reduction in noise levels throughout area of interest

• Determine threshold noise level that results in a change in signaling (e.g., microhabitat choice or signal detection) • Focus conservation and mitigation efforts on noise that exceed threshold levels and overlap in space and time with essential behaviors and communication events • Assess impact of noise mitigation on signaling • Revisit sound levels and mitigation strategies often	• Strategies for noise reduction include natural or artificial berms, vegetation buffers, re-siting of roads • Alternatively, identify quiet natural areas near baseline levels and work to keep them quiet o include quiet regions within larger areas noise • Assess impact of noise mitigation on noise levels • Revisit sound levels and mitigation strategies often

Species-specific Approaches

Noise may not affect each species in the same way. Studies on communities of birds show that vocal traits such as the frequency or pitch of song can predict whether the species will be affected by noise [132]. We see this in the Australian example of one species, the grey shrike-thrush (*Colluricincla harmonica*), which is adversely impacted by traffic noise while its neighbor, the grey fantail (*Rhipidura albiscapa*) appears unaffected [133]. Species-specific signal traits combined with species-specific responses suggest that we may need to develop unique noise management for different species, despite consistent management goals. The goals are two-fold: to preserve high-quality, low-noise sound environments in such a way to allow successful communication as it relates to particular species and to implement mitigation strategies that benefit species for which noise interferes with communication. A species-specific approach may be informed by levels of noise exposure above which signals change or detection is diminished (i.e., threshold effects), followed by habitat conservation, management or exposure mitigation to keep noise below threshold levels. Although this may make intuitive sense, an important caveat is that limited testing of threshold effects exists, especially within the context of communication. Even when considering species, such as greater sage grouse (*Centrocercus urophasianus*), for which considerable research has been conducted to identify appropriate noise thresholds, researchers advocate for continued revision of standards as more information becomes available [130].

Although research supports the general conclusion of noise-induced changes in communication in many species, comparative evidence that identifies the specific noise levels that are problematic is unfortunately limited. Evidence exists for threshold noise levels associated with flushing distance, but researchers state that their recommendations for noise management are relevant only to the species of interest and caution against extrapolating to other species [124, 134]. Such cautions are necessary because noise is measured in different ways across studies and often without calibrations, such that reported noise levels are not comparable. If consistent approaches are used to measure noise, as McKenna et al. [122] advocate, then management plans developed for one species could be evaluated and implemented, if appropriate, across populations and species for greater impact. For details on recommended noise analysis, we refer readers to guidelines presented by McKenna et al. [122].

Ideally, species-level approaches require data about noise, communication, and fitness. As noted earlier, fitness impacts of noise on communication are poorly known. From a conservation perspective, efforts would be best placed in pursuing fitness data for threatened and endangered (hereafter listed) species (e.g., [130]), particularly if listed species avoid habitat with noise above threshold levels [63]. Umbrella species could be useful targets, and if they are more sensitive to noise than other species in the same habitat, management aimed at them (or listed species) protects many others from excessive noise exposure. Such species would additionally act as indicator species, whose presence conveys a healthy sound environment.

To frame our discussion of species-level approaches, we build upon Endler's [10] model of signaling, which includes many aspects of communication discussed earlier, from signal structure to receiver physiology. This model is useful as it provides behavioral targets for noise management, can be reframed as a hierarchy of decisions made by individuals, either signalers and receivers, and can also be extended to include other behaviors affected by noise that are relevant for communication (Fig. 4). For example, management and mitigation can be aimed at protecting habitat if birds avoid noisy areas that are poor quality in terms of signaling or at noise refuges within territories and home ranges if birds avoid noisy locations for certain behaviors (i.e., microhabitat choice). Thus, birds face a series of decisions about signaling in relation to noise, and management and mitigation may be aimed at each successive step in the decision-making pathway or just one decision level, if appropriate. To use such a strategy, managers need to know (a) threshold levels for habitat avoidance, (b) space use within territories by both signalers and receivers, and (c) signal adjustments and detection probabilities, and whether consistent differences exist for threshold levels associated with each decision. We predict that threshold levels for habitat avoidance would be higher than those influencing microhabitat selection and signaling interactions. Threshold levels influencing space use might differ across behaviors, such that noisier areas within territories may be used for particular functions (e.g., singing, if song adjustments occur) and quieter areas for other behaviors (e.g., foraging and nesting). To identify noise thresholds, noise playback experiments that present multiple noise levels [124, 135] should be conducted.

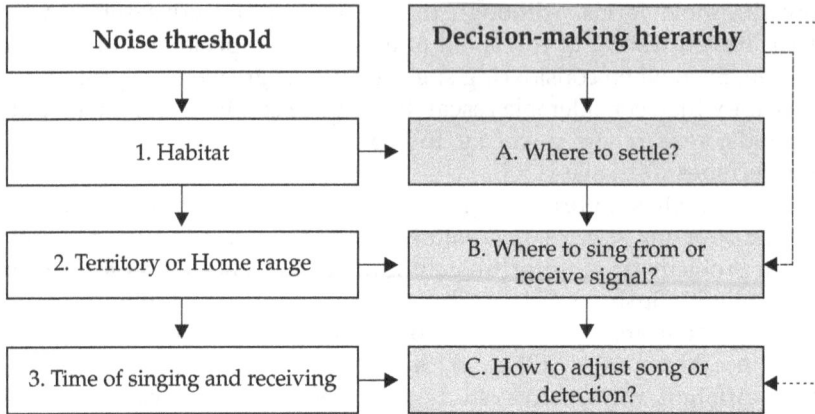

Figure 4 Decision-making framework outlining the influence of noise threshold used in decisions regarding habitat selection and communication. As with other environmental cues, noise varies over time and space (e.g., [61], [136]) and individuals that tune into this variation will benefit via increased efficacy with communication. We suggest that birds use information about noise thresholds at different spatial and temporal scales (1–3) in a series of hierarchical decisions (A-C) to ameliorate the effects of anthropogenic noise on communication. On return from migration or after dispersal (1), individuals use variation in noise thresholds at the scale of habitat to select low-noise habitat. Once on territories (2), individuals tune into within-territory variation to select low-noise areas from which to sing or receive signals. Finally, before signaling, males tune into moment-to-moment variation in the noise thresholds (3) to produce signals that optimize transmission under the condition detected (C). Under this framework, different species may be limited by thresholds at each level (A-C, solid arrows), or one or two levels (dashed lines illustrate two possible examples). Depending on these limitations, multiple management and mitigation strategies to conserve signal transmission would be possible.

Assuming that we know threshold levels for occupancy, next steps would be to understand spatial variation in noise at the level of territory or home range by developing noise maps [61] and assessing whether birds use space differently depending on noise levels. Our narrow concern here is for noise levels encountered when signaling or receiving; will singing birds avoid noisy places within their territories and is there a particular noise value above which birds will not use particular space? Noise maps can also be used in other ways, such as in conjunction with nest monitoring to determine if birds avoid placing nests in noisy parts of their territories. Variation in noise over space may result in noise refuges, which provide the opportunity for spatial release from masking and which could be targeted for heightened protection. Noise refuges are likely to be associated with habitat features (e.g., topography, vegetation structure) that minimize noise exposure. If such features can be identified, then the possibility exists to recreate those features within a management unit at an appropriate scale for the species in question (i.e., the size of the territory or home range) or identify similar refuges in other areas for conservation.

In this framework, the final step in a species-based approach considers the effects of noise on signaling and receiving. As many species adjust their signals in noise and altered signals are detectable and discriminated by signalers, no interventions may be needed. For species that are unable to adjust their signals, noise mitigation, using strategies described above or in the next section, are needed to reduce the risk of losing them from noisy environments. This is relevant for habitats that support numerous species with varying sensitivities to noise [132]. If threshold values are known for listed species, noise introduction and acoustic degradation could be made illegal. Data on noise levels and acoustic quality are needed to define thresholds at which acoustic degradation occurs, which would allow development of similar policy and legislation for listed species.

Area-based Approaches

A second approach to noise management and mitigation focuses on implementation at the scale of the landscape, particularly within protected areas. The goals of area-based management are the same as those for species-based approaches, without reference to species-specific needs. That is, area-based approaches aim to decrease noise exposure within protected spaces and to conserve quiet areas within protected areas or even the overall protected area itself; identifying quiet areas and keeping them quiet may be easier than removing noise sources or mitigating noise exposure. As an added benefit, area-based approaches could protect those spaces for human enjoyment of nature as well. Without reference to the requirements of a particular species, identification of target noise levels could be challenging. Area-based approaches require knowledge of undisturbed baseline noise levels to articulate noise management goals [130], which should address noise variation over time and space [34, 61], influences of stationary and moving sources on noise levels, and strategies to reduce noise exposure throughout the landscape when warranted or in portions of the landscape heavily affected by noise.

In practice, some area-based approaches may require potentially unrealistic large-scale changes in human use and activities to reduce noise exposure to wildlife. As a major source of continuous and intermittent anthropogenic noise, traffic noise presents a prime target for mitigation. Approaches for mitigation may be complex, as sources of traffic noise may originate within or outside protected areas [136, 137] and governance associated with protected areas may have various degrees of monitoring, mitigation, regulatory and enforcement capabilities. Within the context of large national parks, Buxton et al. [138] recommend decreasing traffic through habitat defined as critical as well as the use of noise absorbing asphalt for roads. Patricelli et al. [130] additionally suggest that siting of roads should be considered in management plans, factoring in the distance at which the decay of traffic noise falls below threshold response levels. In addition to vehicular traffic, low-altitude air traffic can also pose challenges for communication. The US National Park Service (NPS) worked with the Federal Aviation Administration to alter flight paths over particular national parks (e.g., Grand Canyon, Yellowstone). Rather than disrupt the most remote and quieter areas of parks, flights are routed to coincide with locations of busy roads [137].

Noise may also be generated by recreation within multi-use protected areas. Keyel et al. [129] used noise propagation models and mapping to inform the placement of trails and paths used for recreation (ATVs or snow mobiles) versus quiet, walking trails through natural areas. These models could also inform decisions regarding siting of potential sources of noise outside of protected areas, such as infrastructure for drilling. Processing noise propagation models may require advanced technical or analytical abilities, and development of online interfaces that allow stakeholders to input data from their management units (e.g., dimensions of protected area, topographic maps, distance to potential noise sources) could benefit many by allowing diverse organizations to develop noise maps for their protected areas. We recommend noise modeling and mapping to inform management decisions and planning, either through collaboration or by the land managers themselves.

Some of our suggestions may require action beyond the scope of many conservation organizations. Changing road surfaces and re-routing flight paths, for example, seem unrealistic for the majority of protected areas, many of which are small and occur within human land-use matrices (e.g., [139]). For regional protected areas as well as privately owned and managed lands (e.g., land conservancies and conservation easements), alternative strategies that can be implemented at the conservation sites appear more feasible (although not without costs). For example, an approach for protected areas situated near major highways or railways could include planting of high-density native vegetation buffers, as has been recommended for cities along busy roadways (e.g., [140]), or construction of natural berms or raised embankments parallel to roads (reviewed in [141]). High traffic areas, such as parking lots, could be targeted for remediation as well. Public outreach may be effective for changing human behavior, with signage to inform visitors about the value of quiet, to reduce speeds through sensitive areas, and to mark protected quiet zones.

Underlying noise mitigation is the idea of a noise reduction target, which may be based on a level of noise that reflects ambient or natural levels combined with threshold response levels. Baseline ambient levels reflect sound levels of natural area in the absence of human activities, but also animal sounds, and local wind and rain. Such values have not been identified and until data are widely available on threshold effects, Patricelli et al. [130] recommends managers use 16–20 dBA (SPL re 20 uPA) as baseline ambient noise, which roughly corresponds to the L_{90} given in an 1971 EPA report (16–22 dBA) as well as the 24 h L_{90} (21.6 dBA), limited to frequencies between 12.5–800 Hz, measured within US National Parks ([137]; recall that L_{90} means the sound pressure level exceeded 90% of the time). With ambient levels identified, the next step investigates the extent to which noise may exceed ambient levels without harming animal communication, increasing stress, or affecting fitness. Determining ambient baseline and threshold levels of a site is challenging and adherence to noise measurement recommendations is needed to be confident in those values (see [122]).

We recommend that acoustic attributes be incorporated into habitat assessments, building a knowledge base regarding noise levels across diverse types of protected areas. Protected areas home to listed species could experience acoustic degradation, yet without baseline data, noise changes would be impossible to document. Of course, without historical records, the data are not necessarily reflective of ambient SPLs (see discussion in [130]), but would allow for monitoring of acoustic habitat quality over time and documentation of degradation of critical acoustic habitat for listed species. With noise data at hand, large-scale analyses of drivers (such as distance from possible noise sources and patterns of surrounding land use) of noise variation could be used to predict areas with high noise levels. Data from US National Parks [138] provide an excellent starting point, but data are needed for smaller protected areas as well. Small protected areas often exist within human-land use matrices and thus they form an important source of protected lands [139].

Unanswered Questions

Anthropogenic noise disrupts communication in diverse ways across bird species. Fitness impacts of noise-induced signal changes are poorly understood, yet such information is critical justification for informing species-level noise management and mitigation.

Noise-induced signal changes, on their own, may not affect fitness, but signal changes could have additive effects with other noise impacts on avian populations (e.g., [131, 142], Beckman et al. Chapter 6). Future studies that combine experimental manipulation of intermittent and continuous noise with measurement of multiple fitness proxies, including pairing success, hatching and fledging success, and stress, are needed. Short-term noise experiments have been instrumental in demonstrating vocal flexibility, but future experiments might also include longer term noise exposures, either over the breeding season or perhaps at particular life history stages, to establish more clearly the link between noise exposure, vocal changes, and fitness.

Conclusions

Across many regions of the world bird species are in decline (e.g., [143, 144]) and all, to varying extents, rely on vocalizations to attract mates, communicate with young, and avoid predators. We know that human activities are the main driver of these declines. However, the role of changed habitat acoustics, and especially noise, is not fully known. Research has shown that anthropogenic noise is likely to act as an additional and widespread stressor on breeding birds, contributing to the declining health of individuals and populations, and even acting as a potential filter for species occupying certain areas. While we have observed many species that adjust their communication behaviors and patterns to increased noise levels, many others appear unable to do so. We are therefore still trying to understand exactly how noise may impact vulnerable populations and species, while simultaneously lacking key data to more fully inform noise management and mitigation for species- and area-based conservation.

New research and discoveries will help guide our understanding on how best to mitigate and manage noise and its effects on songbirds. In the meantime, whether a species-based or area-based approach best suits any given natural area will depend on the natural area itself. Some considerations during the process of deciding between species- and area-based tactics should include: (1) the resources available to the land manager (i.e., is funding and time allocated to conservation?), (2) species composition of natural area (are there birds of conservation concern?), and (3) feasibility of management and mitigation at the landscape level (i.e., does the nature area exist in the middle of an urban center or on the outskirts of a suburban town?). In closing, songbird communication is vital for survival and persistence. However, careful consideration of how human-generated environment change (noise and other factors) affects communication and other behaviors critical to survival must be addressed to make successful (meaningful) conservation gains.

▓ LITERATURE CITED

[1]   Brumm, H. and S.A. Zollinger. 2013. Avian vocal production in noise. pp. 187–227. *In*: H. Brumm (ed.). Animal Communication and Noise. Springer, Berlin, Heidelberg.

[2]   Halfwerk, W., B. Lohr and H. Slabbekoorn. (2018) Impact of man-made sound on birds and their songs. pp. 209–242. *In*: H. Slabbekoorn, R. Dooling, A. Popper and R. Fay. (eds). Effects of Anthropogenic Noise on Animals. Springer Handbook of Auditory Research, vol 66. Springer, New York, NY. https://doi.org/10.1007/978-1-4939-8574-6_8

[3]   Halfwerk, W., S. Bot, J. Buikx, M. van der Velde, J. Komdeur, C. ten Cate, et al. 2011. Low-frequency songs lose their potency in noisy urban conditions. Proc. Nat. Acad. Sci. 108: 14549–14554. doi:10.1073/pnas.1109091108

[4]   Phillips, J.N. and E.P. Derryberry. 2018. Urban sparrows respond to a sexually selected trait with increased aggression in noise. Sci. Rep. 8: 7505. doi:10.1038/s41598-018-25834-6

[5]   Grabarczyk, E.E. and S.A. Gill. 2019. Anthropogenic noise affects male house wren response to but not detection of territorial intruders. PLoS One. 14: e0220576. doi:10.1371/journal. pone.0220576

[6]   Grabarczyk, E.E., M. Araya-Salas, M.J. Vonhof and S.A. Gill. 2020. Anthropogenic noise affects female, not male house wren response to change in signaling network. Ethology. 126: 1069–1078. doi:10.1111/eth.13085

[7]   Rheindt, F.E. 2003. The impact of roads on birds: does song frequency play a role in determining susceptibility to noise pollution? Die Auswirkungen von Straßen auf Vögel: Ist Gesangsfrequenz ein Faktor für Lärmempfindlichkeit? J. Ornithol. 144: 295–306. doi:10.1046/j.1439-0361.2003.03004.x

[8]   Slabbekoorn, H. and M. Peet. 2003. Ecology: birds sing at a higher pitch in urban noise. Nature. 424: 267. doi:10.1038/424267a 424267a

[9]   Searcy, W.A. and S. Nowicki. 2005. The Evolution of Animal Communication. Princeton, USA, Princeton University Press.

[10]   Endler, J.A. 1992. Signals, signal conditions, and the direction of evolution. Am. Nat. 139: S125–S153.

[11]   Catchpole, C.K. and P.J.B. Slater. 2008. Bird Song, 2nd Ed. Cambridge, UK, Cambridge University Press.

[12]   Gill, S.A. and S.G. Sealy. 2004. Functional reference in an alarm signal given during nest defence: seet calls of yellow warblers denote brood-parasitic brown-headed cowbirds. Behav. Ecol. Sociobiol. 56: 71–80. doi:10.1007/s00265-003-0736-7

[13]   Leavesley, A.J. and R.D. Magrath. 2005. Communicating about danger: urgency alarm calling in a bird. Anim. Behav. 70: 365–373. doi:10.1016/j.anbehav.2004.10.017

[14]   Podos, J. 1997. A Performance constraint on the evolution of trilled vocalizations in a songbird family (Passeriformes: Emberizidae). Evolution. 51: 537–551. doi:10.1111/j.1558-5646.1997. tb02441.x

[15]   Henry, K.S., M.D. Gall, A.Vélez and J.R. Lucas. 2016. Avian auditory processing at four different scales: variation among species, seasons, sexes, and individuals. pp. 17–55. *In*: M.A. Bee and C.T. Miller (eds). Psychological Mechanisms in Animal Communication. Springer International Publishing. Cham. doi:10.1007/978-3-319-48690-1_2

[16]   Morton, E.S. 1975. Ecological sources of selection on avian sounds. Am. Nat. 109: 17.

[17]   Marten, K. and P. Marler. 1977. Sound transmission and its significance for animal vocalization. Behav. Ecol. Sociobiol. 2: 271–290. doi:10.1007/bf00299740

[18]   Wiley, R.H. and D. Richards. 1978. Physical constraints on acoustic communication in the atmosphere: implications for the evolution of animal vocalizations. Behav. Ecol. Sociobiol. 3: 69–94. doi:10.1007/BF00300047

[19]   Nicholls, J.A. and A.W. Goldizen. 2006. Habitat type and density influence vocal signal design in satin bowerbirds. J. Anim. Ecol. 75: 549–558. doi:10.1111/j.1365-2656.2006.01075.x

[20]   Naguib, M. and H. Wiley. 2001. Estimating the distance to a source of sound: mechanisms and adaptations for long-range communication. Anim. Behav. 62: 825–837. doi:10.1006/ anbe.2001.1860

[21]   Brittan-Powell, E.F. and R.J. Dooling. 2000. Development of auditory sensitivity in budgerigars. J. Acoust. Soc. Ame. 107: 2785–2785. doi:10.1121/1.428953

[22]   Dooling, R. and S. Blumenrath. 2013. Avian sound perception in noise. pp. 229–250. *In*: H. Brumm (ed.). Animal Communication and Noise. Springer, Berlin, Heidelberg.

[23] Lucas. J., A. Vélez and K. Henry. 2015. Habitat-related differences in auditory processing of complex tones and vocal signal properties in four songbirds. J. Comp. Phys. 201. doi:10.1007/s00359-015-0986-7

[24] Naguib, M., K. Van Oers, A. Braakhuis, M. Griffioen, P. De Goede and J.R. Waas. 2013. Noise annoys: effects of noise on breeding great tits depend on personality but not on noise characteristics. Anim. Behav. 85: 949–956. doi:10.1016/j.anbehav.2013.02.015

[25] Huber, S.K. and J. Podos. 2006. Beak morphology and song features covary in a population of Darwin's finches (*Geospiza fortis*). Biol. J. Linn. Soc. 88: 489–498. doi:10.1111/j.1095-8312.2006.00638.x

[26] Hendry, A.P., P.R. Grant, B.R. Grant, H.A. Ford, M.J. Brewer and J. Podos. 2006. Possible human impacts on adaptive radiation: beak size bimodality in Darwin's finches. Proc. R. Soc. B. 273: 1887–1894. doi:10.1098/rspb.2006.3534

[27] Podos, J. 2001. Correlated evolution of morphology and vocal signal structure in Darwin's finches. Nature. 409: 185–188.

[28] Gardner, J.L., M.R.E. Symonds, L. Joseph, K. Ikin, J. Stein and L.E.B. Kruuk. 2016. Spatial variation in avian bill size is associated with humidity in summer among Australian passerines. Clim Chang Responses. 3: 11. doi:10.1186/s40665-016-0026-z

[29] Tattersall, G.J., B. Arnaout and M.R.E. Symonds. 2017. The evolution of the avian bill as a thermoregulatory organ. Biol. Rev. 92: 1630–1656. https://doi.org/10.1111/brv.12299.

[30] Forseth, I.N. and A.F. Innis. 2004. Kudzu (*Pueraria montana*): history, physiology, and ecology combine to make a major ecosystem threat. Crit. Rev. Plant Sci. 23: 401–413. doi:10.1080/07352680490505150

[31] Warren, P.S., M. Katti, M. Ermann and A. Brazel. 2006. Urban bioacoustics: it's not just noise. Anim. Behav. 71: 491–502.

[32] Slabbekoorn, H., P. Yeh and K. Hunt. 2007. Sound transmission and song divergence: a comparison of urban and forest acoustics. Condor. 109: 67–78. doi:10.1650/0010-5422(2007)109[67:stasda]2.0.co;2

[33] Dowling, J.L., D.A. Luther and P.P. Marra. 2012. Comparative effects of urban development and anthropogenic noise on bird songs. Behav. Ecol. 23: 201–209. doi:10.1093/beheco/arr176

[34] Job, J.R., S.L. Kohler and S.A. Gill. 2016. Song adjustments by an open habitat bird to anthropogenic noise, urban structure, and vegetation. Behav. Ecol. 27(6): 1734–1744. doi:10.1093/beheco/arw105

[35] Potvin, D.A. and K.M. Parris. 2013. Song convergence in multiple urban populations of silvereyes (*Zosterops lateralis*). Ecol. Evol. 2: 1977–1984. doi:10.1002/ece3.320

[36] Boycott, T.J., J. Gao and M.D. Gall. 2019. Deer browsing alters sound propagation in temperate deciduous forests. PLoS One. 14: e0211569. doi:10.1371/journal.pone.0211569

[37] Shannon, G., M.F. McKenna, L.M. Angeloni, K.R. Crooks, K.M. Fristrup, E. Brown, et al. A synthesis of two decades of research documenting the effects of noise on wildlife. Biol. Rev. 91: 982–1005. doi:10.1111/brv.12207

[38] Magrath, R.D., B.J. Pitcher and A.H. Dalziell. 2007. How to be fed but not eaten: nestling responses to parental food calls and the sound of a predator's footsteps. Anim. Behav. 74: 1117–1129.

[39] Parris, K.M. and M.A. McCarthy. 2013. Predicting the effect of urban noise on the active space of avian vocal signals. Am. Nat. 182: 452–464. doi:10.1086/671906

[40] Leonard, M.L., A.G. Horn, K.N. Oswald and E. McIntyre. 2015. Effect of ambient noise on parent–offspring interactions in tree swallows. Anim. Behav. 109: 1–7. doi:10.1016/j.anbehav.2015.07.036

[41] McIntyre, E., M.L. Leonard and A.G. Horn. 2014. Ambient noise and parental communication of predation risk in tree swallows, *Tachycineta bicolor*. Anim. Behav. 87: 85–89. doi:10.1016/j.anbehav.2013.10.013

[42] Templeton, C.N., S.A. Zollinger and H. Brumm. 2016. Traffic noise drowns out great tit alarm calls. Curr. Biol. 26: R1173–R1174. doi:10.1016/j.cub.2016.09.058

[43] Potvin, D.A., M.T. Curcio, J.P. Swaddle and S.A. MacDougall-Shackleton. 2016. Experimental exposure to urban and pink noise affects brain development and song learning in Zebra finches. *Taenopygia guttata*. PeerJ. 4: e2287.

[44] Peters, S., E.P. Derryberry and S. Nowicki. 2012. Songbirds learn songs least degraded by environmental transmission. Biol. Lett. 8: 736–739. doi:10.1098/rsbl.2012.0446

[45] Rosa, P. and N. Koper. 2018. Integrating multiple disciplines to understand effects of anthropogenic noise on animal communication. Ecosphere. 9: e02127. doi:10.1002/ecs2.2127

[46] Chan, A.A.Y.-H., P. Giraldo-Perez, S. Smith and D.T. Blumstein. 2010. Anthropogenic noise affects risk assessment and attention: the distracted prey hypothesis. Biol. Lett. 6: 458–461. doi:10.1098/rsbl.2009.1081

[47] Grade, A.M. and K.E. Sieving. 2016. When the birds go unheard: highway noise disrupts information transfer between bird species. Biol. Lett. 12.

[48] Dominoni, D.M.,. 2020. Why conservation biology can benefit from sensory ecology. Ecol. Evol. 4: 502–511. doi:10.1038/s41559-020-1135-4

[49] Morris-Drake, A., J.M. Kern and A.N. Radford. 2016. Cross-modal impacts of anthropogenic noise on information use. Curr. Biol. CB26: R911–R912. doi:10.1016/j.cub.2016.08.064

[50] Blickley, J.L., K.R. Word, A.H. Krakauer, J.L. Phillips, S.N. Sells, C.C. Taff, et al. 2012. Experimental chronic noise is related to elevated fecal corticosteroid metabolites in lekking male greater sage-grouse. *Centrocercus urophasianus*. PLoS One. 7: e50462. doi:10.1371/journal.pone.0050462

[51] Nowicki, S., W. Searcy and S. Peters. 2002. Brain development, song learning and mate choice in birds: a review and experimental test of the 'nutritional stress hypothesis'. J. Comp. Physiol. A. 188: 1003–1014. doi:10.1007/s00359-002-0361-3

[52] Siemers, B.M. and A. Schaub. 2011. Hunting at the highway: traffic noise reduces foraging efficiency in acoustic predators. Proc. R. Soc. Lond. B. 278: 1646–1652. doi:10.1098/rspb.2010.2262

[53] Cardoso, G.C., Y. Hu and C.D. Francis. 2018. The comparative evidence for urban species sorting by anthropogenic noise. R. Soc. Open Sci. 5: 172059. doi:10.1098/rsos.172059

[54] Cardoso, G.C. 2020. Exposure to noise pollution across North American passerines supports the noise filter hypothesis. Glob. Ecol. Biogeogr. doi/full/10.1111/geb.13085.

[55] Potvin, D.A. 2016. Coping with a changing soundscape: avoidance, adjustments and adaptations. Anim. Cognit. 20: 1–10. doi:10.1007/s10071-016-0999-9

[56] McClure, C.J., H.E. Ware, J. Carlisle, G. Kaltenecker and J.R. Barber. 2013. An experimental investigation into the effects of traffic noise on distributions of birds: avoiding the phantom road. Proc. R. Soc. Lond. B. 280: 20132290.

[57] Kleist, N.J., R.P. Guralnick, A. Cruz and C.D. Francis. 2017. Sound settlement: noise surpasses land cover in explaining breeding habitat selection of secondary cavity-nesting birds. Ecol. Appl. 27: 260–273. doi:10.1002/eap.1437

[58] Mejia, E.C., C.J.W. McClure and J.R. Barber. 2019. Large-scale manipulation of the acoustic environment can alter the abundance of breeding birds: evidence from a phantom natural gas field. J. Appl. Ecol. 56: 2091–2101. doi:10.1111/1365-2664.13449

[59] Vitousek, P.M., C.M. D'antonio, L.L. Loope, M. Rejmánek and R. Westbrooks. 1997. Introduced species: a significant component of human-caused global change. N. Z. J. Ecol. 21: 1–16.

[60] Ibisch, P.L. 2016. A global map of roadless areas and their conservation status. Science. 354: 1423–1427. doi:10.1126/science.aaf7166

[61] Job, J.R., K. Myers, K. Naghshineh and S.A. Gill. 2016. Uncovering spatial variation in acoustic environments using sound mapping. PLoS One. 11: e0159883. doi:10.1371/journal.pone.0159883

[62] McLaughlin, K.E. and H.P. Kunc. 2013. Experimentally increased noise levels change spatial and singing behaviour. Biol. Lett. 9. doi:10.1098/rsbl.2012.0771

[63] Blickley, J.L., D. Blackwood and G.L. Patricelli. 2012. Experimental evidence for the effects of chronic anthropogenic noise on abundance of greater sage-grouse at leks. Conserv. Biol. 26: 461–471. doi:10.1111/j.1523-1739.2012.01840.x

[64] Brumm, H., K.A. Robertson and E. Nemeth. 2011. Singing direction as a tool to investigate the function of birdsong: an experiment on sedge warblers. Anim. Behav. 81: 653–659.

[65] Halfwerk, W., S. Bot and H. Slabbekoorn. 2012. Male great tit song perch selection in response to noise-dependent female feedback. Funct. Ecol. 26: 1339–1347. doi:10.1111/j.1365-2435.2012.02018.x

[66] Montague, M.J., M. Danek-Gontard and H.P. Kunc. 2013. Phenotypic plasticity affects the response of a sexually selected trait to anthropogenic noise. Behav. Ecol. 24: 343–348. doi:10.1093/beheco/ars169

[67] Potvin, D.A. and R.A. Mulder. 2013. Immediate, independent adjustment of call pitch and amplitude in response to varying background noise by silvereyes (*Zosterops lateralis*). Behav. Ecol. 24: 1363–1368. doi:10.1093/beheco/art075

[68] Grabarczyk, E.E., M.J. Vonhof and S.A. Gill. 2020. Social context and noise affect within and between male song adjustments in a common passerine. Behav. Ecol. 31(5): 1150–1158. doi:10.1093/beheco/araa066

[69] Lombard, E. 1911, Le signe de le elevation de la voix. Ann. Malad. l'Oreille, Larynx, Nez, Pharynx. 37: 101–119.

[70] Brumm, H. 2004, The impact of environmental noise on song amplitude in a territorial bird. J. Anim. Ecol. 73: 434–440.

[71] Zollinger, S.A., H. Brumm. 2011. The lombard effect. Curr. Biol. 21: R614–R615.

[72] Hardman, S.I., S.A. Zollinger, K. Koselj, S. Leitner, R.C. Marshall and H. Brumm. 2017. Lombard effect onset times reveal the speed of vocal plasticity in a songbird. J. Exp. Biol. 220: 1065–1071. doi:10.1242/jeb.148734

[73] Dooling, R.J. and A.N. Popper. 2016. Some lessons from the effects of highway noise on birds. Proc. Mtgs. Acoust. 27: 010004. doi:10.1121/2.0000244

[74] Zollinger, S.A., P.J.B. Slater, E. Nemeth and H. Brumm. 2017. Higher songs of city birds may not be an individual response to noise. Proc. R. Soc. B: Biol. Sci. 284.

[75] Francis, C.D., C.P. Ortega and A. Cruz. 2011. Different behavioural responses to anthropogenic noise by two closely related passerine birds. Biol. Lett. 7: 850–852. doi:10.1098/rsbl.2011.0359

[76] Warrington, M.H., C.M. Curry, B. Antze and N. Koper. 2018. Noise from four types of extractive energy infrastructure affects song features of Savannah Sparrows. Condor. 120: 1–15. doi:10.1650/CONDOR-17-69.1

[77] Liu, Q., H. Slabbekoorn and K. Riebel. 2020. Zebra finches show spatial avoidance of near but not far distance traffic noise. Behaviour. 157: 333–362. doi:10.1163/1568539X-bja10004

[78] Mahjoub, G., M.K. Hinders and J.P. Swaddle. 2015. Using a "sonic net" to deter pest bird species: Excluding European starlings from food sources by disrupting their acoustic communication. Wildl. Soc. Bull. 39: 326–333. doi:10.1002/wsb.529

[79] Vargas-Salinas, F., G.M. Cunnington, A. Amézquita and L. Fahrig. 2014. Does traffic noise alter calling time in Frogs and Toads? A case study of anurans in Eastern Ontario, Canada. Urban Ecosyst. 17: 945–953.

[80] Proppe, D.S. and E. Finch. 2017. Vocalizing during gaps in anthropogenic noise is an uncommon trait for enhancing communication in songbirds. J. Ecoacoust. 1: TLP16D. doi:10.22261/JEA.TLP16D

[81] Arroyo-Solís, A., J.M. Castillo, E. Figueroa, J.L. López-Sánchez and H. Slabbekoorn. 2013. Experimental evidence for an impact of anthropogenic noise on dawn chorus timing in urban birds. J. Avian Biol. 44: 288–296. doi:10.1111/j.1600-048X.2012.05796.x

[82] Yang, X.J. and H. Slabbekoorn. 2014. Timing vocal behavior: Lack of temporal overlap avoidance to fluctuating noise levels in singing Eurasian wrens. Behav. Process. 108: 131–137. doi:10.1016/j.beproc.2014.10.002

[83] Fuller, R.A., P.H. Warren and K.J. Gaston. 2007. Daytime noise predicts nocturnal singing in urban robins. Biol. Lett. 3: 368–370. doi:10.1098/rsbl.2007.0134

[84] Da Silva, A., M. Valcu and B. Kempenaers. 2015. Light pollution alters the phenology of dawn and dusk singing in common European songbirds. Philos. Trans. R. Soc. B: Biol. Sci. 370: 20140126. doi:10.1098/rstb.2014.0126

[85] Hennigar, B., J.P. Ethier and D.R. Wilson. 2019. Experimental traffic noise attracts birds during the breeding season. Behav. Ecol. 30: 1591–1601. doi:10.1093/beheco/arz123

[86] Stuart, C.J., E.E. Grabarczyk, M.J. Vonhof and S.A. Gill. 2019. Social factors, not anthropogenic noise or artificial light, influence onset of dawn singing in a common songbird. Auk. 136. doi:10.1093/auk/ukz045

[87] Pohl, N.U., E. Leadbeater, H. Slabbekoorn, G.M. Klump and U. Langemann. 2012. Great tits in urban noise benefit from high frequencies in song detection and discrimination. Anim. Behav. 83: 711–721.

[88] Polak, M. 2014. Relationship between traffic noise levels and song perch height in a common passerine bird. Transp. Res. D: Transp. Environ. 30: 72–75. doi:10.1016/j.trd.2014.05.004

[89] Villain, A.S., M.S.A. Fernandez, C. Bouchut, H. Soula and C. Vignal. 2016. Songbird mates change their call structure and intra-pair communication at the nest in response to environmental noise. Anim. Behav. 116: 113–129. doi:10.1016/j.anbehav.2016.03.009

[90] Dent, M.L., E.M. McClaine, V. Best, E. Ozmeral, R. Narayan, F.J. Gallun, et al. 2009. Spatial unmasking of birdsong in zebra finches (*Taeniopygia guttata*) and Budgerigars (*Melopsittacus undulatus*). J. Comp. Psychol. 123: 357–367. doi:10.1037/a0016898

[91] Klump, G.M. 2000. Sound localization in birds. pp. 249–307. *In*: R.J. Dooling, R.R. Fay and A.N. Popper (eds). Comparative Hearing: Birds and Reptiles. Springer, New York, NY. doi:10.1007/978-1-4612-1182-2_6

[92] Luther, D.A. and E.P. Derryberry. 2012 Birdsongs keep pace with city life: changes in song over time in an urban songbird affects communication. Anim. Behav. 83: 1059–1066.

[93] Read, J., G. Jones and A.N. Radford. 2014. Fitness costs as well as benefits are important when considering responses to anthropogenic noise. Behav. Ecol. 25: 4–7. doi:10.1093/beheco/art102

[94] Slabbekoorn, H. and T.B. Smith. 2002. Bird song, ecology and speciation. Philos. Trans. R. Soc. Lond. B: Biol. Sci. 357: 493–503. doi:10.1098/rstb.2001.1056

[95] Gill, S.A., J.R. Job, K. Myers, K. Naghshineh and M.J. Vonhof. 2015. Toward a broader characterization of anthropogenic noise and its effects on wildlife. Behav. Ecol. 26: 328–333. doi:10.1093/beheco/aru219

[96] Habib, L., E.M. Bayne and S. Boutin. 2007. Chronic industrial noise affects pairing success and age structure of ovenbirds *Seiurus aurocapilla*. J. Appl. Ecol. 44: 176–184. doi:10.1111/j.1365-2664.2006.01234.x

[97] Forman, R.T., B. Reineking and A.M. Hersperger. 2002. Road traffic and nearby grassland bird patterns in a suburbanizing landscape. Environ. Manage. 29: 782–800. doi:10.1007/s00267-001-0065-4

[98] Francis, C.D. and C.P. Ortega, Cruz A. 2009. Noise pollution changes avian communities and species interactions. Curr. Biol. 19: 1415–1419. doi:10.1016/j.cub.2009.06.052

[99] Dale, J., C.J. Dey, K. Delhey, B. Kempenaers and M. Valcu. 2015. The effects of life history and sexual selection on male and female plumage colouration. Nature. 527: 367–370. doi:10.1038/nature15509

[100] Hill, G.E. and R. Montgomerie. 1994. Plumage colour signals nutritional condition in the house finch. Proc. R. Soc. Lond. B: Biol. Sci. 258: 47–52. doi:10.1098/rspb.1994.0140

[101] Bókony, V., L.Z. Garamszegi, K. Hirschenhauser and A. Liker. 2008. Testosterone and melanin-based black plumage coloration: a comparative study. Behav. Ecol. Sociobiol. 62: 1229. doi:10.1007/s00265-008-0551-2

[102] Grindstaff, J.L., M.B. Lovern, J.L. Burtka and A. Hallmark-Sharber. 2012. Structural coloration signals condition, parental investment, and circulating hormone levels in Eastern bluebirds (*Sialia sialis*). J. Comp. Physiol. A. 198(8): 625–637.

[103] Hill, G.E. 2000. Energetic constraints on expression of carotenoid-based plumage coloration. J. Avian Biol. 31: 559–566. doi:10.1034/j.1600-048X.2000.310415.x

[104] Geens, A., T. Dauwe and M. Eens. 2009. Does anthropogenic metal pollution affect carotenoid colouration, antioxidative capacity and physiological condition of great tits (*Parus major*)? Comp. Biochem. Physiol. C: Toxicol. Pharmacol. 150: 155–163. doi:10.1016/j.cbpc.2009.04.007

[105] Krause, E.T., O. Krüger, P. Kohlmeier and B.A. Caspers. 2012. Olfactory kin recognition in a songbird. Biol. Lett. 8: 327–329. doi:10.1098/rsbl.2011.1093

[106] Leclaire, S., M. Strandh, J. Mardon, H. Westerdahl and F. Bonadonna. 2017. Odour-based discrimination of similarity at the major histocompatibility complex in birds. Proc. R. Soc. B: Biol. Sci. 284: 20162466. doi:10.1098/rspb.2016.2466

[107] Whittaker, D.J., H.A. Soini, J.W. Atwell, C. Hollars, M.V. Novotny and E.D. Ketterson. 2010. Songbird chemosignals: volatile compounds in preen gland secretions vary among individuals, sexes, and populations. Behav. Ecol. 21: 608–614. doi:10.1093/beheco/arq033

[108] Corfield, J.R., K. Price, A.N. Iwaniuk, C. Gutierrez-Ibañez, T. Birkhead and D.R. Wylie. 2015. Diversity in olfactory bulb size in birds reflects allometry, ecology, and phylogeny. Front. Neuroanat. 9. doi:10.3389/fnana.2015.00102

[109] McFrederick, Q.S., J.D. Fuentes, T. Roulston, J.C. Kathilankal and M. Lerdau. 2009. Effects of air pollution on biogenic volatiles and ecological interactions. Oecologia. 160: 411–420. doi:10.1007/s00442-009-1318-9

[110] Feng, A.S., P.M. Narins, C.-H. Xu, W.-Y. Lin, Z.-L. Yu, Q. Qiu, et al. 2006. Ultrasonic communication in frogs. Nature. 440: 333–336. doi:10.1038/nature04416

[111] Davidson, B.M., G. Antonova, H. Dlott, J.R. Barber and C.D. Francis. 2017. Natural and anthropogenic sounds reduce song performance: insights from two emberizid species. Behav. Ecol. 28: 974–982. doi:10.1093/beheco/arx036

[112] Wong, B.B.M. and U. Candolin. 2015. Behavioral responses to changing environments. Behav. Ecol. 26: 665–673. doi:10.1093/beheco/aru183

[113] Mulholland, T.I., D.M. Ferraro, K.C. Boland, K.N. Ivey, M.L. Le, C.A. LaRiccia, et al. 2018. Effects of experimental anthropogenic noise exposure on the reproductive success of secondary cavity nesting birds. Integr. Comp. Biol. 58: 967–976. doi:10.1093/icb/icy079

[114] McKinney, M.L. 2006. Urbanization as a major cause of biotic homogenization. Biol. Conserv. 127: 247–260.

[115] Proppe, D.S., C.B. Sturdy and C.C. St. Clair. 2013. Anthropogenic noise decreases urban songbird diversity and may contribute to homogenization. Global Change Ecol. 19: 1075–1084.

[116] Díaz, S., U. Pascual, M. Stenseke, B. Martín-López, R.T. Watson, Z. Molnár, et al. 2018. Assessing nature's contributions to people. Science. 359: 270–272. doi:10.1126/science.aap8826.

[117] Convention on Biological Diversity (CBD). 2020. Strategic Plan for Biodiversity 2011–2020, including Aichi Biodiversity Targets. See https://www.cbd.int/sp/ (accessed on 7 October 2020).

[118] Jones, K.R., O. Venter, R.A. Fuller, J.R. Allan, S.L. Maxwell, P.J. Negret, et al. 2018. One-third of global protected land is under intense human pressure. Science. 360: 788–791. doi:10.1126/science.aap9565

[119] Coad, L., J.E. Watson, J. Geldmann, N.D. Burgess, F. Leverington, M. Hockings, K. Knights and M.D. Marco. 2019. Widespread shortfalls in protected area resourcing undermine efforts to conserve biodiversity. Front. Ecol. Environ. 17: 259–264. doi:10.1002/fee.2042

[120] Gill, S.A., E.E. Grabarczyk, K.M. Baker, K. Naghshineh and M.J. Vonhof. 2017. Decomposing an urban soundscape to reveal patterns and drivers of variation in anthropogenic noise. Sci. Total Environ. 599: 1191–1201.

[121] Merchant, N.D., K.M. Fristrup, M.P. Johnson, P.L. Tyack, M.J. Witt, P. Blondel, et al. 2015. Measuring acoustic habitats. Methods Ecol. Evol. 6: 257–265. doi:10.1111/2041-210X.12330

[122] McKenna, M.F., G. Shannon and K. Fristrup. 2016. Characterizing anthropogenic noise to improve understanding and management of impacts to wildlife. Endanger. Species Res. 31: 279–291. doi:10.3354/esr00760

[123] Williams, R. 2015. Impacts of anthropogenic noise on marine life: Publication patterns, new discoveries and future directions in research and management. Ocean Coastal Manage. 115: 17–24. doi:10.1016/j.ocecoaman.2015.05.021

[124] Karp, D.S. and T.L. Root. 2009. Sound the stressor: how Hoatzins (*Opisthocomus hoazin*) react to ecotourist conversation. Biodivers Conserv. 18: 3733. doi:10.1007/s10531-009-9675-6

[125] Delaney, D.K., T.G. Grubb, P. Beier, L.L. Pater and M.H. Reiser. 1999. Effects of helicopter noise on Mexican spotted owls. J. Wildl. Manage. 63: 60–76. doi:10.2307/3802487

[126] Tyack, P.L. 2009. Human-generated sound and marine mammals. Physics Today. 62: 39–44. doi:10.1063/1.3265235

[127] Patricelli, G.L. and J.L. Blickley. 2006. Avian communication in urban noise: causes and consequences of vocal adjustment. Auk. 123: 639–649. doi:10.1642/0004-8038(2006)123[639:aci unc]2.0.co;2

[128] Buxton, R.T., M.F. McKenna, D. Mennitt, K. Fristrup, K. Crooks, L. Angeloni, et al. 2017. Noise pollution is pervasive in U.S. protected areas. Science. 356: 531–533.

[129] Keyel, A.C., S.E. Reed, K. Nuessly, E. Cinto-Mejia, J.R. Barber and G. Wittemyer. 2018. Modeling anthropogenic noise impacts on animals in natural areas. Landscape Urban Plann. 180: 76–84.

[130] Patricelli, G., J. Blickley and S. Hooper. 2013. Recommended management strategies to limit anthropogenic noise impacts on greater sage-grouse in Wyoming. Hum.-Wildl. Interact. 7(2): 230–249. doi:10.26077/7qfc-6d14

[131] Chan, A.A.Y.H. and D.T. Blumstein. 2011. Attention, noise, and implications for wildlife conservation and management. Appl. Anim. Behav. Sci. 131: 1–7. doi:10.1016/j.applanim. 2011.01.007

[132] Francis, C.D. 2015. Vocal traits and diet explain avian sensitivities to anthropogenic noise. Global Change Biol. 21: 1809–1820. doi:10.1111/gcb.12862

[133] Parris, K.M. and A. Schneider. 2009. Impacts of traffic noise and traffic volume on birds of roadside habitats. Ecol. Soc. 14: 29.

[134] Pater, L.L., T.G. Grubb and D.K. Delaney. 2009. Recommendations for improved assessment of noise impacts on wildlife. J. Wildl. Manage. 73: 788–795. doi:10.2193/2006-235

[135] Grubb, T.G., D.K. Delaney, W.W. Bowerman and M.R. Wierda. 2010. Golden eagle indifference to heli-skiing and military helicopters in northern utah. J. Wildl. Manage. 74: 1275–1285. doi:10.1111/j.1937-2817.2010.tb01248.x

[136] Barber, J.R., C.L. Burdett, S.E. Reed, K.A. Warner, C. Formichella, K.R. Crooks, D.M. Theobald and K.M. Fristrup. 2011. Anthropogenic noise exposure in protected natural areas: estimating the scale of ecological consequences. Landscape Ecol. 26: 1281. doi:10.1007/ s10980-011-9646-7

[137] Lynch, E., D. Joyce and K. Fristrup. 2011. An assessment of noise audibility and sound levels in U.S. National Parks. Landscape Ecol. 26: 1297. doi:10.1007/s10980-011-9643-x

[138] Buxton, R.T., M.F. McKenna, D. Mennitt, E. Brown, K. Fristrup, K.R. Crooks, L.M. Angeloni and G. Wittemyer. 2019. Anthropogenic noise in US national parks—sources and spatial extent. Front. Ecol. Environ. 17: 559–564. doi:10.1002/fee.2112

[139] Volenec, Z.M. and A.P. Dobson. 2020. Conservation value of small reserves. Conserv. Biol. 34: 66–79. doi:10.1111/cobi.13308

[140] Ow, L.F. and S. Ghosh. 2017. Urban cities and road traffic noise: Reduction through vegetation. Appl. Acoust. C: 15–20. doi:10.1016/j.apacoust.2017.01.007

[141] Van Renterghem, T., J. Forssén, K. Attenborough, P. Jean, J. Defrance, M. Hornikx, et al. 2015. Using natural means to reduce surface transport noise during propagation outdoors. Applied Acoustics. 92: 86–101. doi:10.1016/j.apacoust.2015.01.004

[142] Bayne, E.M., L. Habib and S. Boutin. 2008. Impacts of chronic anthropogenic noise from energy-sector activity on abundance of songbirds in the boreal forest. Conserv. Biol. 22: 1186–1193. doi:10.1111/j.1523-1739.2008.00973.x

[143] Inger, R., R. Gregory, J.P. Duffy, I. Stott, P. Voříšek and K.J. Gaston. 2015. Common european birds are declining rapidly while less abundant species' numbers are rising. Ecol. Lett. 18: 28–36. doi:10.1111/ele.12387

[144] Rosenberg, K.V. 2019. Decline of the North American avifauna. Science. 366: 120–124. doi:10.1126/science.aaw1313

Chapter 9

Vocal Learning and Neurobiology in the Anthropocene

Broderick M.B. Parks[1], Andrew G. Horn[2],
Scott A. MacDougall-Shackleton[3,4] and Leslie S. Phillmore[1]

Introduction

Increasing human population and shrinking natural habitat worldwide, largely due to urban expansion, is a conservation concern for many animal species, but songbirds seem particularly susceptible to these changes in the Anthropocene. In North America alone, bird populations have declined by 29%, or 3 billion individuals, since 1970, largely from anthropogenic causes, such as increased habitat loss and pesticide usage [1]. One reason

[1]Department of Psychology and Neuroscience, Dalhousie University, Halifax, NS, Canada
[2]Department of Biology, Dalhousie University, Halifax, NS, Canada
[3]Advanced Facility for Avian Research, University of Western Ontario, London, ON, Canada
[4]Department of Psychology, University of Western Ontario, London, ON, Canada

songbirds might be particularly at risk in urban and industrialized areas is because acoustic communication is especially critical to songbirds, from its early development to its adult use, and changing habitats compromise the acoustic properties of the natural environments to which bird communication is adapted. In addition, changes in the Anthropocene may induce stress that interferes with the ability of offspring to learn song from adults, and compromises the underlying neural architecture that supports song learning. Finally, toxins and chemicals introduced to habitats may interfere directly with the neural machinery underlying vocal learning and production.

In this chapter, after reviewing the ontogeny and neural basis of song learning, we describe evidence from both laboratory and field experiments that examine effects of anthropogenic changes, such as increased noise, stress from noise and food restriction, and environmental contaminants on vocal learning. We also consider why results from both field and laboratory studies on some, but not other stressors associated with the changing Anthropocene, may vary. Despite inconsistent evidence, habitat conservation efforts that allow songbirds to perform critical behaviours during development are required for their reproductive success and, ultimately, species survival.

Vocal Learning—Behaviour

Like humans and a few other taxa (e.g., elephants, Elephantidae and bats, Chiroptera), songbirds (Passeriformes) learn species-specific vocalizations first through hearing, and then imitating a model. In songbirds, this process of vocal learning is well illustrated by zebra finches (*Taeniopygia guttata*), a species studied extensively in the laboratory for vocal communication and song learning [2]. Native to the grasslands of Australia, these birds breed opportunistically when water and food are abundant [3]. Only male zebra finches learn to produce song, and the song development period and its stages are well-defined, occurring over a 90-day period from hatch to adulthood [4].

In the first, sensory phase, the juvenile bird listens to and memorizes the song of an adult tutor, and then in the second, sensorimotor phase, the juvenile reproduces these memorized vocalizations, relying on auditory feedback and modifying the motor output of its vocalizations to match its memorized model [4]. The sensorimotor phase terminates with the crystallization of song, in which song becomes stereotyped and less reliant on auditory feedback [4], although some auditory feedback is still required for adult song maintenance [5, 6]. Over song development, there is a sensitive phase for exposure to conspecific (own species) song, during which disruptions can result in aberrant song. For example, if zebra finches hear heterospecific (different species), but not conspecific, song, then in the absence of the "correct" model, they may incorporate characteristics of that heterospecific song [7]. This phenomenon is not restricted to zebra finches. For example, great tits (*Parus major*) and Eurasian blue tits (*Cyanistes caeruleus*) cross-fostered with adults of the opposite species during song development show varying degrees of adaptation of heterospecific song features [8]. If the sensitive period is disrupted with respect to which songs are heard or when they are heard, then the sensitive period may be extended beyond typical timelines, as if the bird is waiting for conspecific input to develop species-typical song [7]. Some experimental evidence suggests that playback of noise during song memorization interferes with song development (e.g., [9]); however, depending on when and how long noise exposure occurs, the disruption may resolve in adulthood (e.g., [10]).

The sensory acquisition phase in zebra finches occurs between day 15 and day 60, and the sensorimotor phase occurs between day 30 and day 90 post-hatch [4]. Thus, juvenile finches can memorize tutor song and practice their own song simultaneously. Song crystallization occurs at the end of the sensorimotor phase, at approximately 90 days post-hatch. However, stages of song learning and time to crystallization varies across species, often in apparently adaptive ways. For example, many songbird species that breed seasonally in the temperate zone, including white-crowned sparrows (*Zonotrichia leucophrys*) and black-capped chickadees (*Poecile atricapillus*) follow a more protracted timeline for song learning. The sensory acquisition phase tends to occur after hatching but before fledging, and the sensorimotor phase after fledging, with no overlap between the two. If the species is migratory, vocal practice occurs after migration, on non-breeding grounds [4]. Song consolidation, and ultimately song crystallization, occurs the following spring (after species return to their breeding grounds, in migratory species), approximately 1 year post-hatch [4, 11].

A further behavioural variation among songbirds is whether adults can continue to add to, or change, their vocal repertoire after initial crystallization. These species are known as open-ended learners. For example, canaries (*Serinus canaris*) experience cycles of song destabilization when not breeding and re-crystallization when breeding begins again, changing their repertoire each time [12]. Other species, such as European starlings (*Sturnus vulgaris*), add new elements (e.g., syllables and motifs) to their repertoire throughout their entire lives [13] (reviewed in [14]), such that older starlings have a much larger and complex repertoire than younger starlings [13, 15]. Most songbird species studied to date, however, are closed-ended learners like zebra finches, in which song learning is restricted to the sensitive period early in life [16]. However, this does not mean closed-ended learners cannot adjust their crystallized song in response to changes in the acoustic environment: adult Bengalese finches (*Lonchura striata* var. *domesticata*), for example, change the pitch of an individual note in response to white noise delivered when they sing that note [17].

Vocal Learning—Neurobiology

Singing-related behaviour—whether production, perception or learning—is supported by specialized and distinct brain regions, or nuclei, allowing songbirds to respond and adapt to changes in the acoustic environment. Specifically, songbirds have regions critical for song production and learning that non-songbirds, such as pigeons (Columbidae) or chickens (*Gallus gallus domesticus*), do not, mirroring the lack of song learning in non-songbirds [18, 19]. Within songbird species there are sex differences in the volumes and function of nuclei associated with song learning and production that parallel behavioural differences in singing behaviour between males and females (reviewed in [20]).

The nuclei of the vocal-control system make up three interconnected pathways, each with a primary, but not singular, function (Fig. 1). First, the ascending auditory pathway, underlying song perception in both males and females, begins at the level of the hair cells of the avian inner ear, where transduced neural signals travel via the auditory nerve, and after synapsing in several small nuclei of the hindbrain, midbrain, and thalamus (including the nucleus ovoidalis, Ov) terminates in the pallium in Field L, a region analogous to the mammalian primary auditory cortex [21–23]. Field L sends projections to secondary auditory regions, including the caudomedial nidopallium (NCM) and caudomedial mesopallium (CMM), areas crucial for processing auditory information and forming song-related memories [24]. NCM and CMM also project to area HVC (letter-based proper name), a well-studied nucleus implicated in all three pathways, but primarily involved in song production (see below). Field L, NCM, and CMM all show increased neuronal activation when a bird hears song compared to silence [23, 25] and different types of stimuli induce varying levels of activation (e.g., [26, 27]).

The descending motor pathway (also known as the vocal motor pathway) begins in HVC and connects directly to the robust nucleus of the arcopallium (RA), which in turn connects to several midbrain nuclei which directly innervate respiratory motor neurons (e.g., nucleus retroambigualis, RAm; nucleus parambigualis, PAm) and muscles in the trachea and syrinx (e.g., tracheosyringeal part of the hypoglossal nucleus, nXIIts), both required for vocal production [23]. While HVC is connected to nuclei in all three pathways, its primary function relates to vocal production. In the seminal study by Nottebohm et al. [28], adult canaries became permanently mute following bilateral lesions to HVC, although these lesioned birds continued to assume the appropriate posture and movements associated with song production. Further, neurons in HVC that project to RA fire synchronously with song production [29]. HVC displays pronounced seasonal plasticity not only in its size, but also in the number of new neurons incorporated in this area from the subventricular zone [30]. There is also a varying degree of sex difference in both the size and amount of neuronal turnover in HVC, which is linked to the magnitude of difference in singing behaviour (reviewed in [20]).

Finally, the anterior forebrain pathway (AFP) is an indirect cortical-basal ganglia-thalamic loop [23] which connects HVC to RA through Area X (a key component of the avian basal ganglia [31]), the thalamic nucleus dorsolateralis anterior, pars medialis (DLM) and the lateral magnocellular nucleus of the anterior nidopallium (LMAN). The AFP, while not essential for previously crystallized song, is critical for vocal plasticity in both juveniles and adults (reviewed in [32]). Both Area X and LMAN are active during periods of vocal production [23, 33, 34]. In juveniles, lesions to Area X lead to an inability to produce crystallized song, while lesions to LMAN produce aberrant, but stable, song [34]. In adults, lesions to Area X destabilize previously crystalized song [35]. Area X, similar to HVC, displays seasonal plasticity in both size and neuronal density (reviewed in [30, 36]).

A diverse array of neurotransmitters and neuromodulators are required for efficient and effective neural transmission within the vocal-control system, and disturbance at the neurochemical level can drastically alter the songbird brain and singing behaviour, particularly during the sensitive critical period. Glutamate and its receptors (AMPA and NMDA), for example, play a large role in developmental and seasonal plasticity; in particular, NMDA receptor composition changes over development and, in open-ended learners, seasonally, promoting or supressing neural plasticity [37]. Two neuromodulators also play significant roles in the songbird brain: dopamine (DA), typically associated with reward [38] (reviewed in [39]), and acetylcholine (Ach) (reviewed in [40]). As in mammalian brains, dopaminergic inputs originate in the ventral tegmental area (VTA) and connect to the vocal-control system via Area X (e.g., [41]). This connection to the AFP is essential for recognition of errors during the sensorimotor phase and for maintenance of song following crystallization [39, 42]. Recent work has also implicated acetylcholine in song learning and production. Specifically, blockade of cholinergic inputs to nucleus RA in zebra finches impaired song learning [43], and promoting Ach in HVC with agonists increased activity in HVC neurons, which translated to behavioural changes in song that made it higher in pitch, louder, faster, and more stereotyped [44].

In summary, songbird brains have intricate neural architecture and neurochemistry supporting behaviour related to vocal communication. Accurate production and perception of species-typical vocalizations requires exposure to specific stimuli in particular conditions over a critical period during development. The neural infrastructure supports not only vocal learning during development, but supports all aspects of vocal behaviour and perception during adulthood, for either maintenance in closed-ended learners, or for changing or adding songs in open-ended learners. The complexity of the vocal learning process, both behavioural and neural, and continued importance of vocal communication throughout a songbird's life means that there are multiple ways a changing Anthropocene could disrupt vocal behaviour and perception at all ages, and in both sexes. We will explore three of these potential insults to vocal learning in the next section: noise, stress associated with noise and food restriction, and chemical disruption of neural processes.

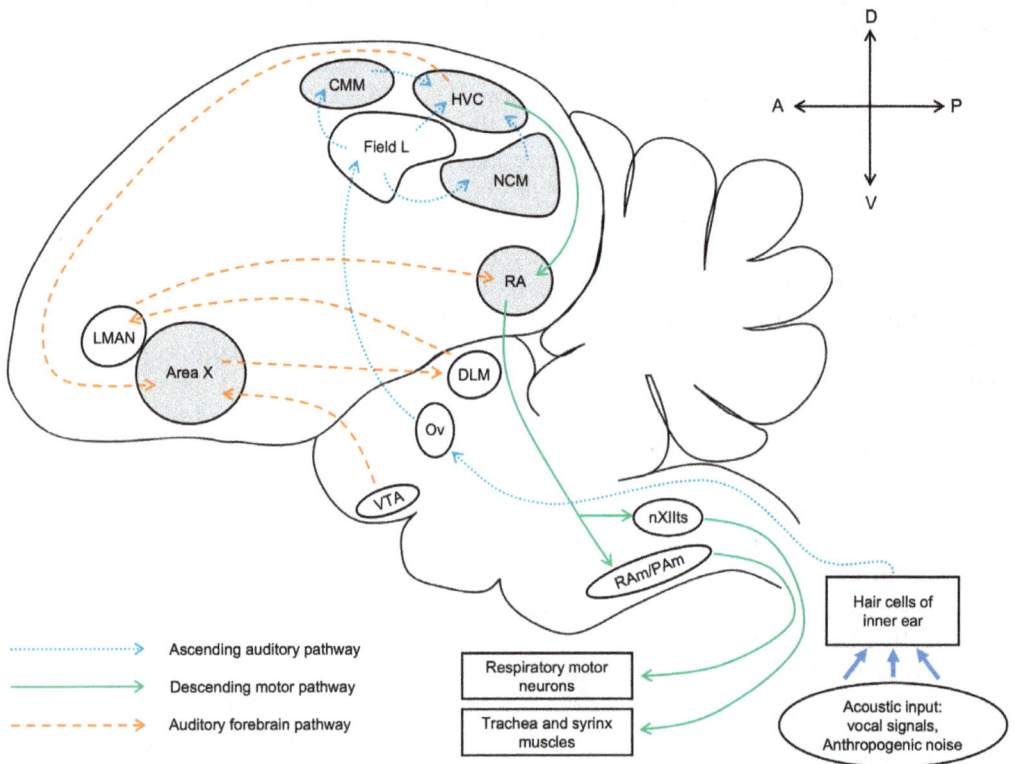

Figure 1 Illustration of the three interconnected pathways of the vocal-control system, brain nuclei important in vocal communication in songbirds. Shaded regions indicate nuclei discussed specifically in this chapter with respect to the Anthropocene.

The Changing Anthropocene and Vocal Learning

In the previous sections we outlined the behavioural conditions that lead to ideal song learning (i.e., hearing an appropriate model at the right time), the neural architecture underlying song learning and perception, and the neurochemicals involved in modulatory signalling within the pathways. Now we discuss three potential ways a changing Anthropocene could disturb vocal communication in songbirds. First, we consider increased noise, which could affect the bioacoustics of the signal itself and interfere with whether other birds can perceive the signal. Second, we consider increased noise and food restriction as stressors, and whether the physiological stress response to noise compromises singing behaviour or the brain of songbirds. Finally, we consider the more direct physiological effects of toxins and pesticides, and how these might disrupt the neurochemistry involved in signalling within the neural pathways important for song behaviour.

Noise—Interfering with the Vocal Signal

Increased noise is one of the primary threats posed by human encroachment on habitats [45]. Increased noise can induce behavioural changes in birds, as evidenced, for example, by birds shifting the start time of their dawn chorus in response to experimental playback of noise (*Sturnus unicolor, Passer domesticus*, [46, 47]), singing at higher frequencies if living in a city rather than in forest (*Turdus merula*, [48]), and singing more in noisier areas (*Serinus serinus*, [49]). Even more convincing is evidence that songbirds sing at higher frequencies (Hz) to mitigate against high anthropogenic (e.g., traffic) noise, and sing better quality songs at lower frequencies when anthropogenic noise abates (*Poecile atricapillus*, [50]; *Zonotrichia leucophrys*, [51]).

During development, increased noise could affect song learning in the sensory phase by making the tutor's song harder to hear, compromising the memory on which the juvenile bases its own song when practicing. Inability to hear an adult vocalizing clearly could also affect females listening to tutors, not for developing own song to sing, but for developing preferences for choosing a mate as an adult [52], in turn compromising the template on which females base assessment of male song quality. In open-ended learners, noise could affect song learning in adults as well. Further, if adults change their song in response to noise, that song, then learned by the next generation, may be judged as lower quality by females, reducing their reproductive success [53].

Playback of noise to nestling tree swallows (*Tachycineta bicolor*) affects the structure of their begging calls, an effect that persisted even after the noise was removed [54]. Given that begging calls develop into adult calls, noise during development can have long term implications for communication. In free living house sparrows (*Passer domesticus*), birds nesting at noisy sites have reduced breeding success, not because of impaired mate choice or impaired survival, but likely because of compromised chick development due to lower parental investment [55]. In an experimental study using zebra finches in the laboratory, juvenile males were reared with their tutor until the memorization phase was over (at 35 days post-hatch) and then moved to isolation chambers where they were then exposed to noise for long or short periods. Although similarity between the tutor and the bird's own song increased over development, indicating that song learning occurred, the songs of experimental birds were in general less similar to that of their tutor compared to control birds who were not tutored in noise [56]. Here the noise was delivered during the sensorimotor phase, indicating interference can occur when birds are practicing and need auditory feedback. Playback of urban noise during the entirety of song development produced similar results [10]: groups exposed to noise showed progressive learning of tutor's song, but certain features of song (e.g., note sequences) were compromised in noise-exposed birds. Noise exposure also affected structure of HVC and Area X: these regions were proportionally smaller compared to males not reared in noise [10].

Noise as a Stressor

While in some cases noise appears to compromise song learning by interfering with the acoustic environment [55], noise could also have indirect effects on song learning through physiological mechanisms, particularly by inducing stress. In birds, response to stressors is, as in mammals, mediated by the HPA-axis, and chronic stressors result in elevated levels of glucocorticoids such as corticosterone [57]. Early-life stress in birds can have especially detrimental effects (reviewed in [58, 59]).

Results from studies examining developmental effects of noise as a stressor on songbirds are mixed, however. In male song sparrows (*Melospiza melodia*), artificially increased levels of corticosterone resulted in lower quality song and decreased volume of RA, a motor nucleus, but did not affect other song vocal-control nuclei [60]. In three other songbird species (*Sialia mexicana, Sialia currucoides, Myiarchus cinerascens*), those that lived near sources of chronic anthropogenic noise (compressors) had higher baseline corticosterone levels compared to birds in the same population living away from the noise [61]. However, noise manipulation does not necessarily increase baseline levels of corticosterone (e.g., [62]), despite being associated with reduced volumes of vocal-control nuclei [10]. Nonetheless, if increased corticosterone is associated with compromised song learning or decreased volume of vocal-control nuclei, then, in turn, the birds living near anthropogenic noise sources may have poor quality songs, as well.

Other studies have used different measures to quantify the physiological effects of stress from noise. Increased corticosterone leads to lower resting metabolic rate (RMR): experimental administration of corticosterone to house sparrows resulted in lower RMR and body mass compared to control birds [63]. Thus lowered RMR may indicate increased stress and be used as a proxy for measuring corticosterone levels. Experimental playback of traffic noise to rural, free-living house sparrows slowed RMR in nestlings compared to those that did not hear playback [64], meaning the noise could have increased corticosterone and stress. Another proposed mechanism for how noise negatively impacts songbirds is through increased oxidative stress. For example, in a study of free-living great tits, Raap et al. [65] quantified haptoglobin (Hp) and nitric oxide (NOx) in plasma as indicators of stress in response to noise and light. Offspring in high-noise level nest boxes did not weigh less than offspring in low-noise level nest boxes, but did show higher levels of Hp, indicating a compensatory response to compromised immunity in these offspring. However, NOx levels were not changed. Telomere length, specifically shortened telomere length, is also an indicator of oxidative stress. In a field study of nestling great tits, nestlings living in noisier areas had shorter telomeres than those living in quieter areas, but this effect was only seen in the smallest nestlings of the brood [66]. However, not all studies find that noise increases indicators of oxidative stress (e.g., [67]). More importantly, these studies did not measure directly whether vocal learning (i.e., song quality as adults), ability to choose high quality mates, or functioning of neural regions associated with song are affected.

Thus, the relationship among noise, stress, and effects on vocal behaviour and vocal-related brain nuclei is complex and inconsistent, varying with free-living and laboratory studies, and possibly whether noise is perceived as a stressor. Research quantifying how these factors directly influence vocal learning, perception, and neuroanatomy should be carried out. Further, we should also consider whether the lack of consistent results in how noise affects singing behaviour, physiology and neurophysiology is due to the ability of songbirds to quickly adapt to the changing Anthropocene. As mentioned earlier, some birds can shift the timing of their dawn chorus to avoid increased environmental noise [46, 47], meaning tutor song may not be masked for juveniles memorizing the song for templates. In fact, nestling tree swallows can discriminate among vocalizations of conspecifics despite experimentally introduced noise [68], indicating that noise may not interfere with ability to distinguish conspecifics. Further, juvenile white-crowned sparrows can selectively learn tutor songs that are less masked by traffic noise [69]. Although their song learning overall was compromised, given the choice among playback of both high and low frequency tutor songs, birds who heard the tutor songs at the same time as traffic noise with masking frequencies chose to copy the songs that overlapped least with the traffic frequencies [69]. In other words, these young birds were able to adapt and make the best of a noisy situation.

In addition to these examples of resilience in behaviour, mixed results of studies on whether noise induces stress in songbirds may also indicate some degree of physiological resilience. Specifically, birds may adapt to chronic noise disruptions by habituating to background noise rather than maintaining an elevated baseline stress response. Unpredictable, or a sudden onset of acute noise is perhaps a greater issue than chronic noise, and the timing associated with experiencing the stressor may also temper effects of noise: stress during a critical period of learning may have greater effects than once song is learned, especially in closed-ended learners. Further, it may be easier to compensate for stress in a laboratory setting where, typically, only one variable is manipulated (e.g., noise) and other potential stressors (e.g., food unreliability and the need to forage) are eliminated. Thus, a combination of both field and laboratory studies are essential to elucidate not only how the changing Anthropocene negatively affects songbirds, their singing, and their neurophysiology, but perhaps how these diverse species are adapting to their habitats to maintain survival.

Food Restriction as a Stressor

Human urbanization and the resulting encroachment on songbird habitat results not only in increased noise levels in closer proximity to songbirds, but also rapid loss of appropriate habitat in which songbirds live and breed [1]. This decrease in habitat results in both less breeding territory, and a potential decrease in overall food availability and altered foraging strategies within that habitat. Compromised access to food during the critical period of vocal development can have lasting impacts on neurobiology and, ultimately, vocal behaviour, resulting in altered breeding success. Food restriction, as a developmental stressor, has been well-studied and its impacts on both songbird behaviour (i.e., birdsong) and physiology have been well-documented (reviewed in [58]). Beginning in the late 1990s, Nowicki and colleagues proposed the developmental-stress hypothesis (originally the nutritional-stress hypothesis), positing that birdsong can act as an indicator of mate quality in males, given that the neural architecture responsible for vocal-control develops during a period when birds can experience significant developmental stress, and that this underlying neural architecture is susceptible to these stressors [70, 71]. In this section, we discuss how food restriction, lower food quality, and alterations to foraging opportunities during development, which may be a result of the changing Anthropocene, affect both brain and behaviour in male and female songbirds. Specifically, we look at whether volume of vocal-control nuclei (HVC, RA), neurogenesis, and response of auditory perceptual regions (NCM, CMM), are affected by various methods of food restriction.

The effects of food restriction on the songbird brain have been studied in several songbird species in a variety of field and laboratory experiments. These studies have primarily focused on the effects of food restriction or food unpredictability on HVC and RA, given the role of these two nuclei in vocal production. One of the first laboratory experiments [71] restricted food access in nestling swamp sparrows (*Melospiza georgiana*) from 7 to 28 days post-hatch by feeding them only 70% of the average food volume consumed by nestlings in the control group. Sparrows with restricted access to food during this period of development (corresponding to the time parents would be feeding them until independence) had significantly smaller HVC and RA by volume in adulthood, compared to controls. These birds also spent longer in the subsong and plastic song phases of song development and made poorer copies of their tutors' songs [71]. This seminal study shows that even brief periods of developmental stress early in life (and early in the song memorization phase) can disrupt the vocal learning process and have major consequences on the songbird brain and vocal behaviour through adulthood [71].

Subsequent studies have had mixed results. For example, in juvenile zebra finches where food stress was created from 5 to 30 days post-hatch using food with filler, one study found reduced HVC volume [72] but not another [73]. Juvenile song sparrows fed 65% of the weight of food consumed by the control group from 3 to approximately 25 days post-hatch had smaller HVC, but not RA as juveniles (sacrificed at approximately 25 days; [74]). However, a different study in song sparrows that used the same method of restriction (65% food weight of controls) from 7 to 25 days post-hatch combined with random, restricted access to food from 25 to 60 days post-hatch (reflecting the change from dependent to independent feeding) did not result in a reduction in adult HVC volume, but did result in reduced RA volume [60]. Additionally, although HVC was not different between food-stressed and non-stressed groups, adult males that had experienced this extended food restriction had fewer song types and syllables in their vocal repertoire and made poorer copies of their tutors' songs [60].

One mechanism through which volume of vocal-control nuclei may be affected as a result of food stress is through a reduction in neurogenesis (e.g., [74]). However, the relationship between brain region volume and rate of neuronal recruitment and survival is tenuous: zebra finches who received a low-quality diet from hatching to 35 days post-hatch did not have a smaller HVC by volume, but did have fewer new neurons in HVC (i.e., reduced neurogenesis) compared to finches who received a higher quality, supplemented diet [73]. Although a recent study in non-songbirds has shown that early-life food restriction leads to fewer new neurons specifically in the avian hippocampal formation [75], the relationship between food restriction, volume of vocal-control nuclei, and neurogenesis needs to be studied explicitly and in more species.

As outlined above, the effects of food restriction on vocal-control nuclei have been studied in detail, especially in males who produce song. However, stress from food restriction may also affect the auditory perceptual regions, NCM and CMM. This is perhaps especially critical to consider in female songbirds, where assessment of male song is integral for mate choice. Again, results are varied. Female European starlings exposed to unpredictable food access (food removed for a random three hour interval per day) between 35 and 115 days post-hatch (i.e., after being able to feed independently) had reduced response in NCM and CMM (as measured by number of cells expressing ZENK, an immediate-early gene marking neuronal activity) after hearing playback of male conspecific song [76]. However, female song sparrows exposed to 65% food weight of controls from 7 to 25 days post-hatch combined with random, restricted access to food from 25 to 60 days post-hatch did not show reduced ZENK response in NCM and CMM after playback of male conspecific song [77]. Furthermore, in males, song sparrows exposed to a low quality diet from 5 to 30 days post-hatch had decreased ZENK response in NCM and CMM in response to playback of tutor song compared to control males [78].

While the results of these studies examining vocal-control brain region volume, neurogenesis, and ZENK response may be inconsistent this may be explained partly by the variability in the method of food restriction used to create developmental stress (e.g., percentage consumed of controls, food quality), the timing of food restriction (e.g., before memorization, during sensorimotor practice), and species studied. Clearly, the potential for altered food availability to affect songbird development depends critically on when the young birds experience these effects. As with results from studies on noise-induced stress, this may indicate songbirds have some level of resilience and ability to adapt to early-life stress from decreased food availability and quality, and, perhaps, resilience to stress in general.

Environmental Contaminants and their Effects on the Avian Brain

Another major issue associated with the onset of the Anthropocene is contamination and pollution [79]. This includes presence of heavy metal and organohalide (e.g., flame retardant) pollution, as well as increased use of pesticides (including neonicotinoids, organophosphates, and carbamates); all of which have detrimental effects on songbird abundance [80–82]. In the birds that survive, these pollutants also have detrimental effects on their physiology and, by extension, their neurophysiology.

Birds are often used as bioindicators of pollution within an ecosystem (e.g., [83, 84]), particularly heavy metal pollution [83]. Much of the research to date on the physiological consequences of heavy metal pollution in birds have been studied in non-songbirds, including waterfowl, raptors, and seabirds (e.g., [85–89]), which are especially likely to consume food containing bioaccumulated heavy metals. Nonetheless, once studies showed changes in birdsong between non-polluted and heavy metal-polluted sites (e.g., [90–92]), researchers began examining whether the underlying neural architecture in songbirds was also affected. Male zebra finches treated with methylmercury (MeHg) at two different environmentally relevant concentrations (low- and high-dose) *in ovo* did not show any significant differences in the size of vocal-control nuclei (HVC, RA, Area X) in adulthood (90 days post-hatch) compared to non-treated finches [93]. However, birds in the high-treatment group had smaller overall brain volumes than the other groups, attributed to a chronic neuroinflammatory response produced by the MeHg treatment [93]. This suggests that the changes in birdsong observed near sites of heavy metal pollution are not directly caused by metal-induced alterations to the vocal-control system, but rather result likely from a combination of sub-lethal changes in both songbird physiology (reviewed in [94]) and behaviour.

In addition to heavy metals, organohalide pollution also has significant effects on avian physiology. Specifically, the neurotoxic effects of polybrominated diphenyl ethers (PBDEs), a family of commonly-used flame retardants present in a variety of household and industrial products, have been well-studied across a variety of species, including humans (e.g., [95]). Although widely restricted or banned in most jurisdictions, these chemicals continue to persist and accumulate in the environment [96], posing a severe threat to organisms near PBDE deposits, including at waste disposal sites and landfills. The toxic effects of flame retardants, including the pentabromodiphenyl ether, BDE-99, have also been studied in songbirds, given their tendency to congregate and nest near disposal sites and landfills (reviewed in [84, 97]). In zebra finches, *in ovo* exposure to increasing concentrations of BDE-99 led to a dose-dependent decrease in the volume of HVC, RA and Area X, all components of the avian vocal-control system integral for production, learning, and maintenance [98]. Interestingly, exposing nestling finches to increasing concentrations of BDE-99 had no effect on the same nuclei, but did result in behavioural changes later in adulthood [99]. Furthermore, while the well-known organohalide pesticide dichlorodiphenyltrichloroethane (DDT) has been banned in many jurisdictions since the 1970s, the accumulation of this chemical around the world continues to have major neurotoxic effects in many species, including songbirds (reviewed in [100, 101]). For example, American robins (*Turdus migratorius*) exposed to DDT *in ovo* had significantly smaller brains, and within the vocal-control system, smaller HVC and RA [102].

Increased use of pesticides and herbicides during the Anthropocene has been explicitly named as a leading cause of the significant decline in songbird populations in recent years [1, 103, 104]. Most common pesticides (including organophosphates, carbamates, and more recently, neonicotinoids) are intended to disrupt peripheral neural transmission in insects by inhibiting acetylcholinesterase (AchE) activity and overstimulating muscarinic and nicotinic acetylcholine (Ach) receptors, ultimately leading to paralysis and death (reviewed in [105]). Thus it should not be surprising that these chemicals have secondary physiological and behavioural effects on other organisms within an ecosystem, including songbirds (reviewed in [106]). These pesticides also interfere with AchE inhibition in songbirds, causing detrimental effects, not only on songbird brain function, but also behaviour. As described previously, proper Ach functioning is necessary for song learning and production [40]. Widespread spraying of the organophosphate Fenitrothion (210 g/ha) over eight days near a large population of white-throated sparrows (*Zonotrichia albicollis*) in New Brunswick, Canada, not only caused adults to abandon territory and nests, but also, among adults that remained, reduced their ability to defend their territory [107]. While the Fenitrothion spraying led to one-third reduction in overall sparrow population within the survey area [107], a follow-up study found that over 50% of the surviving sparrows sprayed with this pesticide had AchE activity inhibited by greater than 40% [108]. Cholinesterase activity inhibition also occurs following application of organophosphate or carbamate spraying in other species, such as New World warblers (Parulidae), blue jays (*Cyanocitta cristata*), and red-bellied woodpeckers (*Melanerpes carolinus*) [109, 110]. Furthermore, recent studies of both songbirds (American goldfinches, *Spinus tristis*) and non-songbirds (Japanese quail, *Coturnix japonica*) found that an environmentally relevant application of the neonicotinoid pesticide imidacloprid led to acute toxicity and death after rapid absorption and accumulation within the brain and other tissues [111, 112]. Although not directly studied in songbirds, a recent study using domesticated chickens found that exposure to imidacloprid *in ovo* interfered with the differentiation of the avian neural tube, as well as disrupted neurogenesis by increasing apoptosis [113]. Adult neurogenesis is ongoing in many adult songbirds, so this is yet another pathway for chemicals to disturb critical neural processes related to singing behaviour.

Management Implications

There are multiple pathways through which the Anthropocene could disturb vocal learning in songbirds, both directly and indirectly. These include disruption of the acoustic environment from increased noise, elevated stress, and interference with neural transmission required for species-typical behaviour. Disruption during development turns into long-lasting effects in adulthood, not only for male and female singers, but for also for the sex (usually females) that chooses singers based on preference. In open-ended learners, disruptions to vocal learning continues past development and throughout adulthood.

That songbirds seem particularly susceptible to the changing Anthropocene may be one reason why they have been used as sentinels for the effects of human spread into their habitats. Nonetheless, songbirds also appear to have a remarkable degree of resilience to acute and chronic changes in their environment. Both tutor and learner can adapt behaviour to maximize song learning in a noisy habitat. Further, noise does not always induce a reliable stress response in birds, whether measured via quantification of changes in the HPA-axis or in oxidative stress, nor does it always affect the neural regions and connections that underlie vocal production and perception. However, this does not mean that mitigating measures to reduce noise, such as erecting noise barriers, building roads using alternative materials which produce less noise, or limiting the time and frequency of noisy anthropogenic activities (reviewed in [45, 114, 115]) should not be implemented.

Results from studies on food restriction are also inconsistent, however this could be attributed to the fact that methods to induce food-related stress vary by quantity or quality, as well as when during development they are applied, and even species used in the experiment. Some variability in results may also be due to using both field and laboratory studies to ask questions about how the changing Anthropocene may affect songbird neurobiology and behaviour. However, using both approaches should be viewed as a strength, rather than weakness, and is why we have included both in our review of the literature. Field studies lend a measure of ecological validity to results, while laboratory studies may show detrimental effects of specific factors only revealed under controlled conditions. Laboratory studies may be especially important in identifying specific targets for mitigating conservation efforts that, if studied in the field would remain obscured, because birds may compensate for stressors when they have access to uncontrolled resources. Further, studying effects of stressors on the vocal-control system and the songbird brain necessitates a laboratory component. We argue that using a combination of field- and laboratory-based techniques are required not only to reveal subtleties of effects not clear in field studies alone, but also to test in the field the efficacy of mitigation strategies informed by laboratory results. A nice example of this approach, albeit not from songbirds, is a suite of experiments on juvenile Ambon damselfish (*Pomacentrus amboinensis*), a coral reef fish. Laboratory studies showed that boat noise hampered learning of predator avoidance, then field studies showed that fish whose learning was compromised in this way had lower survival. Regulation of boat visits to reefs, which are increasing in the study area, thus must account for long-lasting harm caused by even brief periods of traffic, a consideration less likely to be revealed by just laboratory or field studies alone [116]. Unfortunately, such mitigation studies for birds are currently lacking, and should be a priority for future research.

Box 1   Vocal Learning and Neurobiology in the Anthropocene

Background
- Songbirds learn species-specific vocalizations through imitation
- Vocal behaviour supported by distinct brain regions in three pathways:
  - *Ascending auditory pathway:* auditory perception (nuclei include: NCM, CMM)
  - *Descending motor pathway:* vocal production (nuclei: HVC, RA)
  - *Anterior forebrain pathway:* song learning and plasticity (nuclei: HVC, RA, Area X)

How Does the Changing Anthropocene Affect Vocal Learning and Neurobiology in Songbirds?

Noise Interferes with Vocal Signals

- Leads to behavioural changes, including dawn chorus timing, song duration and frequency
- Masks tutor song and compromises juvenile's song template
- Difficult for females to judge male song quality
- Urban noise playback during vocal development reduces volume of HVC and Area X

Noise is a Stressor

- Noise may induce stress via the HPA-axis or by increasing oxidative stress
- Chronic anthropogenic noise leads to higher baseline corticosterone levels (in turn, affecting the brain)
- Relationship among noise, stress and the brain is complex and inconsistent

Decreased Food Availability

- Effects of food restriction on the brain and vocal behaviour are mixed, but suggest:
  - Food restriction early in life may affect HVC and RA volume
  - Food restriction may reduce neurogenesis in HVC
  - Food restriction may reduce NCM and CMM response to playback in adult females

Contaminants Interfere with Neural Signaling

- Heavy metal pollution has indirect neuroinflammatory affects on the songbird brain
- *In ovo* exposure to organohalides (BDE-99, a flame retardant; DDT, a pesticide) cause significant reductions in HVC, RA, Area X
- Pesticides (organophosphates, carbamates, neonicotinoids) directly interfere with cholinergic transmission; may also affect neurogenesis

Management Implications and Conclusions
- **Multiple pathways** through which the Anthropocene can compromise songbird neurobiology and vocal learning
- Songbirds have a **remarkable degree of resilience** to acute and chronic Anthropogenic-induced changes in the environment
- Evidence and approaches **from both field and laboratory studies** are valuable
- **Habitat conservation efforts should still be made** to allow birds to perform critical vocal behaviour without interference; and such changes may help mitigate the effects of the Anthropocene

While songbirds may appear, at least in part, to be able to adapt to noise through behavioural change, resident songbirds cannot avoid the introduction of pollutants, toxins, and pesticides directly into their habitat. These agents, if they do not kill the bird outright, can severely disrupt and alter the neural transmission and neural plasticity necessary for vocal learning and perception, compromising vocal communication in birds from nest to adult. The measures to prevent this are much clearer cut: avoid administration of such chemicals in critical songbird habitats. It was research linking the use of DDT to adverse physiological effects *in ovo* (e.g., [117, 118], reviewed in [119]) which raised awareness on the detrimental effects of DDT on songbirds, and which ultimately led to the ban of DDT by the United States in 1972. This ban is widely considered one of the major causes of the revival in the large raptor populations in North America, including bald eagles (*Haliaeetus leucocephalus*) and peregrine falcons (*Falco peregrinus*) [1, 120]. While some jurisdictions, such as the European Union, have recently banned or significantly restricted the use of neonicotinoid pesticides due their harmful secondary effects on wildlife, including songbirds [121, 122], the use of these pesticides is still common across North America, despite an increasing body of evidence showing that their usage continues to have detrimental effects on songbirds (e.g., [123]). In areas where pesticide usage continues to be prevalent, it is anticipated that songbird populations will continue to decline, thus longer-term studies on whether populations of songbirds have recovered in areas where these pesticides have been banned should be used to assess effectiveness of these measures, as well as to raise awareness and inform future conservation efforts and policies in areas where such practices continue.

Conclusions

In this chapter, we reviewed the ontogeny and neural basis of vocal learning and described evidence from both laboratory and field experiments examining how factors associated with the changing Anthropocene, such as increased noise, stress from noise and food restriction, and environmental contaminants affect songbirds (Box 1). Increased anthropogenic noise appears to directly interfere with songbird communication, impeding song learning by making tutor song more difficult to hear during the sensory phase, and consequently weakening the memory on which the juvenile bases its own song during the sensorimotor phase. Increased anthropogenic noise also appears to affect vocal learning and the underlying vocal-control system via traditional physiological mechanisms associated with stress, including both increased HPA-axis activation and increased oxidative stress. Food availability and food quality, changes to which are likely to occur as a result of the decreased habitat availability in the changing Anthropocene, also appear to have detrimental effects on the structure and function of the songbird brain, particularly those nuclei involved in vocal production (e.g., HVC and RA, critical for males) and vocal perception (e.g., NCM and CMM, critical for females). Furthermore, increased heavy metal and organohalide pollution, as well as increased use of pesticides such as neonicotinoids, have direct and detrimental effects on songbird populations worldwide, in large part due to their harmful effects on the songbird brain and neural transmission.

While both field and laboratory studies examining the effects of some (but not other) stressors associated with the changing Anthropocene produced inconsistent results, this is likely a result of variability in study design (e.g., timeline differences in exposure to noise, food restriction) and the particular species under study. However, these results may also indicate that songbirds display a degree of resilience in the face of the changing Anthropocene, and can rapidly adapt to changes in their environment. Even so, purported resilience should not overshadow the evidence indicating that Anthropocene-induced environmental changes, such as increased noise, food restriction, or increased pollution, even in small measures and for short periods, can in fact drastically alter songbird neurobiology and vocal behaviour. In short, these mixed results should not be interpreted as an affirmation that the Anthropocene has little-to-no effect on the songbird brain and behaviour.

Despite evidence that the changing Anthropocene negatively affects the songbird brain and behaviour, to date, there are no studies which have directly tested how conservation measures mitigate these effects. This is one area where research on songbirds lags behind research on marine mammals, much of which is mandated by regulations that have long placed more emphasis on noise in marine, rather than terrestrial, environments [124]. Therefore, we need to conduct experiments to determine which habitat conservation efforts are most effective at mitigating the effects of the changing Anthropocene on songbird survival. Given that songbird populations (as well as overall global biodiversity) are expected to continue declining over the next several years and decades (e.g., [125–127]), the time is for significant and impactful changes to manage and mitigate the effects of the changing Anthropocene on songbirds is now.

LITERATURE CITED

[1]   Rosenberg, K.V., A.M. Dokter, P.J. Blancher, J.R. Sauer, A.C. Smith, P.A. Smith, et al. 2019. Decline of the North American avifauna. Science. 366: 120–124.

[2]   Derégnaucourt, S. 2011. Birdsong learning in the laboratory, with especial reference to the song of the Zebra Finch (*Taeniopygia guttata*). Interact. Stud. 12: 324–350.

[3]   Zann, R.A. 1996. The Zebra Finch: A Synthesis of Field and Laboratory Studies. Oxford University Press, Oxford, UK.

[4]   Mooney, R. 2009. Neural mechanisms for learned birdsong. Learn. Mem. 16: 655–669.

[5]   Nordeen, K.W. and E.J. Nordeen. 1992. Auditory feedback is necessary for the maintenance of stereotyped song in adult Zebra Finches. Behav. Neural Biol. 57: 58–66.

[6]   Sober, S.J. and M.S. Brainard. 2009. Adult birdsong is actively maintained by error correction. Nat. Neurosci. 12: 927–931.

[7]   Gobes, S.M.H., R.B. Jennings and R.K. Maeda. 2019. The sensitive period for auditory-vocal learning in the Zebra Finch: Consequences of limited-model availability and multiple-tutor paradigms on song imitation. Behav. Processes. 163: 5–12.

[8]   Johannessen, L.E., T. Slagsvold and B.T Hansen. 2006. Effects of social rearing conditions on song structure and repertoire size: experimental evidence from the field. Anim. Behav. 72: 83–95.

[9]   Funabiki, Y. and K. Funabiki. 2009. Factors limiting song acquisition in adult Zebra Finches. Dev. Neurobio. 69: 752–759.

[10]   Potvin, D.A., M.T. Curcio, J.P. Swaddle and S.A. MacDougall-Shackleton. 2016. Experimental exposure to urban and pink noise affects brain development and song learning in Zebra Finches (*Taenopygia guttata*). PeerJ. 4: e2287.

[11]   Plamondon, S.L., G.J. Rose and F. Goller. 2010. Roles of syntax information in directing song development in White-crowned Sparrows (*Zonotrichia leucophrys*). J. Comp. Psychol. 124: 117–32.

[12]   Leitner, S., C. Voigt and M. Gahr. 2001. Seasonal changes in the song pattern of the non-domesticated Island Canary (*Serinus canaria*), a field study. Behaviour. 138: 885–904.

[13]   Eens, M., R. Pinxten and R.F. Verheyen. 1992. Song learning in captive European starlings, *Sturnus vulgaris*. Anim. Behav. 44: 1131–43.

[14]   Brenowitz, E.A. and M.D. Beecher. 2005. Song learning in birds: diversity and plasticity, opportunities and challenges. Trends Neurosci. 28: 127–132.

[15]   Mountjoy, J.D. and R.E Lemon. 1995. Extended song learning in wild European starlings. Anim. Behav. 49: 357–366.

[16]   Araya-Salas, M. and T. Wright. 2013. Open-ended song learning in a hummingbird. Biol. Lett. 9: 20130625.

[17]   Tumer, E.C. and M.S. Brainard. 2007. Performance variability enables adaptive plasticity of 'crystallized' adult birdsong. Nature. 450: 1240–1244.

[18]   Brenowitz, E.A. 1991. Evolution of the vocal control system in the avian brain. Sem. Neurosci. 3: 399–407.

[19]   Nottebohm, F. 2005. The neural basis of birdsong. PLoS Biology. 3: e164.

[20]   Ball, G.F. and S.A. Macdougall-Shackleton. 2001. Sex differences in songbirds 25 years later: what have we learned and where do we go? Microsc. Res. Tech. 54: 327–334.

[21]   Gentner, T.Q. and D. Margoliash. 2003. Neuronal populations and single cells representing learned auditory objects. Nature. 424: 669–674.

[22]   Nagel, K.I. and A.J. Doupe. 2006. Temporal processing and adaptation in the songbird auditory forebrain. Neuron. 51: 845–859.

[23]   Bolhuis, J.J. and S. Moorman. 2015. Birdsong memory and the brain: in search of the template. Neurosci. Biobehav. Rev. 50: 41–55.

[24]   Bolhuis, J.J. and M. Gahr. 2006. Neural mechanisms of birdsong memory. Nat. Rev. Neurosci. 7: 347–357.

[25]   Moorman, S., C.V. Mello and J.J. Bolhuis. 2011. From songs to synapses: molecular mechanisms of birdsong memory. Bioessays. 33: 377–385.

[26]   Duffy, D.L., G.E. Bentley and G.F. Ball. 1999. Does sex or photoperiodic condition influence ZENK induction in response to song in European starlings? Brain Res. 844: 78–82.

[27]   Phillmore, L.S., A.S. Veysey and S.P. Roach. 2011. Zenk expression in auditory regions changes with breeding condition in male Black-capped chickadees (*Poecile atricapillus*). Behav. Brain Res. 225: 464–472.

[28]   Nottebohm, F., T.M. Stokes and C.M. Leonard. 1976. Central control of song in the Canary, *Serinus canarius*. J. Comp. Neurol. 165: 457–486.

[29]   Kozhevnikov, A.A. and M.S. Fee. 2007. Singing-related activity of identified HVC neurons in the Zebra Finch. J. Neurophysiol. 97: 4271–4283.

[30]   Brenowitz, E.A. 2004. Plasticity of the adult avian song control system. Ann. N. Y. Acad. Sci. 1016: 560–585.

[31]   Doupe, A.J., D.J. Perkel, A. Reiner and E.A. Stern. 2005. Birdbrains could teach basal ganglia research a new song. Trends Neurosci. 28: 353–363.

[32]   Mooney, R., J. Prather and T. Roberts. Neurophysiology of birdsong learning. 2008. pp. 441–474. *In:* J.H. Byrne (ed.). Learning and Memory: A Comprehensive Reference. Academic Press, Oxford, UK.

[33]   White, S.A. 2001. Learning to communicate. Curr. Opin. Neurobiol. 11: 510–520.

[34]   Brenowitz, E.A. and S.M.N. Woolley. 2004. The avian song control system: a model for understanding changes in neural structure and function. pp. 228–284. *In:* T.N. Parks, E.W. Rubel A.N. Popper and R.R. Fay (eds). Plasticity of the Auditory System. Springer, New York, USA.

[35] Scharff, C. and F. Nottebohm. 1991. A comparative study of the behavioral deficits following lesions of various parts of the zebra finch song system: implications for vocal learning. J. Neurosci. 11: 2896–2913.

[36] Tramontin, A.D. and E.A. Brenowitz. 2000. Seasonal plasticity in the adult brain. Trends Neurosci. 23: 251–258.

[37] Nordeen, K.W. and E.J. Nordeen. 2004. Synaptic and molecular mechanisms regulating plasticity during early learning. Ann. N. Y. Acad. Sci. 1016: 416–437.

[38] Berridge, K.C. 2007. The debate over dopamine's role in reward: the case for incentive salience. Psychopharmacology. 191: 391–431.

[39] Kubikova, L. and L Koštál. 2010. Dopaminergic system in birdsong learning and maintenance. J. Chem. Neuroanat. 39: 112–123.

[40] Shea, S.D. and D. Margoliash. 2010. Behavioral state-dependent reconfiguration of song-related network activity and cholinergic systems. J. Chem. Neuroanat. 39: 132–140.

[41] Saravanan, V., L.A. Hoffmann, A.L. Jacob, G.J. Berman and S.J. Sober. 2019. Dopamine Depletion affects vocal acoustics and disrupts sensorimotor adaptation in songbirds. eNeuro. 6: ENEURO.0190-19.2019.

[42] Hoffmann, L.A., V. Saravanan, A.N. Wood, L. He and S.J. Sober. 2016. Dopaminergic contributions to vocal learning. J. Neurosci. 36: 2176–2189.

[43] Puzerey, P.A., K. Maher, N. Prasad and J.H. Goldberg. 2018. Vocal learning in songbirds requires cholinergic signaling in a motor cortex-like nucleus. J. Neurophysiol. 120: 1796–1806.

[44] Jaffe, P.I. and M.S. Brainard. 2020. Acetylcholine acts on songbird premotor circuitry to invigorate vocal output. eLife. 9: e53288.

[45] Shannon, G., M.F. McKenna, L.M. Angeloni, K.R. Crooks, K.M. Fristrup, E. Brown, et al. 2016. A synthesis of two decades of research documenting the effects of noise on wildlife. Biol. Rev. 91: 982–1005.

[46] Arroyo-Solís, A., J.M. Castillo, E. Figueroa, J.L. López-Sánchez and H. Slabbekoorn. 2013. Experimental evidence for an impact of anthropogenic noise on dawn chorus timing in urban birds. J. Avian Biol. 44: 288–296.

[47] Gil, D., M. Honarmand, J. Pascual, E. Pérez-Mena and C. Macías Garcia. 2015. Birds living near airports advance their dawn chorus and reduce overlap with aircraft noise. Behav. Ecol. 26: 435–443.

[48] Nemeth, E., N. Pieretti, S.A. Zollinger, N. Geberzahn, J. Partecke and A.C. Miranda. 2013. Bird song and anthropogenic noise: vocal constraints may explain why birds sing higher-frequency songs in cities. Proc. Royal Soc. B. 280: 20122798.

[49] Díaz, M., A. Parra and C. Gallardo. 2011. Serins respond to anthropogenic noise by increasing vocal activity. Behav. Ecol. 22: 332–336.

[50] Proppe, D.S., M.T. Avey, M. Hoeschele, M.K. Moscicki, T. Farrell, C.C. St Clair, et al. 2012. Black-capped chickadees *Poecile atricapillus* sing at higher pitches with elevated anthropogenic noise, but not with decreasing canopy cover. J. Avian Biol. 43: 325–332.

[51] Derryberry, E.P., J.N. Phillips, G.E. Derryberry, M.J. Blum and D. Luther. 2020. Singing in a silent spring: Birds respond to a half-century soundscape reversion during the COVID-19 shutdown. Science. 370: 575–579.

[52] Lauay, C., N.M. Gerlach, E. Adkins-Regan and T.J. DeVoogd. 2004. Female Zebra Finches require early song exposure to prefer high-quality song as adults. Anim. Behav. 68: 1249–1255.

[53] Halfwerk, W., L.J.M. Holleman, C.M. Lessells and H. Slabbekoorn. 2011. Negative impact of traffic noise on avian reproductive success. J. Appl. Ecol. 48: 210–219.

[54] Leonard, M.L. and A.G. Horn. 2008. Does ambient noise affect growth and begging call structure in nestling birds? Behav. Ecol. 19: 502–507.

[55] Schroeder, J., S. Nakagawa, I.R. Cleasby, and T. Burke. 2012. Passerine birds breeding under chronic noise experience reduced fitness. PLoS One. 7: e39200.

[56] Funabiki, Y. and M. Konishi. 2003. Long memory in song learning by Zebra Finches. J. Neurosci. 23: 6928–6935.

[57] Kriengwatana, B., H. Wada, K.L. Schmidt, M.D. Taves, K.K. Soma and S.A. MacDougall-Shackleton. 2014. Effects of nutritional stress during different developmental periods on song and the hypothalamic-pituitary-adrenal axis in Zebra Finches. Horm. Behav. 65: 285–293.

[58] MacDougall-Shackleton, S.A. and K.A. Spencer. 2012. Developmental stress and birdsong: current evidence and future directions. J. Ornithol. 153: 105–117.

[59] Schmidt, K.L., E.A. MacDougall-Shackleton, S.P. Kubli and S.A. MacDougall-Shackleton. 2014. Developmental stress, condition and birdsong: a case study in song sparrows. Integr. Comp. Biol. 54: 568–577.

[60] Schmidt, K.L., S.D. Moore, E.A. MacDougall-Shackleton and S.A MacDougall-Shackleton. 2013. Early-life stress affects song complexity, song learning and volume of the brain nucleus RA in adult male song sparrows. Anim. Behav. 86: 25–35.

[61] Kleist, N.J., R.P. Guralnick, A. Cruz, C.A. Lowry and C.D. Francis. 2018. Chronic anthropogenic noise disrupts glucocorticoid signaling and has multiple effects on fitness in an avian community. PNAS. 115: E648–E657.

[62] Zollinger, S.A., A. Dorado-Correa, W. Goymann, W. Forstmeier, U. Knief, A.M. Bastidas-Urrutia, et al. 2020. Traffic noise exposure depresses plasma corticosterone and delays offspring growth in breeding zebra finches. Conserv. Physiol. 7: coz056.

[63] Dupont, S.M., J.K. Grace O. Lourdais, F. Brischoux and F. Angelier. 2019. Slowing down the metabolic engine: impact of early-life corticosterone exposure on adult metabolism in house sparrows (*Passer domesticus*). J. Exp. Biol. 222: jeb211771.

[64] Brischoux, F., A. Meillère, A. Dupoué, O. Lourdais and F. Angelier. 2017. Traffic noise decreases nestlings' metabolic rates in an urban exploiter. J. Avian Biol. 48: 905–909.

[65] Raap, T., R. Pinxten, G. Casasole, N. Dehnhard and M. Eens. 2017. Ambient anthropogenic noise but not light is associated with the ecophysiology of free-living songbird nestlings. Sci. Rep. 7: 2754.

[66] Grunst, M.L., A.S. Grunst, R. Pinxten and M. Eens. 2020. Anthropogenic noise is associated with telomere length and carotenoid-based coloration in free-living nestling songbirds. Environ. Pollut. 260: 114032.

[67] Casasole, G., T. Raap, D. Costantini, H. AbdElgawad, H. Asard, R. Pinxten, et al. 2017. Neither artificial light at night, anthropogenic noise nor distance from roads are associated with oxidative status of nestlings in an urban population of songbirds. Comp. Biochem. Physiol. Part A Mol. Integr. Physiol. 210: 14–21.

[68] Horn, A.G., M. Aikens, E. Jamieson, K. Kingdon and M.L. Leonard. 2020. Effect of noise on development of call discrimination by nestling tree swallows, *Tachycineta bicolor*. Anim. Behav. 164: 143–8.

[69] Moseley, D.L., G.E. Derryberry, J.N. Phillips, J.E. Danner, R.M. Danner, D.A. Luther, et al. 2018. Acoustic adaptation to city noise through vocal learning by a songbird. Proc. Royal Soc. B. 285: 20181356.

[70] Nowicki, S., S. Peters and J. Podos. 1998. Song learning, early nutrition and sexual selection in songbirds. Integr. Comp. Biol. 38: 179–90.

[71] Nowicki, S., W.A. Searcy and S. Peters. 2002. Brain development, song learning and mate choice in birds: a review and experimental test of the "nutritional stress hypothesis. J. Comp. Physiol. A. 188: 1003–1014.

[72] Buchanan, K.L., S. Leitner, K.A. Spencer, A.R. Goldsmith and C.K. Catchpole. 2004. Developmental stress selectively affects the song control nucleus HVC in the zebra finch. Proc. Royal Soc. B. 271: 2381–2386.

[73] Honarmand, M., C.K. Thompson, A. Schatton, S. Kipper and C Scharff. 2016. Early developmental stress negatively affects neuronal recruitment to avian song system nucleus HVC. Dev. Neurobiol. 76: 107–118.

[74] MacDonald, I.F., B. Kempster, L. Zanette and S.A. MacDougall-Shackleton. 2006. Early nutritional stress impairs development of a song-control brain region in both male and female juvenile song sparrows (*Melospiza melodia*) at the onset of song learning. Proc. Royal Soc. B. 273: 2559–2564.

[75] Robertson, B.-A., L. Rathbone, G. Cirillo, R.B. D'Eath, M. Bateson, T. Boswell, et al. 2017. Food restriction reduces neurogenesis in the avian hippocampal formation. PLoS One. 12: e0189158.

[76] Farrell, T.M., M.A.C. Neuert, A. Cui and S.A. MacDougall-Shackleton. 2015. Developmental stress impairs a female songbird's behavioural and neural response to a sexually selected signal. Anim. Behav. 102: 157–167.

[77] Schmidt, K.L., E.S. McCallum, E.A. MacDougall-Shackleton and S.A. MacDougall-Shackleton. 2013. Early-life stress affects the behavioural and neural response of female song sparrows to conspecific song. Anim. Behav. 85: 825–837.

[78] Bell, B.A., M.L. Phan, A. Meillère, J.K. Evans, S. Leitner, D.S. Vicario, et al. 2018. Influence of early-life nutritional stress on songbird memory formation. Proc. Royal Soc. B. 285: 20181270.

[79] Caro, T., J. Darwin, T. Forrester, C. Ledoux-Bloom and C. Wells. 2014. Conservation in the anthropocene. pp. 185–188. *In:* G. Wuerthner, E. Crist and T. Butler (eds). Keeping the Wild: Against the Domestication of Earth. Foundation for Deep Ecology, San Francisco, USA.

[80] Kekkonen, J., I.K. Hanski, R.A. Vaisanen and J.E. Brommer. 2012. Levels of heavy metals in House Sparrows (*Passer domesticus*) from urban and rural habitats of southern Finland. Ornis Fenn. 89: 91–99.

[81] Terborgh, J. 1992. Why American songbirds are vanishing. Sci. Am. 266: 98–105.

[82] Mineau, P. and M. Whiteside. 2013. Pesticide acute toxicity is a better correlate of U.S. grassland bird declines than agricultural intensification. PLoS One. 8: e57457.

[83] Dauwe, T., L. Bervoets, R. Blust, R. Pinxten and M Eens. 2000. Can excrement and feathers of nestling songbirds be used as biomonitors for heavy metal pollution? Arch. Environ. Contam. Toxicol. 39: 541–546.

[84] Guigueno, M.F. and K.J. Fernie. 2017. Birds and flame retardants: a review of the toxic effects on birds of historical and novel flame retardants. Environ. Res. 154: 398–424.

[85] Sileo, L., W. Nelson Beyer and R. Mateo. 2003. Pancreatitis in wild zinc-poisoned waterfowl. Avian Pathol. 32: 655–660.

[86] Vallverdú-Coll, N., R. Mateo, F. Mougeot and M.E. Ortiz-Santaliestra. 2019. Immunotoxic effects of lead on birds. Sci. Total Environ. 689: 505–515.

[87] Gil-Sánchez, J.M., S. Molleda, J.A. Sánchez-Zapata, J. Bautista, I. Navas, R. Godinho, et al. 2018. From sport hunting to breeding success: patterns of lead ammunition ingestion and its effects on an endangered raptor. Sci. Total Environ. 613-614: 483–491.

[88] Stewart, F.M., R.W. Furness and L.R. Monteiro. 1996. Relationships between heavy metal and metallothionein concentrations in lesser Black-backed Gulls, *Larus fuscus*, and Cory's shearwater, *Calonectris diomedea*. Arch. Environ. Contam. Toxicol. 30: 299–305.

[89] Eckbo, N., C. Le Bohec, V. Planas-Bielsa, N.A. Warner, Q. Schull, D. Herzke, et al. 2019. Individual variability in contaminants and physiological status in a resident Arctic seabird species. Environ. Pollut. 249: 191–199.

[90] Gorissen, L., T. Snoeijs, E. Van Duyse and M. Eens. 2005. Heavy metal pollution affects dawn singing behaviour in a small passerine bird. Oecologia. 145: 504–509.

[91] Hallinger, K.K., D.J. Zabransky, K.A. Kazmer and D.A. Cristol. 2010. Birdsong differs between mercury-polluted and reference sites. Auk. 127: 156–161.

[92]   McKay, J.L. and C.R. Maher. 2012. Relationship between blood mercury levels and components of male song in Nelson's sparrows (*Ammodramus nelsoni*). Ecotoxicology. 21: 2391–2397.

[93]   Yu, M.S., M.L. Eng, T.D. Williams, M.F. Guigueno and J.E. Elliott. 2017. Assessment of neuroanatomical and behavioural effects of *in ovo* methylmercury exposure in Zebra Finches (*Taeniopygia guttata*). Neurotoxicology. 59: 33–39.

[94]   Whitney, M.C. and D.A. Cristol. 2018. Impacts of sublethal mercury exposure on birds: a detailed review. Rev. Environ. Contam. Toxicol. 244: 113–163.

[95]   Grandjean, P. and P.J. Landrigan. 2014. Neurobehavioural effects of developmental toxicity. Lancet Neurol. 13: 330–338.

[96]   Law, R.J., A. Covaci, S. Harrad, D. Herzke, M.A.-E. Abdallah, K. Fernie, et al. 2014. Levels and trends of PBDEs and HBCDs in the global environment: Status at the end of 2012. Environ. Int. 65: 147–158.

[97]   Tongue, A.D.W., S.J. Reynolds, K.J. Fernie and S. Harrad. 2019. Flame retardant concentrations and profiles in wild birds associated with landfill: a critical review. Environ. Pollut. 248: 646–658.

[98]   Eng, M.L., V. Winter, J.E. Elliott, S.A. MacDougall-Shackleton and T.D. Williams. 2018. Embryonic exposure to environmentally relevant concentrations of a brominated flame retardant reduces the size of song-control nuclei in a songbird. Dev. Neurobiol. 78: 799–806.

[99]   Eng, M.L., J.E. Elliott, S.A. MacDougall-Shackleton, R.J. Letcher and T.D. Williams. 2012. Early exposure to 2,2-,4,4-,5-pentabromodiphenyl ether (BDE-99) affects mating behavior of Zebra Finches. Toxicol. Sci. 127: 269–276.

[100] Walker, C.H. 2003. Neurotoxic pesticides and behavioural effects upon birds. Ecotoxicology. 12: 307–316.

[101] Richardson, J.R., V. Fitsanakis, R.H.S. Westerink and A.G. Kanthasamy. 2019. Neurotoxicity of pesticides. Acta Neuropathol. 138: 343–362.

[102] Iwaniuk, A.N., D.T. Koperski, K.M. Cheng, J.E. Elliott, L.K. Smith, L.K. Wilson, et al. 2006. The effects of environmental exposure to DDT on the brain of a songbird: changes in structures associated with mating and song. Behav. Brain Res. 173: 1–10.

[103] Hallmann, C.A., R.P.B. Foppen, C.A.M. van Turnhout, H. de Kroon and E. Jongejans. 2014. Declines in insectivorous birds are associated with high neonicotinoid concentrations. Nature. 511: 341–343.

[104] Ertl, H.M.H., M.A. Mora, D.J. Brightsmith and J.A. Navarro-Alberto. 2018. Potential impact of neonicotinoid use on Northern bobwhite (*Colinus virginianus*) in Texas: A historical analysis. PLoS One. 13: e0191100.

[105] Costa, L.G., G. Giordano, M. Guizzetti and A. Vitalone. 2008. Neurotoxicity of pesticides: a brief review. Front. Biosci. 13: 1240–1249.

[106] Grue, C.E., P.L. Gibert and M.E. Seeley. 1997. Neurophysiological and behavioral changes in non-target wildlife exposed to organophosphate and carbamate pesticides: thermoregulation, food consumption and reproduction. Integr. Comp. Biol. 37: 369–388.

[107] Busby, D.G., L.M. White and P.A. Pearce. 1990. Effects of aerial spraying of fenitrothion on breeding White-throated Sparrows. J. Appl. Ecol. 27: 743–755.

[108] Busby, D.G., L.M. White and P.A. Pearce. 1991. Brain acetylcholinesterase activity in forest songbirds exposed to a new method of UULV fenitrothion spraying. Arch. Environ. Contam. Toxicol. 20: 25–31.

[109] Busby, D.G., S.B. Holmes, P.A. Pearce and R.A. Fleming. 1987. The effect of aerial application of Zectran® on brain cholinesterase activity in forest songbirds. Arch. Environ. Contam. Toxicol. 16: 623–629.

[110] White, D.H. and J.T. Seginak. 1990. Brain cholinesterase inhibition in songbirds from pecan groves sprayed with phosalone and disulfoton. J. Wildl. Dis. 26: 103–106.

[111] Rogers, K.H., S. McMillin, K.J. Olstad and R.H. Poppenga. 2019. Imidacloprid poisoning of songbirds following a Drench application of trees in a residential neighborhood in California, USA. Environ. Toxicol. Chem. 38: 1724–1727.

[112] Bean, T.G., M.S. Gross, N.K. Karouna-Renier, P.F.P. Henry, S.L. Schultz, M.L. Hladik, et al. 2019. Toxicokinetics of imidacloprid-coated wheat seeds in Japanese Quail (*Coturnix japonica*) and an evaluation of hazard. Environ. Sci. Technol. 53: 3888–3897.

[113] Liu, M., G. Wang, S.-Y. Zhang, S. Zhong, G.-L. Qi, C.-J. Wang, et al. 2016. Exposing imidacloprid interferes with neurogenesis through impacting on chick neural tube cell survival. Toxicol. Sci. 153: 137–48.

[114] Slabbekoorn, H. and E.A.P. Ripmeester. 2008. Birdsong and anthropogenic noise: implications and applications for conservation. Mol. Ecol. 17: 72–83.

[115] Blickley, J.L. and G.L. Patricelli. 2010. Impacts of anthropogenic noise on wildlife: research priorities for the development of standards and mitigation. J. Int. Wildl. Law Policy. 13: 274–292.

[116] Ferrari, M.C.O., M.I. McCormick, M.G. Meekan, S.D. Simpson, S.L. Nedelec and D.P. Chivers. 2018. School is out on noisy reefs: the effect of boat noise on predator learning and survival of juvenile coral reef fishes. Proc. Royal Soc. B. 285: 2018033.

[117] Carson, R. 2002. Silent Spring: Fortieth Anniversary Edition. Houghton Mifflin, Boston, USA.

[118] Ratcliffe, D.A. 1970. Changes attributable to pesticides in egg breakage frequency and eggshell thickness in some British birds. J. Appl. Ecol. 7: 67–115.

[119] Hellou, J., M. Lebeuf and M. Rudi. 2012. Review on DDT and metabolites in birds and mammals of aquatic ecosystems. Environ. Rev. 21: 53–69.

[120] Farmer, C.J., L.J. Goodrich, E.R. Inzunza and J.P. Smith. 2008. Conservation status of North America's birds of prey. pp. 303–420. In: K.L. Bildstein, J.P. Smith, I.E. Ruelas, R.R. Veit (eds). The State of North America's Birds of Prey. Series in Ornithology 3. Nuttall Ornithological Club and American Ornithologists' Union, Washington, D.C., USA.

[121] de Montaigu, C.T. and D. Goulson. 2020. Identifying agricultural pesticides that may pose a risk for birds. PeerJ. 8: e9526.

[122] Roy, C.L. and P.L. Coy. 2020. Wildlife consumption of neonicotinoid-treated seeds at simulated seed spills. Environ. Res. 190: 109830.

[123] Eng, M.L., B.J.M. Stutchbury and C.A. Morrissey. 2019. A neonicotinoid insecticide reduces fueling and delays migration in songbirds. Science. 365: 1177–1180.

[124] Verfuss, U.K., C.E. Sparling, C. Arnot, A. Judd and M. Coyle. 2016. Review of offshore wind farm impact monitoring and mitigation with regard to marine mammals. pp. 1175–1182. In: A.N. Popper, A. Hawkins (eds.). The Effects of Noise on Aquatic Life II. Springer, New York, USA.

[125] Jetz, W., D.S. Wilcove and A.P. Dobson. 2007. Projected impacts of climate and land-use change on the global diversity of birds. PLoS Biol. 5: e157.

[126] Pimm, S.L., C.N. Jenkins, R. Abell, T.M. Brooks, J.L. Gittleman, L.N. Joppa, et al. 2014. The biodiversity of species and their rates of extinction, distribution and protection. Science. 344: 1246752.

[127] Bonnot, T.W., W.A. Cox, F.R. Thompson and J.J. Millspaugh. 2018. Threat of climate change on a songbird population through its impacts on breeding. Nat. Clim. Change. 8: 718–722.

Chapter **10**

Personality and Behavioural Syndromes

Kimberley J. Mathot[1], Sue Anne Zollinger[2]
and Todd M. Freeberg[3]

Introduction

Landscapes have been, and continue to be, dramatically altered by anthropogenic activity. Pervasive and obvious structural changes are clear in urbanized areas, but even many rural areas are profoundly altered by agriculture, transportation networks, and other industrialized infrastructure. These human alterations not only result in structural disturbance of the environment, including habitat fragmentation, but are

[1]Department of Biological Sciences, University of Alberta
[2]Department of Natural Sciences, Manchester Metropolitan University
[3]Department of Psychology, University of Tennessee Knoxville

often accompanied by other factors such as chemical, noise or light pollution, and severe losses in biodiversity, among others. Avian urban ecology examines the ways in which birds adapt to anthropogenic impacts associated with urbanization. Though now a well-established field, studies of urban ecology are still largely focused on population-level investigations. For example, do populations respond to anthropogenic change via avoidance/attraction, adjustment (e.g., through reversible behavioural plasticity or developmental plasticity), or adaptation? This chapter begins by providing examples of these types of population-level studies. However, within populations, individuals often show repeatable differences in their average behaviour (i.e., animal personality), or repeatable differences in suites of correlated traits (i.e., behavioural syndromes; see Box 1 for a glossary of key terms). The chapter describes these phenomena in detail before moving on to discuss the ways in which explicit recognition of animal personality and behavioural syndromes can enrich understanding of the ecological and evolutionary impacts of anthropogenic change on songbirds. The chapter ends with a discussion of questions related to the ecology and evolution of animal personality in the Anthropocene and suggestions for important areas of future research on this topic.

Box 1   Glossary

Among-individual variation: Differences in the average phenotype of individuals across repeated observations.

Animal personality: Term used to describe the presence of repeatable among-individual differences in behaviour in a population. An individual's expressed level of the behaviour is sometimes referred to as its "personality", but is better described as its 'behavioural type'. Statistically, measured as the proportion of phenotypic (behavioural) variation in a population that exists at the among-individual level.

Behavioural type: An individual's average expression of a behaviour which shows repeatable among-individual variation (see Animal personality). For example, exploration may show repeatable among-individual variation (i.e., animal personality), and an individual can be characterized as having high or low aggression (behavioural type).

Behavioural plasticity: Changes in behaviour within individuals across an environmental gradient. Also referred to as phenotypic plasticity or within-individual variation.

Behavioural syndrome: Describes covariation between behaviours at the among-individual level, such as when individuals with relatively high expressions of exploration also have relatively high expressions of aggression.

"Big 5":   Group of five commonly measured behavioural traits in animal personality research (exploration, activity, aggressiveness, shyness-boldness, and sociability). See, Box 2 for a brief history of their use in studies of animal personality and common ways of assaying these behaviours.

Within-individual variation: Differences in the observed phenotype of an individual across repeated observations. These differences are assumed to be the result of non-permanent (i.e., reversible) responses by the individual to changing environmental conditions. Also referred to as phenotypic plasticity or behavioural plasticity when the observed phenotype is a behaviour.

Anthropogenic Disturbance: Population-level Impacts

The past decades have seen a steady rise in published studies of the impact of anthropogenic activities, particularly urbanization, on songbird behaviour, health, and fitness. Three broad categories of response are possible (see also Chapter 8); avoidance/ attraction, adjustment, or adaptation [1]. Many species avoid anthropogenically altered habitats, or may avoid specific features of those habitats. For example, species with high neophobia, low innovation, and low learning ability are less likely to inhabit urban habitats [2]. Species may be sensitive to multiple features of the altered landscape, or particularly sensitive to key features. For example, although human alteration of landscapes often involves suites of changes including increased human activity, anthropogenic noise, and modifications of infrastructure, several studies have now shown that noise alone is sufficient to reduce the abundance of breeding birds, with some species showing complete avoidance of areas affected by anthropogenic noise [3, 4]. Other species do well in urban environments, by taking advantage of anthropogenic resources (urban adaptors) or even depending on them (urban exploiters) [5]. These species often share a range of traits that facilitate exploiting the novel opportunities that urban habitats can provide. For example, urban species often exhibit innovative behaviours, high learning ability, and low neophobia; traits that presumably aid in the exploitation of novel food sources [2].

Avoidance and attraction to anthropogenically altered landscapes are rarely absolute, and when species are found in habitats across gradients of anthropogenic disturbance, they often exhibit suites of behavioural differences across these gradients. For example, urban songbirds commence daily activity earlier than their rural counterparts [6], and songbirds in areas with artificial light at night (ALAN) have earlier dawn songs [7]. Urban birds tend to exhibit less fear towards humans [8, 9], employ different predator escape behaviours [10], are less neophobic [11, 12], and in some cases are less aggressive and less territorial [13]. Except for a few studies, we do not know the extent to which these patterns are due to reversible behavioural plasticity, developmental plasticity, microevolution, or to a combination of these processes.

One of the most widely observed patterns when comparing songbirds across different anthropogenic impact gradients is that songbirds in noise-polluted territories show alterations to song features compared with birds in less noisy territories. As this topic has received perhaps the longest and most diverse attention, it makes a good starting point from which to consider how songbird populations respond to anthropogenic change, and more specifically, the types of studies that are required to tease these alternative mechanisms apart. Of the now hundreds of published studies investigating avian vocal behaviour in noise, the majority report associations between average song features, such as frequency, amplitude, or temporal patterns and levels of background noise pollution levels (reviewed in [14]). Correlations between the minimum or peak frequency of songs and high anthropogenic noise levels have been reported now in scores of species, and on every continent where there are cities. Numerous bird species increase the frequency of their songs to move above the noise floor of frequencies caused by anthropogenic factors like industry or traffic noise. The first species for which a shift in frequency was reported was great tits, *Parus major* [15], an observation that was followed up with additional reports of the same phenomenon in great tits in cities across Europe and in Japan [16, 17]. But are these population differences the result of immediate behavioural plasticity, effects of noise on earlier song learning or development, or selection for certain song types that are less masked?

While there are still relatively few studies that investigate this question, the evidence for one mechanism over another is not consistent. For zebra finches (*Taeniopygia guttata*) and great tits, higher frequencies of songs in noise do not seem to result from developmental plasticity, because juveniles exposed to noise during their song learning phase did not learn to sing higher frequency songs than their tutors or than control groups [18, 19]. Nor does reversible behavioural plasticity appear to be a key mechanism mediating the observed population-level vocal shifts. In great tits, adult birds exposed to changing noise conditions did not exhibit behavioural plasticity in the frequency of their songs and calls, but did immediately increase vocal amplitude [3]. Thus, the widely observed pattern of higher song frequencies in urban great tits is not the result of irreversible developmental plasticity or reversible within-individual plasticity, but is likely happening at the community level over the course of several generations, consistent with micro-evolutionary change. Further evidence for micro-evolutionary responses in song features to increased urbanization comes from studies in white-crowned sparrows (*Zonotrichia leucophrys*). Using historical records of song, Luther and Derryberry [20] showed that song frequencies have changed over a 36-year period as noise levels have risen. The population-level shifts in song frequency may be a result of selection for certain song types by receivers, or may be the outcome of generations of individuals selectively learning higher songs, until certain low-pitched tutor songs are lost from the collective repertoire. Experimental exposure of young white crowned sparrows to traffic noise during ontogeny does result in individuals with higher frequency song types, through developmental plasticity during the learning process [21]. Young sparrows exposed to low frequency noise during development preferentially learned higher frequency song types, which may provide a mechanism explaining the gradual population shift observed in this species [20]. Still other studies show evidence that immediate and reversible plasticity enables songbirds to adjust song features to current levels of noise [22, 23–26]. Thus, the role of individual plasticity in frequency as a strategy employed by songbirds to cope with fluctuating anthropogenic noise levels appears variable.

These examples illustrate that adaptation to novel anthropogenic factors can and does occur via micro-evolution, developmental plasticity, and reversible phenotypic plasticity (also referred to as within-individual flexibility). These examples further demonstrate that the relative importance of these processes varies across species and the specific behavioural trait in question. How does consideration of animal personality enrich understanding of how these mechanisms might impact the ways in which songbirds adapt in the Anthropocene? To answer that, it is crucial to understand what is meant by "animal personality".

What is Animal Personality?

Animal personality generally relates to internal mechanisms within individuals that lead to repeatable behaviour patterns over time and over a wide range of contexts [27]. The notion of personality-like influences has gone by different terms over time—e.g., temperament, behavioural predispositions, reaction norms, coping styles, behavioural type and behavioural syndromes—and has a long history in the fields of animal and human behaviour. In Lamarck's discussions of inner feelings, temperaments, inclinations, and habits [28], and in Darwin's writings on inherited tendencies and associated habitual movements [29], there are arguments about individual behaviour being constrained by factors internal to the individual. Constraints could come from evolutionary or developmental factors, and could be based on anatomical, physiological, or behavioral factors. Importantly, these early notions of personality and constraint were generally thought to be consistent across individuals within a species, views that had a profound influence on the field of ethology [30]. Overlapping in time with early ethology, research on human psychology in the early and middle 1900s [31] pointed to the importance of individual differences and repeatable variations in tendencies to behave in certain ways. This view extended into research on non-human primates as well [32].

More recently, the study of among-individual differences in behavioural tendencies has been expanded to include all animal taxa [33, 34]. Researchers have often held the view that behaviours are labile or flexible traits and individuals can quickly and reversibly adjust their level of expression of a given behaviour in response to current conditions, which at first may seem inconsistent with the notion of constraints in behavioural expression. Given the often high degree of flexibility observed in behavioural expression, it is not surprising that traditionally, many studies in behavioural ecology focused on how behaviour changes *on average* in response to changing conditions. These studies generally assumed that any observed changes in behaviour in response to a particular set of conditions reflected behavioural plasticity, a within-individual process [35]. While it has long been appreciated that individuals in populations differ, quantification of the extent and repeatability of among-individual differences in behaviour has only emerged as a major area of research in behavioural ecology in the last few decades [36]. There is now a large body of work showing that in organisms ranging from invertebrates to fish and mammals, individuals often exhibit what is referred to as repeatable among-individual variation, or animal personality [36]. Importantly, within-individual plasticity does not preclude the presence of repeatable among-individual variation [37], as seen below, and songbirds have been model systems for much of the work in this field [38].

What does it mean to have "repeatable among-individual variation"? This simply means that part of the total variation that exists in a population of animals exists because individuals differ in their average expression of behaviour. In statistical terms, individuals have different intercepts for behaviour. Fig. 1 shows how the same total amount of variance in a trait may or may not be associated with repeatable among-individual differences. In both panels in Fig. 1, there is an overall shift in the behaviour (trait A) across the environmental gradient, but in panel (a) there are no repeatable among-individual differences. Individuals do not differ from each other in predictable ways, and they show the same overall response. In panel (b) however, individuals do show repeatable variation. The individual shown with a triangle has consistently higher expression of trait A compared to the individual shown with a square, who itself has a constantly higher expression of trait A compared to the individual shown with a circle. As an example, a captive study of female-male pairs of Carolina chickadees (*Poecile carolinensis*) involved swapping of individuals among pairs midway through the study. In this case, the environmental gradient is the social partner identity. Individuals that called, ate, and drank more and that were more aggressive with their first social partner, also tended to call, eat, and drink more, and be more aggressive with their second social partner [39].

In the example illustrated in Fig. 1, the behaviour is labelled generically as 'Trait A', but the empirical example here describes several behaviours for which Carolina chickadees exhibit "personality". What kind of behavioural traits normally exhibit repeatable among-individual variation? The answer is, all of them [36]. In a meta-analysis summarizing the 759 estimates of repeatability of behavioural traits, Bell and colleagues found that the global average repeatability of behavioural traits was $r = 0.37$ [36]. This means that 37% of the variation that exists in the expression of a given behaviour within a population of animals exists because individuals differ on average in their expression of those traits (i.e., at the among-individual level). Although the estimated repeatability varied for different types of behaviours, all behaviours that were considered in the analysis showed significant repeatability, ranging from an average repeatability of $r \approx 0.20$ to $r > 0.8$ [36].

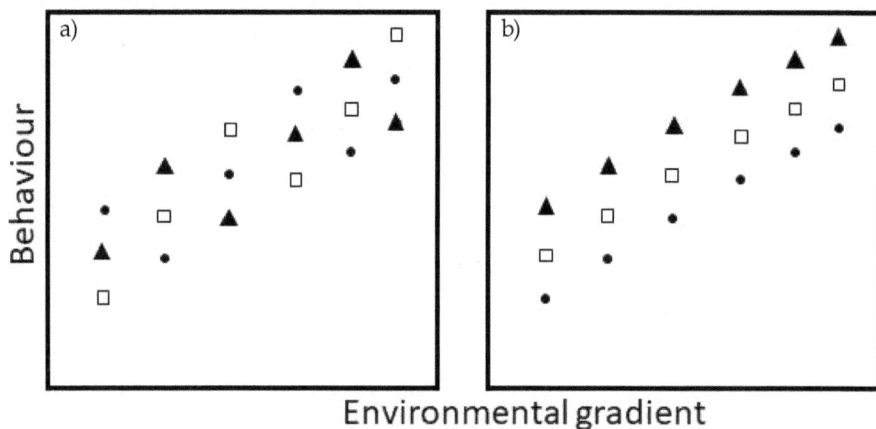

Figure 1 Illustration of how the same amount of phenotypic variation (i.e., total variance in behaviour), can either (a) have no associated among-individual differences, or (b) significant repeatable among-individual differences (e.g., triangle consistently has a higher expression than square or circle, right panel). In each panel, individuals are represented by different shapes, and in both panels, individuals show the same average increase in the expression of behaviour across the environment.

Although any behaviour can be studied in an animal personality framework, an overwhelming majority of studies of animal personality attempt to classify behaviours along one of five behavioural axes; shyness-boldness, exploration-avoidance, activity, aggressiveness, and sociability [34] (see Box 2). More recently, researchers have begun to study a wider range of behaviours under the animal personality framework, and include more observation of behaviours expressed in the wild. As long as data are collected in a way that allows within- and among-individual level variation to be decomposed, any behavioural trait can be studied within the framework of animal personality (see Box 3). The recent shifts in animal personality research to include a broader range of behavioural categories have come about for several reasons. Concerns have been raised about what many of the standardized assays traditionally used in this research field are actually measuring, and whether the same assay can be interpreted in the same way for different species or even different populations of the same species [40] (see also Box 2). Additionally, many behaviours that are of interest to behavioural ecologists and are clearly functionally important do not always fit neatly into one of these five categories. For example, should parent birds provisioning their young at high rates be considered highly active, bold, or both? Does seeking extra-pair matings equate to novelty seeking (i.e., unfamiliar mate), aggressiveness (i.e., willingness to intrude on another male's territory), activity (i.e., moving across larger areas), or some combination of these?

Box 2 Approaches for assaying 'The Big 5' in songbirds

In 2007, Réale and colleagues proposed five broad categories of behaviour for studies of animal personality: exploration-avoidance, activity, aggressiveness, shyness-boldness, and sociability [34], which were subsequently adopted en masse by many researchers in the field. These five categories were proposed as a working tool, and were never meant to be an exhaustive list [34]. However, they shaped much of the empirical work on animal personality over the following decade. The popularity of these five categories of behaviour stemmed in part from the ease with which they could be assessed in standardized ways, facilitating repeated measures in the same context, as required to estimate among-individual variance (see Box 3).

Perhaps the most widely studied behaviour under the umbrella of animal personality is what was termed **"exploration"** or "exploration-avoidance", which is a measure of an individual's reaction to a novel situation. This may include reactions to novel food, new habitat, new objects, etc., and is sometimes referred to as neophilia-neophobia. The first widely used assay in the context of animal personality research involved bringing songbirds into a novel room fitted with "artificial trees", and scoring the extent of movement in the first two minutes after being introduced to the room [38] (Fig. 2(a)). More recently, a field compatible version of this assay was developed, which uses a bird cage as the novel environment [92] (Fig. 2(b)). This method allows assays to be performed quickly in the field, which may be useful when the field site is either far from lab infrastructure, or during the breeding season, when removal of individuals from their territories should be kept as short as possible. **Activity** refers to the degree of movement exhibited by an individual in a non-risky and non-novel environment. This is often assessed using the same testing environments as for exploration, but after individuals have had an opportunity to become familiar with them. **Aggressiveness** is a measure of the extent and/or rate of agonistic reactions towards conspecifics and/or heterospecifics. In captivity, this may be measured using the rate of agonistic interactions between individuals, and in the field, it is commonly measured using standardized territorial intrusions (STIs). STIs can be performed either by experimentally broadcasting the call of conspecifics on the territory, presenting a model of a conspecific intruder, or both [93] (Fig. 2(c)).

Another major axis of behavioural variation studied in the animal personality framework is shyness-boldness. **Shyness-boldness** is a measure of an individual's reaction to risky situations. This can include reactions to predators or humans, but should not include reactions to novel situations (see exploration-avoidance). For example, individuals may be startled at a feeder, and their boldness scored as the latency to resume feeding (Fig. 2(d)). The fifth behavioural category proposed by Réale and colleagues is **sociability**, which is a measure of an individual's response to conspecifics, excluding agonistic responses, which are captured under the trait of aggression. Sociable individuals seek the presence of conspecifics and unsociable individuals avoid conspecifics. This behavioural category has probably received the least attention of the "Big 5" proposed by Réale and colleagues. However, the use of PIT tags has enabled several recent studies to construct "social" networks in free-living birds based on patterns of association between pairs of individuals at feeders equipped with RFID antennas [94, 95]. Such RFID data, unfortunately, cannot capture the nature of the interactions between individual (e.g., agonistic or pro-social).

These types of standardized assays have many advantages, including being relatively easy to implement, and in some cases, removing subjects from their environment and

Box 2   Approaches for assaying 'The Big 5' in songbirds (Contd....)

minimizing the risk of assaying pseudo-personality (i.e., when environmental features shape behavioural expressions, and individuals are assayed in different environments) [96]. However, it is also important to note that no behaviour is by definition a 'personality trait', and repeated behavioural observations per individual are necessary to properly quantify animal personality (see Box 3). Further, the functional significance of these assays is not always clear, and may differ for different species or contexts [40]. For example, high exploration may equate to high information gathering when the cues being assessed are conspicuous, but low (sometimes called slow, or thorough) exploration may equate to higher information gathering when cues are cryptic [97]. Thus, in addition to the importance of quantifying among-individual differences by properly decomposing variance (Box 3), validating behavioural assays is also critical [40]. The most powerful approach for validating behaviours involves using a combination of standardized (but potentially artificial assays), and measures of natural behaviours assayed in free-living animals (which may be noisy due to uncontrolled variables).

Figure 2   Images of some common standardized behavioural assays used in songbird research on animal personality. (a) The standard novel environment test is often a white room with artificial 'trees' used to assay exploration. Movement between the five artificial "trees" is used to assess 'exploration'. The dark squares on the walls are the sliding doors through which birds can enter the novel environment without being handled by observers. (b) A cage version of the novel environment test used to assay exploration in the field. A great tit can be seen in the cage, and movement between perches is used to assess 'exploration'. (c) An example of a standardized territorial intrusion test in the field used to assay aggression. In this photo, a stuffed male great tit was placed a few meters in front of a nest-box. The resident male can be seen perched on the protective wire mesh around the mount, attacking it. (d) Example of a field experiment of boldness in black-capped chickadees. A stuffed merlin (*Falco culumbarius*) was placed in the vicinity of a feeder together with a speaker broadcasting chickadee alarm calls, and latency to resume feeding was tracked with RFID antennas. An observer can be seen inserting their own PIT tag into the feeder antenna to automatically log the beginning of the predator presentation. All photos are by Jan Wijmenga and used with permission.

Box 3 Contemporary approaches to studying animal personality

The strong emphasis in earlier studies of animal personality on the behavioural categories outlined by Réale and colleagues [34] led to a common misconception that specific behaviours are synonymous with animal personality, and therefore, that a single measure of that behaviour revealed an individual's "personality". For example, it is not uncommon to find papers that read something like "we measured two personality traits: exploration and activity", or some other combination of the big five described in Box 2, but then which proceed to take only a single measure per individual per trait. In fact, a recent review by Niemelä and Dingemanse [98] found that >60% of studies which purported to address questions related to animal personality relied on a single measurement per individual to quantify "personality". Why is this a problem? Using a single measure per individual makes the assumption that the phenotypic correlation (i.e., the correlation between pairs of observations of traits) accurately captures the among-individual correlation, an assumption that has been termed the "individual gambit" [99]. While it is possible for phenotypic correlations to accurately estimate among-individual correlations (Fig. 3(a)), more often they will yield biased estimates of the true relationship. That is because the estimate will be biased whenever the within-individual association (i.e., due to plasticity) and among-individual association (i.e., animal personality) are not identical (Fig. 3(b) and 3(c)), which can reasonably be assumed to be the norm.

Contemporary research in animal personality recognizes that a given trait is not synonymous with animal personality, and that as long as there is repeatable among-individual variation in the expression of the behaviour, there are important questions to consider about the causes and consequences of that among-individual variation. This requires that repeated measures of the behaviour can be obtained on the same individuals to properly decompose variation to the among-individual level [98]. Given that the appropriate data can be collected, different statistical approaches are possible to evaluate both the repeatability of the trait [100], and associations between multiple traits at the among-individual level [101, 102].

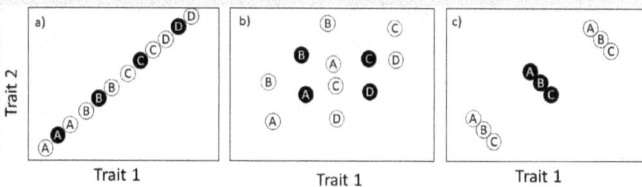

Figure 3 Illustration of how a single measure per individual may accurately capture the among-individual correlation (a), over-estimate (b), or under-estimate (c) the among-individual correlation. In all three panels, individuals (denoted by different letters) show repeatable among-individual differences in behaviour, as indicated by the difference in mean trait values (black circles) across three repeated measures of each trait. In panel (a) the within-individual correlation is identical to the among-individual correlation, such that a single measure per individual (dashed circles) would provide an unbiased estimate of the among-individual correlation. In panels (b) and (c), the within and among-individual correlations are not the same, and single measures per individual lead to biased estimates of the among-individual correlation. In panel (b) there is no among-individual correlation, but a strong within-individual correlation between traits 1 and 2. A single measure per individual yields an overestimate of the among-individual correlation. In panel (c) within- and among-individual correlations are in the opposite direction, and a single observation per individual would tend to incorrectly estimate a positive correlation between traits.

Progress towards scoring functionally relevant behaviour in free-living populations has been aided greatly by recent technological advances that have made it possible to automatically collect repeated behavioural data on free-living animals. For example, passive integrated transponder (PIT) tags make it possible to automatically record the presence of animals at regularly visited sites that are equipped with radio frequency identification (RFID) antennas, such as feeders or nest-boxes [41]. Use of this technology in free-living populations has revealed that great tits show repeatable among-individual variation in feeding rates [42], willingness to resume feeding after an experimental presentation of a model predator [43], time required to solve novel foraging tasks [44, 45], and ability to locate new feeding opportunities [46]. In blue tits (*Cyanistes caeruleus*), males differ in how often they visit off-territory nest-boxes, and this extraterritorial behaviour predicts rates of extra-pair paternity [47]. RFID technology and inexpensive nest-box infrared cameras have also aided in establishing that in songbirds, individuals often show repeatable variation in provisioning effort [48, 49]. These technological advances have increased the ability to track behaviour in free-living birds, and have enabled the study of many more natural behaviours under the umbrella of animal personality research.

An Among-individual Perspective on Songbirds in the Anthropocene

Why does animal personality matter when considering songbirds in the Anthropocene? The presence of animal personality is clearly important for micro-evolution, which by definition requires variation in the phenotype expressed among individuals in a population (and that the variation is heritable and association with differences in fitness). However, the presence of animal personality also has implications for the understanding of the roles of developmental and reversible plasticity as means of coping with a changing world. In the remaining sections, the chapter provides conceptual and empirical examples to illustrate how explicit consideration of among-individual differences can enrich understanding of the ecological and evolutionary impacts of anthropogenic change on songbirds. Additionally, these sections point to some of the ways that greater recognition of animal personality can improve our ability to evaluate anthropogenic impacts on populations.

Animal Personality in the Anthropocene

The previous section revealed that individuals within populations often show repeatable variation in a wide range of behavioural traits. How does this matter in the context of understanding anthropogenic impacts on songbirds? First, variation in the distribution of behaviours across environmental gradients is often assumed to reflect behaviourally plastic responses within individuals, whether reversible and flexible behavioural changes or irreversible developmental changes as described above. However, these patterns could also arise over ecological timescales through personality-related differences in habitat selection. If certain behavioural types are more likely to encounter or tolerate novel conditions brought about through human activities, those behavioural types may disproportionately settle in anthropogenically-impacted landscapes (also called founder effects), or disperse out of them. For example, more exploratory great tits show different spatial responses to changes in food availability. When previously rewarding feeders suddenly ceased to be rewarding, more exploratory individuals travelled longer distances in search of new food [46], suggesting they may be more likely to encounter novel habitats via their greater dispersal propensity. This mechanism could account for the common observation that exploration behaviour is higher in urban compared with rural great tits [50–52].

Compelling evidence for personality-related settlement patterns comes from another study in great tits [53]. After accounting for the magnitude of behaviorally plastic responses (i.e., changes in aggression across observations with differing levels of disturbance), among-individual differences in mean behaviour remained strongly associated with among-territory differences in average level of urbanization (as measured by the number of pedestrians, cyclists, and cars) [53]. More aggressive great tits were more likely to settle in areas with more disturbances. The spatial scale of the associations between aggression and urbanization were too small to be the result of micro-evolution. A similar result was found in a study of dunnocks (*Prunella modularis*) [54]. Boldness was assessed based on the flight initiation distance in response to an approaching human observer; bolder individuals took flight at shorter distances. Boldness was found to be highly repeatable, and bolder individuals were more likely to occupy high disturbance habitats [54]. Importantly, behavioural adjustment over time contributed very little to the observed personality-related habitat choice. The presence of personality-based habitat matching implies that observed, even genetically underpinned, differences in behaviours across urban/rural habitats can arise in the absence of micro-evolutionary change.

When personality-related differences in settlement into novel habitats have genetic underpinnings, as has been shown in great tits [51], population divergence can proceed more rapidly, but at the cost of a loss of genetic diversity [55]. Even in the absence of personality-related habitat choice, human altered landscapes can generate population variation through (micro-) evolutionary change if behavioural type is heritable and mediates fitness in anthropogenically-disturbed landscapes. For example, using the same measure of boldness as described above (flight initiation distance in response to human approach), Møller [8] showed that across 44 common species of European birds, urban birds were consistently bolder than rural birds, but further, that the magnitude of this effect was predicted by the number of generations since urbanization [8], strongly suggestive of micro-evolutionary adaptation.

The recolonization of historic ranges by western bluebirds (*Sialia mexicana*), demonstrates non-random habitat selection and micro-evolutionary processes simultaneously. Western bluebirds are obligate secondary cavity nesters whose range contracted dramatically in the early 20th century because logging reduced the availability of old growth trees [56]. Mountain bluebirds (*Sialia currocoides*) were similarly affected by a lack of nesting opportunities due to logging. However, because mountain bluebirds can inhabit higher elevation habitats, they persisted in pockets of habitat that were unaffected by logging [57]. Over the last 40 years, nest-boxes have been provided to facilitate recolonization of the historic ranges of these two species. The differences in habitat refugia for the western bluebirds and the mountain bluebirds described above created important differences in the temporal dynamics of recolonization between these species [57].

Mountain bluebirds recolonized the eastern portion of the historic range more quickly than western bluebirds did, because the high elevation refugia for mountain bluebirds meant they had shorter distances to travel. As a result, as western bluebirds re-established their historic ranges, they encountered already-established mountain bluebird populations, often at high densities, resulting in competition for nest-boxes [57]. In order to successfully re-establish, western bluebirds needed both to arrive in these areas (i.e., have a high dispersal propensity), and be able to outcompete mountain bluebirds when they arrived (i.e., be aggressive). In a large-scale study investigating the temporal dynamics of re-establishment of western bluebirds, Duckworth and Badyaev [57] showed that there was a biased dispersal of highly aggressive males on the invasion front of western bluebirds. Following establishment, aggression decreased over the next several generations due to selection against both dispersal and aggression. Thus, human impacts on the landscape, first through logging and later through the provisioning of nest-boxes, combined with heritable among-individual differences in aggression and dispersal propensity, have directed microevolution in this species.

Personality-related differences in habitat selection can also have important implications for the understanding and quantification of the consequences of different types of anthropogenic disturbance. For example, in many songbirds, parents show repeatable variation in the level of parental care that they provide, even across multiple breeding attempts over multiple years [e.g., 58]. Some individuals are "good" parents, consistently provisioning their young at relatively high rates, while others are "bad" parents. These differences in parental care may reflect differences in individual quality whereby high-quality individuals can afford to pay the costs of high levels of parental care. If high-quality individuals preferentially settle in high-quality territories, then the fitness consequences of disturbance may be overestimated (Fig. 4). For example, chestnut-collared longspurs (*Clacarius ornatus*) breeding in closer proximity to oil and gas developments and to roads have lower levels of parental care [59]. This is interpreted as evidence of a negative effect of these disturbances on birds. However, to remove any confounding effect of differential settlement of birds of different qualities in disturbed versus undisturbed areas would require either translocation experiments to document within-individual changes, experimental application of disturbance treatments after territory settlement, or detailed measurements of morphological correlates of individual quality. Indeed, there was some evidence for non-random territory settlement in this study. Structurally smaller females were more likely to breed near roads, and older and heavier females were more likely to settle near oil and gas infrastructure [59], suggesting potential individual quality-related differences in territory settlement.

Another possibility is that novel anthropogenic conditions generate an ecological trap [60], whereby high-quality individuals preferentially settle in disturbed landscapes (Fig. 4d). In such cases, failure to account for among-individual differences may actually lead to an underestimate of the ultimate effect of the anthropogenic impact. For example, differences observed in parental care in noise-polluted areas may result from individual plasticity in provisioning behaviour, or may instead be the result of acoustic masking of important offspring-parent signals. In house sparrows (*Passer domesticus*), individual females provisioned chicks significantly less when they nested in noise-exposed nest-boxes than when they nested in quieter territories, and even provisioned less during noisy periods during the same season than during quieter moments [61]. Nests from the noisy habitats in this study produced fewer and smaller young than nests in quiet areas, which argues for interpreting the poor provisioning by females as the result of acoustic masking rather than an example of an adaptive change in provisioning strategy by the individuals in response to the change in noise levels (e.g., such as if noisy areas have greater prey available, favouring lower provisioning rates). This study highlights the importance of considering personality and individual plasticity when conducting studies in anthropogenically-disturbed areas. Without longitudinal data, it might be concluded that poor-quality mothers are more likely to settle in the noisy territories, or that the overall negative impact of noise pollution is low. However, by tracking individual female behaviour across breeding seasons, Schroeder and colleagues were able to show that noise-polluted sites may be ecological traps: attracting high-quality parents that do not perceive the noise as a threat to reproductive success [61].

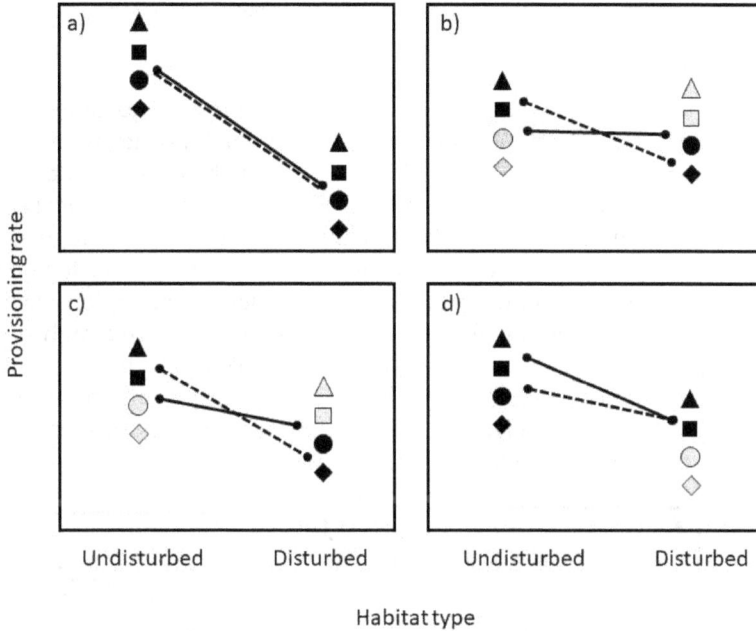

Figure 4 Illustration of how recognition of among-individual differences in habitat selection may alter the interpretation of the impacts of human disturbance on songbirds. Different individuals are represented by different shapes, and shading indicates selection probability (black = high probability, grey = low probability). The solid black line represents the average within-individual change across habitat conditions, and the dotted line represents the observed habitat-related effect when among-individual differences are not accounted for.

In panel (a) all individuals show the same decrease in provisioning in response to human disturbance and there are no personality-related differences in habitat selection. In this case, the observed habitat related differences in provisioning are reflective of the actual within-individual effect. In panel (b), there are no within-individual changes in provisioning across habitats. However, when measuring provisioning rates across habitat types, there is still an apparent habitat effect that arises from the fact that high quality birds are more likely to occupy undisturbed habitats (their absence from disturbed habitats is indicated by their light grey colouring), and vice versa for low quality individuals. Failure to recognize personality-related differences in habitat selection overestimates the habitat-related consequences of disturbance. Panel (c) shows the scenario where both within-individual plasticity and personality-related differences in habitat selection occur simultaneously. Failure to account for among-individual differences will tend to overestimate disturbance-related effects on songbirds. Panel (d) illustrates an ecological trap, where high quality individuals (triangle and square) are more likely to settle in disturbed habitats, but there is a fitness cost of settling there. Failure to recognize the differential settlement patterns of high versus low quality individuals results in an underestimation of the true impact of the disturbance.

Behavioural Syndromes in the Anthropocene

The bluebird example above actually encompassed two types of among-individual differences: among-individual differences in average behaviour (i.e., animal personality), but also, covariation between two behaviours at the among-individual level (i.e., behavioural syndromes) [62]. Individuals showed repeatable, and heritable, variation in both dispersal propensity and aggression. Further, aggression and dispersal propensity were correlated such that individuals that were more aggressive on average also had greater dispersal tendencies [57]. In fact, there are many examples of suites of behavioural traits being correlated in behavioural syndromes [62], and these often reflect underlying genetic correlations [63]. Understanding syndrome structure (i.e., the strength and direction of correlations between behavioural traits at the among-individual level) that exists in a population *before* changing the selection regime is important for understanding the genetic constraints that are present in a population. Knowing these constraints is key to being able to predict the capacity for populations to evolve in response to new selection regimes. In the bluebird example, selection aligned with the direction of the trait correlation; conditions that favoured high dispersal also favoured high aggression, and vice versa. This alignment of selection with the direction of the trait correlation allows for rapid evolution following population re-establishment [64]. On the other hand, when selection does not align with the behavioural syndrome, evolution will be much slower (Fig. 5).

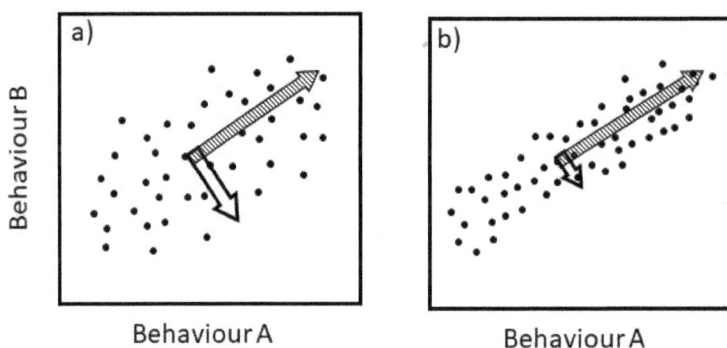

Figure 5 Illustration of how both the presence and strength of a behavioural syndrome constrain evolution. When selection acts in the same direction as the correlation (hatched arrow), evolution can be very rapid. In contrast, when selection acts counter to the correlation (unfilled arrow), for example, high values of trait A are favoured, but low values of trait B are favoured, evolution will be slower. The rate of evolution will be further affected by the strength of the syndrome correlation. When the correlation is weak, as in panel a), evolution will be relatively rapid compared to when the correlation is strong, as in b), particularly when selection does not align with the syndrome correlation.

Understanding syndrome structure *before* changing the selection regime can yield critical insights regarding the functional significance of differences in trait means across different levels of anthropogenic disturbance. For example, exploration and boldness scores in great tits have been linked to singing behaviours, such as level of singing activity during the dawn chorus [65], or level of song activity prior to breeding in early spring [66], as well as song parameters on a smaller temporal scale such as delivery rate or bout length [67]. These findings raise the intriguing possibility that some of the differences in song features observed across rural and urban populations may be artifacts of urban habitats favouring bolder, more aggressive individuals, with song features being "brought along for the ride" due to their inherent association with boldness and aggression. Recent years have seen an increase in experimental investigations into the mechanisms underlying differences recorded in songbird vocal behaviour in anthropogenically-disturbed areas. However, the large diversity in behavioural responses across different species and the difficulty in measuring and interpreting individual responses to fluctuating and diverse urban challenges in the field mean there is still much that is unknown about the interaction of animal personality, urbanization and birdsong.

Existing behavioural syndromes can facilitate or constrain responses to novel selection pressures. Similarly, novel selection pressures originating from human-produced changes in the landscape can generate or break down behavioural syndromes. There are a handful of studies in birds that have considered how syndrome structure varies as a function of anthropogenic factors. Some studies have found no evidence for changes in syndrome structure across urbanization gradients. For example, in great tits, although reaction to handling stress and exploration behaviour were both higher in urban populations, there was no among-individual covariance in these traits in either forest or urban populations [50]. Other studies have shown breakdowns of syndrome structure in urban populations. In house sparrows, food neophobia is part of a syndrome structure with risk-taking and activity in rural birds, but not urban birds [68], possibly because selection against food neophobia would be particularly strong in urban settings, where sparrows might regularly encounter new potential food sources.

In song sparrows (*Melospiza melodia*), behavioural syndromes exist between aggression and boldness in rural populations [69]. Individuals that are bold allow human observers to approach more closely before they flush (i.e., they have shorter flight initiation distances). Bold individuals are also more aggressive, as measured by their approach distance to a speaker that was broadcasting a playback song of an unknown male conspecific. When song sparrows move into urban settings, they were predicted to show increased boldness (i.e., shorter flight initiation distances), because the regular frequency with which urban song sparrows experience human disturbances would make low flight initiation distance (i.e., easily flushed) too costly. Additionally, because of the syndrome structure observed in rural populations, the urban population was expected to show a coincident shift in aggression, and this is what the authors found. However, the syndrome structure actually broke down in the urban population. Even though both boldness and aggression increased in urban birds, they were no longer correlated among individuals. In other words, individuals with higher than average aggression did not have higher than average boldness. This pattern of syndrome breakdown in urban populations has been observed in other studies in song sparrows [70, 71].

Work in other taxa has shown that syndrome structure is often strongest under harsh conditions. For example, a behavioural syndrome between exploration and boldness is present in rural, but not urban populations of great tits [72], and while correlations between object neophobia, risk taking, and activity were similar in both urban and rural populations of house sparrows, food neophobia was also part of the behavioural syndrome in rural populations, but not urban populations [68]. Such observations have led to the interpretation that for some species, including song sparrows, urban settings may reflect higher quality habitat than rural settings, at least in terms of food availability. Such an interpretation implies that at least to some degree, trait expression varies in response to habitat quality through behavioural plasticity. In a follow up study, Foltz et al. [71] experimentally manipulated territory quality in breeding song sparrows by provisioning the area around some nests with supplemental food, and leaving other nests untreated. They found that in the rural birds, providing supplemental food increased aggression to levels similar to the control birds in urban populations, while urban birds showed a small but non-significant increase in aggression in response to food supplementation. These results are consistent with the idea that the breakdown in syndrome structure in urban populations arises via high quality habitats relaxing constraints on trait expression, allowing for greater within-individual plasticity.

Among-individual Differences in Behavioural Plasticity

Importantly, the presence of animal personality does not imply lack of behavioural plasticity. In the right panel of Fig. 1, individuals differ consistently from each other while also showing a predictable response to changes in the environmental gradient. In this example, all individuals show the same level of behavioural plasticity. In statistical terms, they have the same reaction norm slope. Interestingly, among-individual differences in average behaviour are often also associated with among-individual differences in behavioural plasticity [73] (Fig. 6). Understanding how and why individuals differ in their degree of behavioural plasticity can yield important insights for predicting response to human induced change.

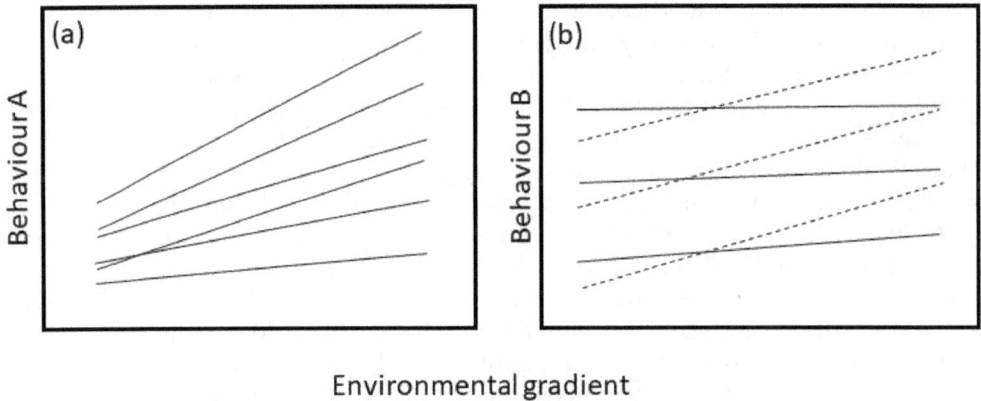

Environmental gradient

Figure 6 Examples of two patterns of personality-related differences in behavioural plasticity. In panel (a), individuals differ in intercept (personality) and slope (plasticity), with intercept-slope covariance for the same behaviour. Individuals with higher average expression of behaviour A are also the most behavioural plastic for behaviour A across the environmental gradient. In panel (b), individuals differ in intercept (personality) for behaviour A, as indicated by the type of line (solid line = high intercept for behaviour A, dashed line = low intercept for behaviour A). The lines indicate individual differences in behavioural plasticity for behaviour B. Individual differences in behaviour A are correlated with individual differences in slope (plasticity) in behaviour B.

When average behaviour and behavioural plasticity are correlated, any anthropogenic impacts that drive shifts in trait means, either through differential settlement, differential survival, or differential reproduction, can indirectly select for higher or lower behavioural plasticity. This may have consequences for the ability of populations to cope with future change. For example, in great tits, lay date and plasticity in lay date are genetically correlated [74]. Over the last several decades, spring temperatures have warmed, resulting in earlier emergence of caterpillar prey, thus selecting for earlier timing of breeding [74]. Because of the genetic correlation between average lay date and plasticity in lay date, climate-change induced selection on lay date has resulted in increased plasticity in timing of laying in great tits. Urbanization has also been shown to impact phenology of songbirds, most often by advancing laying dates and nest building compared with matched populations in nearby more traditional habitats, although in few studies an opposite effect was found with urban birds breeding later than rural counterparts [75]. However, it is less certain how inter-individual differences factor into these population-level divergences, and if phenological shifts in urban populations are the result of variation in individual phenotypic plasticity or genetic drift between populations.

Anthropogenic impacts may also reduce plasticity in traits—potentially limiting the ability of species to cope with future changes. For example, high levels of exploration would be expected to facilitate dispersal into novel environments. Indeed, in great tits, there is a significant and positive genetic correlation between dispersal distance and exploration [76]. At the same time, faster exploring great tits perform worse at reversal learning [77], suggesting that the types of individuals that are more likely to disperse into novel habitats may also be less able to plastically adjust their behaviour to changing contingencies once they have become established. In black-capped chickadees (*Poecile atricapillus*), urban birds have also been shown to be faster explorers, though the mechanisms generating this pattern are unclear (differential settlement, developmental plasticity, or micro-evolution) [78]. As seen in the great tit, fast exploring black-capped chickadees have also been shown to be slower at reversal learning [79]. If urban adapted individuals are generally poor at reversal learning, this leads to another interesting hypothesis that the individuals that thrive in urban habitats might be particularly prone to becoming pests. In urban adapted individuals, once an undesirable association is formed from the perspective of humans, such as tearing open garbage to access food subsidies, these individuals may be less likely to respond to aversive conditioning.

Ecological or Evolutionary Consequences?

This chapter has provided examples of cases of among-individual differences in behaviour predicting responses to anthropogenic change. For example, western bluebirds with high dispersal propensity and high aggressiveness are faster to re-establish in their historical range when suitable nesting sites are provided. In more general cases of species and range expansions, are bolder or fast-exploring individuals typically the ones at the leading edge of the expansion, as suggested in fish [80] and frogs [81]? The chapter has also discussed examples of anthropogenic changes exerting selection pressures in songbird populations. In song sparrows, higher food availability in urban habitats allows for greater behavioural plasticity, leading to a breakdown in the behavioural syndrome structure. Ultimately, understanding how such changes will impact songbird populations on ecological versus evolutionary timescales requires an understanding of the proximate mechanisms underlying among-individual differences in behaviour.

Animal personality, behavioural syndromes, and among-individual differences can all exist in the absence of underlying genetic constraints. Long-term repeatability in behaviour can arise when individuals show repeatable among-individual differences in the state variables that guide behaviour, such as through permanent environment effects [82]. Long-term behavioural repeatability can also arise from processes such as positive feedbacks, where differences in state favour differences in the expression of a second behaviour, which in turn reinforce the initial difference in state [83]. For example, initially small, even stochastic differences in body condition among individuals may lead to differences in the ability to gain priority access to food. If individuals in better condition are better able to monopolize food, this will allow them to further build up their condition, which in turn, will allow them to continue to monopolize the resource. When differences in state and behaviour reinforce each other in this way, this is referred to as positive feedback [83]. Conceptual models show that when there are positive feedbacks, even initially small and random among-individual differences can be maintained over time. State-dependence can also lead to the emergence of behavioural syndromes and among-individual differences in behavioural plasticity. For example, if higher individual quality allows for higher levels of both provisioning behaviour and aggression, as has been shown in house sparrows [84], there will be a syndrome between aggression and parental care. Similarly, differences in behavioural plasticity can be shaped by prior experience [85]. When among-individual differences in the ability to cope with, or in the strength of responses to, environmental change are based on non-heritable variation, resulting changes in behavioural variation within populations will not lead to evolutionary change.

Conversely, when animal personality, behavioural syndromes, or behavioural plasticity have a genetic basis, differences in the propensity to settle, establish, and/or thrive in response to anthropogenic changes as a function of any of these three types of among-individual differences will lead to (micro-) evolutionary change. Indeed, there are many examples of strong heritability of among-individual differences. For example, the chronotype of great tits, which shapes sleep/wake cycles and daily activity patterns, has a heritability of 0.86 [86]. Such high heritability means that changes in environmental conditions, such as increased light at night, that favour certain chronotypes over others can lead to rapid evolution. When behavioural syndromes are the result of an underlying genetic correlation between the two traits, the direction of selection relative to the direction of the genetic correlation underlying the syndrome will influence the rate at which populations can adapt. When the two align, as in the western bluebird example, evolution will be swift [57, 64]. When they do not, evolution will proceed more slowly [64]. Behavioural plasticity, too, can have an underlying genetic basis [74].

Conclusions and Future Directions

Ultimately, understanding global impacts of increased anthropogenic change on songbirds will require a better understanding of the proximate mechanisms driving observed changes. Avian urban ecology, while now a well-established field in its own right, has still primarily focused on population-level investigations, and, more recently, within-individual responses such as behavioural or developmental plasticity. The integration of among-individual variation (i.e., personality) into avian urban ecology is still in its infancy. As this chapter has illustrated, such among-individual differences can have profound consequences for the understanding of anthropogenic impacts.

When the types of individuals that settle or establish in anthropogenically altered environments are not random with respect to behavioural phenotype, this fundamentally affects our ability to estimate the impacts of anthropogenic change. If lower quality individuals are more likely to settle in anthropogenically altered landscapes, negative impacts of anthropogenic effects may be overestimated, while if higher quality individuals are more likely to settle in these landscapes, negative impacts may be underestimated. When among-individual differences in behaviour and/ or behavioural plasticity are integrated in syndromes, understanding the syndrome structure is essential for predicting the scope for future adaptation. When selection is aligned with the syndrome structure this will facilitate future adaptation. However, when selection is perpendicular to the syndrome structure, future adaptation will be constrained. Thus, taking among-individual differences and syndrome structure into account both for estimating short-term anthropogenic impacts on songbirds, as well as for predicting long-term impacts that can be mediated through microevolution. More longitudinal studies on marked populations of songbirds combined with before-and-after intervention studies (also on marked songbirds) are needed to develop a stronger integration between urban ecology and animal personality.

A recent review revealed the extent to which personality and conservation biology are intertwined in amphibians [87]. These authors found that work on personality in amphibians had largely exposed three broad dimensions (that we discussed above): shyness-boldness, exploration-avoidance, and activity. All three dimensions relate to fundamental fitness-related outcomes in conservation-oriented contexts. For example, captive breeding programs could benefit from knowing the extent to which a bold-aggressive syndrome might negatively impact breeding efforts, and whether aligned or complementary personality types result in greater reproductive output. The goals of such captive breeding programs are typically to reintroduce those populations to the wild. Are some behavioural types better than others at re-establishing home ranges and territories in those areas? Are some behavioural types better able to learn to avoid predators in captivity prior to release [88]? Should reintroduction programs seek to maximize diversity of behavioural types in released individuals to try to maximize diversity in behavioral responses to environmental change, as well as to maximize underlying genetic variation [89]?

Although the review just discussed centered on amphibians, the questions and concerns raised clearly pertain to avian work. The behavioral choices birds make stem from a wide range of proximate factors including sensory systems, attention and perceptual processing, histories of experience and conditioning, and current physical and social environmental contexts [90]. As we hope we have made clear here, one of these major proximate factors is the individual's behavioural type (i.e., animal personality). Personality has been one of the behavioral dimensions of least focus for conservation biologists [91] and we hope the discussion in our chapter will help serve as a call to arms to change this.

LITERATURE CITED

[1]   Potvin, D.A. 2017. Coping with a changing soundscape: avoidance, adjustments and adaptations. Anim. Cogn. 20: 9–18.

[2]   Wong, B.B.M. and U. Candolin. 2015. Behavioral responses to changing environments. Behav. Ecol. 26: 665–673.

[3]   McClure, C.J.W., H.E. Ware, J. Carlisle, G. Kaltenecker and J.R. Barber. 2013. An experimental investigation into the effects of traffic noise on distributions of birds: avoiding the phantom road. Proc. R. Soc. B. 280: 20132290.

[4] Cinto Mejia, E., C.J.W. Mcclure and J.R. Barber. 2019. Large-scale manipulation of the acoustic environment can alter the abundance of breeding birds: evidence from a phantom natural gas field. J. Appl. Ecol. 56: 2091–2101.

[5] McKinney, M.L. 2002. Urbanization, biodiversity, and conservation: the impacts of urbanization on native species are poorly studied, but educating a highly urbanized human population about these impacts can greatly improve species conservation in all ecosystems. BioScience. 52: 883–890.

[6] Dominoni, D.M., B. Helm, M. Lehmann, H.B. Dowse and J. Partecke. 2013. Clocks for the city: circadian differences between forest and city songbirds. Proc. R. Soc. B. 280: 20130593.

[7] Kempenaers, B., P. Borgström, P. Loës, E. Schlicht and M. Valcu. 2010. Artificial night lighting affects dawn song, extra-pair siring success, and lay date in songbirds. Curr. Biol. 20: 1735–1739.

[8] Møller, A.P. 2008. Flight distance of urban birds, predation, and selection for urban life. Behav. Ecol. Sociobiol. 6: 63–75.

[9] Cooke, A.S. 1980. Observations on how close certain passerine species will tolerate an approaching human in rural and suburban areas. Biol. Conserv. 18: 85–88.

[10] Møller, A.P. and J.D. Ibáñez-Álamo. 2012. Escape behaviour of birds provides evidence of predation being involved in urbanization. Anim. Behav. 84: 341–348.

[11] Miranda, A.C., H. Schielzeth, T. Sonntag and J. Partecke. 2013. Urbanization and its effects on personality traits: a result of microevolution or phenotypic plasticity? Glob. Chang. Biol. 19: 2634–2644.

[12] Lowry, H., A. Lill and B.B.M. Wong. 2013. Behavioural responses of wildlife to urban environments. Biol. Rev. 88: 537–549.

[13] Carlier, P. and Lefebvre, L. 1997. Ecological differences in social learning between adjacent, mixing, populations of zenaida doves. Ethology. 103: 772–784.

[14] Brumm, H. and S.A. Zollinger. 2014. Avian vocal production in noise. pp. 187–227. *In:* H. Brumm (ed.). Animal Communication and Noise. Springer, Berlin.

[15] Slabbekoorn, H. and M. Peet. 2003. Birds sing at a higher pitch in urban noise. Nature. 424: 267–267.

[16] Hamao, S., M. Watanabe and Y. Mori. 2011. Urban noise and male density affect songs in the great tit *Parus major*. Ethol. Ecol. Evol. 23: 111–119.

[17] Slabbekoorn, H. and A. den Boer-Visser. 2006. Cities change the songs of birds. Curr. Biol. 16: 2326–2331.

[18] Zollinger, S.A., P.J.B. Slater, E. Nemeth and H. Brumm. 2017. Higher songs of city birds may not be an individual response to noise. Proc. R. Soc. B. 284: 20170602.

[19] Liu, Y., S.A. Zollinger and H. Brumm. 2021. Chronic exposure to urban noise during the vocal learning period does not lead to increased song frequencies in zebra finches. Behav. Ecol. Sociobiol. 75: 3.

[20] Luther, D.A. and E.P. Derryberry. 2012. Birdsongs keep pace with city life: changes in song over time in an urban songbird affects communication. Anim. Behav. 83: 1059–1066.

[21] Moseley, D.L., G.E. Derryberry, J.N. Phillips, J.E. Danner, R.M. Danner, D.A. Luther, et al. 2018. Acoustic adaptation to city noise through vocal learning by a songbird. Proc. R. Soc. B. 285: 20181356.

[22] Bermudez-Cuamatzin, E., A.A. Rios-Chelen, D. Gil and C.M. Garcia. 2012. Experimental evidence for real-time song frequency shift in response to urban noise in a passerine bird. Biol. Lett. 8: 320–320.

[23] Zollinger, S.A., J. Podos, E. Nemeth, F. Goller and H. Brumm. 2012. On the relationship between, and measurement of, amplitude and frequency in birdsong. Anim. Behav. 84: E1–E9.

[24] Brumm, H. 2004. Causes and consequences of song amplitude adjustment in a territorial bird: a case study in nightingales. An. Acad. Bras. Ciênc. 76: 289–295.

[25] Potvin, D.A. and R.A. Mulder. 2013. Immediate, independent adjustment of call pitch and amplitude in response to varying background noise by silvereyes (*Zosterops lateralis*). Behav. Ecol. 24: 1363–1368.

[26] Verzijden, M.N., E.A.P. Ripmeester, V.R. Ohms, P. Snelderwaard and H. Slabbekoorn. 2010. Immediate spectral flexibility in singing chiffchaffs during experimental exposure to highway noise. J. Exp. Biol. 213: 2575–2581.

[27] Weiss, A. and D.M. Altschul. 2017. Methods and applications of animal personality research. pp. 179–200. *In*: J. Call (ed.). Handbook of Comparative Psychology. Vol. 1: Basic Concepts, Methods, Neural Substrate, and Behavior. American Psychological Association, Washington D.C.

[28] Lamarck, J.B. 1809/1984. Zoological Philosophy. University of Chicago Press, Chicago.

[29] Darwin, C. 1872. The Expression of the Emotions in Man and Animals. John Murray, London.

[30] Tinbergen, N. 1951. The Study of Instinct. Oxford University Press, Oxford.

[31] Murphy, G. 1947. Personality: a Biosocial Approach to Origins and Structure. Harper & Brothers Publishers, New York.

[32] Yerkes, R.M. 1939. The life history and personality of the chimpanzee. Am. Nat. 73: 97–112.

[33] Carere, C. and D. Maestripieri. 2013. Animal Personalities: Behavior, Physiology, and Evolution. University of Chicago Press, Chicago.

[34] Réale, D., S.M. Reader, D. Sol, P.T. Mcdougall and N.J. Dingemanse. 2007. Integrating animal temperament within ecology and evolution. Biol. Rev. 82: 291–318.

[35] Davies, N.B., J.R. Krebs and S.A. West. 2012. An Introduction to Behavioural Ecology. John Wiley & Sons, Oxford.

[36] Bell, A.M., S.J. Hankison and K.L. Laskowski. 2009. The repeatability of behaviour: a meta-analysis. Anim. Behav. 77(4): 771–783.

[37] Dingemanse, N.J., A.J.N. Kazem, D. Réale and J. Wright. 2010. Behavioural reaction norms: animal personality meets individual plasticity. Trends Ecol. Evolut. 25: 81–89.

[38] Dingemanse, N.J., C. Both, P.J. Drent, K. Van Oers and A.J. Van Noordwijk. 2002. Repeatability and heritability of exploratory behaviour in great tits from the wild. Anim. Behav. 64: 929–938.

[39] Harvey, E.M. and T.M. Freeberg. 2007. Behavioral consistency in a changed social context in carolina chickadees. J. Gen. Psychol. 134(2): 229–245.

[40] Carter, A.J., W.E. Feeney, H.H. Marshall, G. Cowlishaw and R. Heinsohn. 2013. Animal personality: what are behavioural ecologists measuring? Biol. Rev. 88: 465–475.

[41] Bonter, D.N. and E.S. Bridge. 2011. Applications of radio frequency identification (RFID) in ornithological research: a review. J. Field Ornithol. 82: 1–10.

[42] Moiron, M., K.J. Mathot and N.J. Dingemanse. 2018. To eat and not be eaten: Diurnal mass gain and foraging strategies in wintering great tits. Proc. R. Soc. B. 285(1874): 20172868.

[43] Mathot, K.J., M. Nicolaus, Y.G. Araya-Ajoy, N.J. Dingemanse and B. Kempenaers. 2015. Does metabolic rate predict risk-taking behaviour? A field experiment in a wild passerine bird. Funct. Ecol. 29: 239–249.

[44] Cole, E.F., D.L. Cram and J.L. Quinn. 2011. Individual variation in spontaneous problem-solving performance among wild great tits. Anim. Behav. 81(2): 491–498.

[45] Morand-Ferron, J., E.F. Cole, J.E.C. Rawles and J.L. Quinn. 2011. Who are the innovators? A field experiment with 2 passerine species. Behav. Ecol. 22: 1241–1248.

[46] van Overveld, T. and E. Matthysen. 2010. Personality predicts spatial responses to food manipulations in free-ranging great tits (*Parus major*). Biol. Lett. 6: 187–190.

[47] Schlicht, L., M. Valcu and B. Kempenaers. 2015. Male extraterritorial behavior predicts extrapair paternity pattern in blue tits, *Cyanistes caeruleus*. Behav. Ecol. 26: 1404–1413.

[48] Wilkin, T.A., L.E. King and B.C. Sheldon. 2009. Habitat quality, nestling diet, and provisioning behaviour in great tits *Parus major*. J. Avian Biol. 40: 135–145.

[49] Mutzel, A., N.J. Dingemanse, Y.G. Araya-Ajoy and B. Kempenaers. 2013. Parental provisioning behaviour plays a key role in linking personality with reproductive success. Proc. R. Soc. B. 280: 20131019.

[50] Charmantier, A., V. Demeyrier, M. Lambrechts, S. Perret and A. Grégoire. 2017. Urbanization is associated with divergence in pace-of-life in great tits. Front. Ecol. Evol. 5(53).

[51] Riyahi, S., M. Sánchez-Delgado, F. Calafell, D. Monk and J.C. Senar. 2015. Combined epigenetic and intraspecific variation of the DRD_4 and SERT genes influence novelty seeking behavior in great tit *Parus major*. Epigenetics. 10: 516–525.

[52] Tryjanowski, P., A.P. Møller, F. Morelli, W. Biaduń, T. Brauze, M. Ciach, et al. 2016. Urbanization affects neophilia and risk-taking at bird-feeders. Sci. Rep. 6: 28575–28575.

[53] Sprau, P. and N.J. Dingemanse. 2017. An approach to distinguish between pasticity and non-random distributions of behavioral types along urban gradients in a wild passerine bird. Front. Ecol. Evol. 5: 92.

[54] Holtmann, B., E.S.A. Santos, C.E. Lara and S. Nakagawa. 2017. Personality-matching habitat choice, rather than behavioural plasticity, is a likely driver of a phenotype-environment covariance. Proc. R. Soc. B. 284(1864): 20170943.

[55] Isaksson, C. 2018. Impact of urbanization on birds: how they arise, modify and vanish. pp. 235–257. *In*: D.T. Tietze (ed.). Bird Species. Springer International Publishing, Cham.

[56] Brawn, J.D. and R.P. Balda. 1988. Population biology of cavity nesters in northern arizona: do nest sites limit breeding densities? Condor. 90: 61–71.

[57] Duckworth, R.A. and A.V. Badyaev. 2007. Coupling of dispersal and aggression facilitates the rapid range expansion of a passerine bird. Proc. Natl. Acad. Sci. U.S.A. 104: 15017–15022.

[58] Westneat, D.F., M.I. Hatch, D.P. Wetzel and A.L. Ensminger. 2011. Individual variation in parental care reaction norms: Integration of personality and plasticity. Am. Nat. 178: 652–667.

[59] Ng, C.S., P.G. Des Brisay and N. Koper. 2019. Chestnut-collared longspurs reduce parental care in the presence of conventional oil and gas development and roads. Anim. Behav. 148: 71–80.

[60] Hale, R. and S.E. Swearer. 2016. Ecological traps: current evidence and future directions. Proc. R. Soc. B. 283(1824): 8.

[61] Schroeder, J., S. Nakagawa, I.R. Cleasby and T. Burke. 2012. Passerine birds breeding under chronic noise experience reduced fitness. PLoS One. 7(7): e39200.

[62] Sih, A., A.M. Bell, J.C. Johnson and R.E. Ziemba. 2004. Behavioral syndromes: an integrative overview. Q. Rev. Biol. 79: 241–277.

[63] Dochtermann, N.A. 2011. Testing Cheverud's conjecture for behavioral correlations and behavioral syndromes. Evolution. 65: 1814–1820.

[64] Saltz, J.B., F.C. Hessel and M.W. Kelly. 2017. Trait correlations in the genomics era. Trends Ecol. Evolut. 32: 279–290.

[65] Snijders, L., E.P. Van Rooij, M.F.A. Henskens, K. Van Oers and M. Naguib. 2015. Dawn song predicts behaviour during territory conflicts in personality-typed great tits. Anim. Behav. 109: 45–52.

[66] Naguib, M., A. Kazek, S.V. Schaper, K. van Oers and M.E. Visser. 2010. Singing activity reveals personality traits in great tits. Ethology. 116: 763–769.

[67] Amy, M., P. Sprau, P. De Goede and M. Naguib. 2010. Effects of personality on territory defence in communication networks: a playback experiment with radio-tagged great tits. Proc. R. Soc. B. 277(1700): 3685–3692.

[68] Bokony, V., A. Kulcsar, Z. Toth and A. Liker. 2012. Personality traits and behavioral syndromes in differently urbanized populations of house sparrows (*Passer domesticus*). PLoS One. 7(5): 11.

[69] Scales, J., J. Hyman and M. Hughes. 2011. Behavioral syndromes break down in urban song sparrow populations. Ethology. 117: 887–895.

[70] Evans, J., K. Boudreau and J. Hyman. 2010. Behavioural syndromes in urban and rural populations of song sparrows. Ethology. 116: 588–595.

[71] Foltz, S.L., A.E. Ross, B.T. Laing, R.P. Rock, K.E. Battle and I.T. Moore. 2015. Get off my lawn: Increased aggression in urban song sparrows is related to resource availability. Behav. Ecol. 26: 1548–1557.

[72] Riyahi, S., M. Björklund, F. Mateos-Gonzalez and J.C. Senar. 2017. Personality and urbanization: behavioural traits and DRD4 SNP830 polymorphisms in great tits in Barcelona city. J. Ethol. 35: 101–108.

[73] Mathot, K.J. and N.J. Dingemanse. 2015. Plasticity and personality. pp. 55–69. *In*: L.B. Martin and C.K. Ghalambor (eds). Integrative Organismal Biology. Wiley Scientific, Hoboken, NJ.

[74] Nussey, D.H. and E. Postma, P. Gienapp and M.E. Visser. 2005. Selection on heritable phenotypic plasticity in a wild bird population. Science. 310: 304–306.

[75] Deviche, P. and S. Davies. 2013. Reproductive phenology of urban birds: environmental cues and mechanisms. pp. 98–115. *In*: D. Gil and H. Brumm (eds). Avian Urban Ecology. Oxford University Press, Oxford.

[76] Korsten, P., T. Van Overveld, F. Adriaensen and E. Matthysen. 2013. Genetic integration of local dispersal and exploratory behaviour in a wild bird. Nat. Commun. 4: 2362.

[77] Verbeek, M.E.M., P.J. Drent and P.R. Wiepkema. 1994. Consistent individual differences in early exploratory behaviour of male great tits. Anim. Behav. 48: 1113–1121.

[78] Thompson, M.J. and J.C. Evans, S. Parsons and J. Morand-Ferron. 2018. Urbanization and individual differences in exploration and plasticity. Behav. Ecol. 29: 1415–1425.

[79] Guillette, L.M., A.R. Reddon, M. Hoeschele and C.B. Sturdy. 2011. Sometimes slower is better: slow-exploring birds are more sensitive to changes in a vocal discrimination task. Proc. R. Soc. B. 278: 767–773.

[80] Michelangeli, M., J. Cote, D.G. Chapple, A. Sih, T. Brodin, S. Fogarty, et al. 2020. Sex-dependent personality in two invasive species of mosquitofish. Biol. Invasions. 22: 1353–1364.

[81] Brodin, T., M.I. Lind, M.K. Wiberg and F. Johansson. 2013. Personality trait differences between mainland and island populations in the common frog (*Rana temporaria*). Behav. Ecol. Sociobiol. 67: 135–143.

[82] Wolf, M. and F.J. Weissing. 2010. An explanatory framework for adaptive personality differences. Philos. Trans. R. Soc. B. 365: 3959–3968.

[83] Sih, A., K.J. Mathot, M. Moirón, P.-O. Montiglio, M. Wolf and N.J. Dingemanse. 2015. Animal personality and state–behaviour feedbacks: a review and guide for empiricists. Trends Ecol. Evolut. 30: 50–60.

[84] Wetzel, D.P. and D.F. Westneat. 2014. Parental care syndromes in house sparrows: Positive covariance between provisioning and defense linked to parent identity. Ethology. 120: 249–257.

[85] Rojas-Ferrer, I. and J. Morand-Ferron. 2020. The impact of learning opportunities on the development of learning and decision-making: an experiment with passerine birds. Philos. Trans. R. Soc. B. 375(1803): 20190496.

[86] Helm, B. and M.E. Visser. 2010. Heritable circadian period length in a wild bird population. Proc. R. Soc. B. 277: 3335–3342.

[87] Kelleher, S.R., A.J. Silla and P.G. Byrne. 2018. Animal personality and behavioral syndromes in amphibians: A review of the evidence, experimental approaches, and implications for conservation. Behav. Ecol. Sociobiol. 72: 79.

[88] Schakner, Z.A., M.B. Petelle, O. Berger-Tal, M.A. Owen and D.T. Blumstein. 2014. Developing effective tools for conservation behaviorists: reply to Greggor et al. Trends Ecol. Evolut. 29: 651–652.

[89] Watters, J.V. and C.L. Meehan. 2007. Different strokes: can managing behavioral types increase post-release success? Appl. Anim. Behav. Sci. 102: 364–379

[90] Owen, M.A., R.R. Swaisgood and D.T. Blumstein. 2017. Contextual influences on animal decision-making: significance for behavior-based wildlife conservation and management. Integrative Zool. 12: 32–48.

[91] Berger-Tal, O., T. Blumstein D, S. Carroll, R.N. Fisher, S.L. Mesnick, M.A. Owen, et al. 2015. A systematic survey of the integration of animal behavior into conservation. Conserv. Biol. 30: 744–753.

[92] Herborn, K.A., R. Macleod, W.T.S. Miles, A.N.B. Schofield, L. Alexander and K.E. Arnold. 2010. Personality in captivity reflects personlity in the wild. Anim. Behav. 79: 835–843.

[93] Araya-Ajoy, Y.G. and N.J. Dingemanse. 2014. Characterizing behavioural 'characters': a conceptual and statistical framework. Proc. R. Soc. B. 281(1776): 20132645.

[94] Aplin, L.M., J.A. Firth, D.R. Farine, B. Voelkl, R.A. Crates, A. Culina, et al. 2015. Consistent individual differences in the social phenotypes of wild great tits, *Parus major*. Anim. Behav. 108: 117–127.

[95] Evans, J.C. and J. Morand-Ferron. 2019. The importance of preferential associations and group cohesion: constraint or optimality. Behav. Ecol. Sociobiol. 73: 109.

[96] Niemelä, P.T. and N.J. Dingemanse. 2017. Individual versus pseudo-repeatability in behaviour: lessons from translocation experiments in a wild insect. J. Anim. Ecol. 86: 1033–1043.

[97] Mathot, K.J., J. Wright, B. Kempenaers and N.J. Dingemanse. 2012. Adaptive strategies for managing uncertainty may explain personality-related differences in behavioural plasticity. Oikos. 121: 1009–1020.

[98] Niemelä, P. and N.J. Dingemanse. 2018. On the usage of single measurements in behavioural ecology research on indivdiual differences. Anim. Behav. 145: 99–105.

[99] Brommer, J. 2013. On between-individual and residual (co)variances in the study of animal personality: are you willing to take the "individual gambit"? Behav. Ecol. Sociobiol. 67: 1027–1032.

[100] Nakagawa, S. and H. Schielzeth. 2010. Repeatability for gaussian and non-gaussian data: A practical guide for biologists. Biol. Rev. 85: 935–956.

[101] Dingemanse, N.J. and N.A. Dochtermann. 2013. Quantifying individual variation in behaviour: mixed-effect modelling approaches. J. Anim. Ecol. 82: 39–54.

[102] van de Pol, M. and J. Wright. 2009. A simple method for distinguishing within-versus between-subject effects using mixed models. Anim. Behav. 77: 753–758.

Chapter 11

Conclusion: The Role of Human Behavior in Songbird Conservation

Darren S. Proppe[1]

As I pen the concluding chapter of this book, the year 2020 is also coming to close. It has been a year of exceptional hardship, and extraordinary change in human behavior. At the time of writing, the novel COVID-19 virus has infected more than 75 million people worldwide leading to death in > 2% of reported cases. Face coverings (e.g., masks) have become standard apparel for public interaction, although not without substantial disagreement and dissention in some regions. The term 'social distancing' is now a household term, and is generally interpreted as maintaining six feet of separation from individuals not living within one's own household. In Texas, the author's home, shelter-in-place orders shuttered most businesses for several months during the spring.

---

[1]Research Director, Wild Basin Creative Research Center, St. Edward's University

While the impacts on the human population have been dramatic and devastating, the 'anthropause' has created unprecedented opportunities for scientist who study the natural world [1]. For example, noise abatement, a strategy recommended by many authors in this text, was accomplished on an unprecedented scale. At least one species of songbird responded by singing at lower frequencies that were more typical of their pre-industrialization songs [2]. Reduced human activity also brought widespread changes in habitat use demographics across the animal kingdom, with reports of many species using previously vacant sites and increased reproductive success in urban habitats [3, 4]. Greater use of facemasks even reduced fear of humans in a study on Eurasian tree sparrows (*Passer montanus*) [5]. Several parts of the world experienced improvements in water quality [6, 7], and air quality increased in some [8], but not all regions [9]. Yet, the longer term outcomes of this pandemic remain relatively unknown, likely harboring both positive and negative effects on wildlife [10]. The push towards global research collaborations serves as one silver lining to the international COVID-related lockdowns [11], and an increased commitment to 'work together' may bolster conservation objectives [12]. But, perhaps the most compelling message resulting from the pandemic is that 'change is possible'.

Change is Possible

As champions for wildlife and their habitats, ecologists and conservation biologists often make recommendations in the scientific literature and the public arena that seem unlikely to be put into practice due to larger economic forces and societal resistance to change. This disparity has led to detectable levels of hopelessness and depression in the conservation community [13]. The question as to whether society can make the larger changes needed to sustain our planet under imminent ecological crises like climate change looms large [14, 15]. Lifestyle changes, such as reducing one's carbon footprint, are not easily accomplished [16]. Yet, both individual and institutional behavioral changes were immediate and dramatic under COVID-19 restrictions. Entire learning environments converted to virtual platforms in a matter weeks [17, 18]. A vast number of employed adults embraced, if not by choice, working remotely [19, 20]. Online retail soared [21], and wearing facemasks became the norm in many places [22]. The 'typical' lifestyle in 2020 would hardly have been predicted or thought possible in 2019. Taken together, this is a strong indicator that entrenched human behavior can change, albeit not without difficulty.

Throughout the pages of this text, authors repeatedly recommend that changes in the current direction of human behavior are needed—native habitats should be left intact, entire species assemblages should be monitored and managed for stable ecosystem function, noise and light pollution should be reduced, air and water quality should be improved, the pace of urban expansion should be slowed, etc. In many cases, these management recommendations align with those put forth by landscape ecologists. But, adding an understanding of animal behavior, and specifically songbird behavior in this context, to our toolkit will allow managers to detect less perceptible changes in our environments before it is too late [23]. Regardless, detection of a problem is only half of the battle. Implementation of the solutions critical for sustaining diverse populations of songbirds may be even more challenging because they typically require a corresponding change in human behavior.

Human Behavior is Critical

As someone who studies animal behavior on a regular basis, it is easy to see how many of the rules regulating decision making in the non-human world also direct processes in the human realm. This should not be surprising since we too are animals. Yet, the chasm between those who study human behavior and cognition, and those who study the same topics in animals is still larger than should be expected. In addition to the excellent recommendations provided by songbird experts throughout this book, I would like to add that we all *must* possess an understanding of human behavior. That is not say that we should be versed in manipulating human behavior—this approach will likely be perceived as devious and ultimately drive stronger opposition to the lifestyle and societal changes necessary to sustain the healthy ecosystems that support our valued songbirds. Rather, ecologists and conservation biologists should;

1. Recognize the broader human context within which we work
2. Explore the ways that songbirds are meaningful within that context (beyond ourselves)
3. Find ways to distribute our findings to the greater community in a meaningful way

A recent publication by Ferraro et al. demonstrates elements of this process nicely [24]. The authors investigate the impact of birdsong on human satisfaction through a serious of phantom song playback experiments in parks, followed by hiker surveys. Their results indicated conclusively that hearing birdsong was restorative for park users. Media attention has successfully put this documented societal benefit of maintaining songbird diversity into the public sphere for a moment. But we must ask ourselves how we can continue to expand this conversation with the public. And further, how do we ask members of society to make lifestyle changes based on these demonstrated benefits. Indeed, a growing field of study, known collectively as *Human Dimensions of Wildlife*, is explicitly investigating the interactions between wildlife and humans [25]. Historically, coyote conflict and deer-vehicle collisions might have filled the spotlight, but modern perspectives are looking deeper at both the positive and negative interactions between humans and animals [26]. A rather poignant example of human-wildlife interaction is the illegal sale of animals that was a potential vector in the origination of the COVID-19 virus [27, 28].

If our goal is to bring about necessary changes in human behavior that support songbird persistence, it may be helpful to look at what aspects of the 2020 COVID-19 pandemic drove dramatic changes in human behavior. Perhaps the most pertinent trait of the pandemic was its immediate and mortal threat to life. Model predictions presented to the public generally indicated what may happen within the next few days or months, rather than years. This short-term outlook indicated that immediate action was necessary. This is not to suggest that longer-term modelling is not vitally important to our understanding of ecosystems and our ability to plan for the future. Rather, it is a call to consider focusing our dissemination of information to the public on the current situation for songbirds. A good example of this was the 2019 *Science* paper that reported the loss of 3 billion birds since 1970 [29]. The information provided was immediately pertinent, and served as an indicator of a current, rather than future, crisis. I personally know of many people that were impacted by this information, despite being rather oblivious to birds otherwise.

The importance of perceived impact on personal wellbeing for driving behavioral change was also demonstrated by people's willingness to wear a mask during the

COVID-19 pandemic. Mask wearing increased with age (an indicator of risk) and was higher in urban areas (higher incidence rates) [30]. However, mask wearing in the United States also correlated with political affiliation [31]. The politically-driven 'us vs them' mentality that become detrimental to public health during the COVID-19 pandemic also influences conservation-oriented strategies and policies. [32] Like politicians, we must be willing to 'reach across the aisle' and engage stakeholders whose priorities and worldviews differ from our own. Dr. David Lodge, past president of the Ecological Society of America (ESA) sums this up well.

> "...First, what binds ESA members together is our respect for science, commitment to rigorous peer review and publication of research, and a desire to see our science interpreted and used appropriately. We must continue to advocate – more strongly than ever – that representatives of science and rigorous scientific analysis are essential to policy-making... Second, we must not allow ourselves to be arrogant or make it easy for others to perceive us that way. Science must be at the policy-making table, but in a democracy, many diverse considerations belong at the decision-making table. We must be more aggressive promoters of science, but we must simultaneously be humble in recognition that our unique role is not solely important..."

We all live and work among a community of individuals with unique values and goals. The more we can work collaboratively with those who share our objectives, the more likely we are able to achieve our goals. Equally important, the more we are able to work alongside those who have a different set of priorities, the more likely we are to achieve our goals. This book was written to introduce the reader to behavioral concepts that can guide conservation and management, but it is also a means to connect with colleagues that share an interest in developing data-driven solutions for the conservation and management of songbirds. While each author's particular expertise varies, they all maintain a value for understanding songbird behavior. I invite you, the reader, to reach out to your colleagues as you pursue solutions to the complex challenges that the Anthropocene presents. But I also implore you to reach out to your community, so that they too might value the role of songbirds in the nature world, and for human wellbeing.

LITERATURE CITED

[1]   Rutz, C., M.C. Loretto, A.E. Bates, S.C. Davidson, C.M. Duarte, W. Jetz, et al. 2020. COVID-19 lockdown allows researchers to quantify the effects of human activity on wildlife. Nat. Ecol. Evol. 4(9): 1156–1159. https://doi.org/10.1038/s41559-020-1237-z

[2]   Derryberry, E.P., J.N. Phillips, G.E. Derryberry, M.J. Blum and D. Luther. 2020. Singing in a silent spring: Birds respond to a half-century soundscape reversion during the COVID-19 shutdown. Science. 370(6516): 575–579. https://doi.org/10.1126/science.abd5777

[3]   Zellmer, A.J., E.M. Wood, T. Surasinghe, B.J. Putman, G.B. Pauly, S.B. Magle, et al. 2020. What can we learn from wildlife sightings during the COVID-19 global shutdown? Ecosphere. 11(8): e03215. https://doi.org/10.1002/ecs2.3215

[4]   Manenti, R., E. Mori, V. Di Canio, S. Mercurio, M. Picone, M. Caffi, et al. 2020. The good, the bad and the ugly of COVID-19 lockdown effects on wildlife conservation: insights from the first European locked down country. Biol. Conserv. 249: 108728. https://doi.org/10.1016/j.biocon.2020.108728

[5] Jiang, X., J. Liu, C. Zhang and W. Liang. 2020. Face masks matter: Eurasian tree sparrows show reduced fear responses to people wearing face masks during the COVID-19 pandemic. Global Ecol. Conserv. 24: e01277. https://doi.org/10.1016/j.gecco.2020.e01277

[6] Yunus, A.P., Y. Masago and Y. Hijioka. 2020. COVID-19 and surface water quality: Improved lake water quality during the lockdown. Sci. Total Environ. 731: 139012. https://doi.org/10.1016/j.scitotenv.2020.139012

[7] Lokhandwala, S. and P. Gautam. 2020. Indirect impact of COVID-19 on environment: A brief study in Indian context. Environ. Res. 188: 109807. https://doi.org/10.1016/j.envres.2020.109807

[8] Lian, X., J. Huang, R. Huang, C. Liu, L. Wang and T. Zhang. 2020. Impact of city lockdown on the air quality of COVID-19-hit of Wuhan city. Sci. Total Environ. 742: 140556. https://doi.org/10.1016/j.scitotenv.2020.140556

[9] Zangari, S., D.T. Hill, A.T. Charette and J.E. Mirowsky. 2020. Air quality changes in New York City during the COVID-19 pandemic. Sci. Total Environ. 742: 140496. https://doi.org/10.1016/j.scitotenv.2020.140496

[10] Buck, J.C. and S.B. Weinstein. 2020. The ecological consequences of a pandemic. Biol. Lett. 16(11): 20200641. https://doi.org/10.1098/rsbl.2020.0641

[11] Bates, A.E., R.B. Primack, P. Moraga and C.M. Duarte. 2020. COVID-19 pandemic and associated lockdown as a "Global Human Confinement Experiment" to investigate biodiversity conservation. Biol. Conserv. 248: 108665. https://doi.org/10.1016/j.biocon.2020.108665

[12] Bang, A. and S. Khadakkar. 2020. Opinion: biodiversity conservation during a global crisis: consequences and the way forward. P. Natl. Acad. Sci. USA. 117(48): 29995–29999. https://doi.org/10.1073/pnas.2021460117

[13] Park, A., E. Williams and M. Zurba. 2020. Understanding hope and what it means for the future of conservation. Biol. Conserv. 244: 108507. https://doi.org/10.1016/j.biocon.2020.108507

[14] Biesbroek, G.R., J.E.M. Klostermann, C.J.A.M. Termeer and P. Kabat. 2013. On the nature of barriers to climate change adaptation. Reg. Environ. Change. 13(5): 1119–1129. https://doi.org/10.1007/s10113-013-0421-y

[15] Eisenack, K., S.C. Moser, E. Hoffmann, R.J.T. Klein, C. Oberlack, A. Pechan, et al. 2014. Explaining and overcoming barriers to climate change adaptation. Nat. Clim. Change. 4(10): 867–872. https://doi.org/10.1038/nclimate2350

[16] Capstick, S., I. Lorenzoni, A. Corner and L. Whitmarsh. 2014. Prospects for radical emissions reduction through behavior and lifestyle change. Carbon Manage. 5(4): 429–445. https://doi.org/10.1080/17583004.2015.1020011

[17] Lashley, M.A., M. Acevedo, S. Cotner and C.J. Lortie. 2020. How the ecology and evolution of the COVID-19 pandemic changed learning. Ecol. Evol. 10(22): 12412–12417. https://doi.org/10.1002/ece3.6937

[18] Adedoyin, O.B. and E. Soykan. 2020. Covid-19 pandemic and online learning: the challenges and opportunities. Interactive Learning Environments. 1–13. https://doi.org/10.1080/10494820.2020.1813180

[19] Donthu, N. and A. Gustafsson. 2020. Effects of COVID-19 on business and research. J. Bus. Res. 117, 284–289. https://doi.org/10.1016/j.jbusres.2020.06.008

[20] Kniffin, K.M., J. Narayanan, F. Anseel, J. Antonakis, S.P. Ashford, A.B. Bakker, et al. 2021. COVID-19 and the workplace: Implications, issues, and insights for future research and action. Am. Psychol. 76(1): 63-77. http://dx.doi.org/10.1037/amp0000716

[21] Dannenberg, P., M. Fuchs, T. Riedler and C. Wiedemann. 2020. Digital Transition by COVID-19 Pandemic? The German Food Online Retail. Tijdschrift voor Economische en Sociale Geografie. 111(3): 543–560. https://doi.org/10.1111/tesg.12453

[22] Eikenberry, S.E., M. Mancuso, E. Iboi, T. Phan, K. Eikenberry, Y. Kuang, et al. 2020. To mask or not to mask: modeling the potential for face mask use by the general public to curtail the COVID-19 pandemic. Infect. Dis Modell. 5: 293–308. https://doi.org/10.1016/j.idm.2020.04.001

[23] Berger-Tal, O., T. Polak, A. Oron, Y. Lubin, B.P. Kotler and D. Saltz. 2011. Integrating animal behavior and conservation biology: a conceptual framework. Behav. Ecol. 22(2): 236–239. https://doi.org/10.1093/beheco/arq224

[24] Ferraro, D.M., Z.D. Miller, L.A. Ferguson, B.D. Taff, J.R. Barber, P. Newman, et al. 2020. The phantom chorus: birdsong boosts human well-being in protected areas. Proc. R. Soc. B: Biol. Sci. 287(1941): 20201811. https://doi.org/10.1098/rspb.2020.1811

[25] Decker, D.J., S.J. Riley and W.F. Siemer. 2012. Human Dimensions of Wildlife Management. JHU Press.

[26] Morzillo, A.T., K.M. de Beurs and C.J. Martin-Mikle. 2014. A conceptual framework to evaluate human-wildlife interactions within coupled human and natural systems. Ecol. Soc. 19(3): 44. Retrieved from https://www.jstor.org/stable/26269619

[27] Shereen, M.A., S. Khan, A. Kazmi, N. Bashir and R. Siddique. 2020. COVID-19 infection: origin, transmission, and characteristics of human coronaviruses. J. Adv. Res. 24: 91–98. https://doi.org/10.1016/j.jare.2020.03.005

[28] Aguirre, A.A., R. Catherina, H. Frye and L. Shelley. 2020. Illicit wildlife trade, wet markets, and COVID-19: preventing future pandemics. World Medical & Health Policy. 12(3): 256–265. https://doi.org/10.1002/wmh3.348

[29] Rosenberg, K.V., A.M. Dokter, P.J. Blancher, J.R. Sauer, A.C. Smith, P.A. Smith, et al. 2019. Decline of the North American avifauna. Science. 366(6461): 120–124. https://doi.org/10.1126/science.aaw1313

[30] Haischer, M.H., R. Beilfuss, M.R. Hart, L. Opielinski, D, Wrucke, G. Zirgaitis, et al. 2020. Who is wearing a mask? Gender-, age-, and location-related differences during the COVID-19 pandemic. PLoS One. 15(10): e0240785. https://doi.org/10.1371/journal.pone.0240785

[31] Makridis, C. and J.T. Rothwell. 2020. The Real Cost of Political Polarization: Evidence from the COVID-19 Pandemic (SSRN Scholarly Paper No. ID 3638373). Rochester, NY: Social Science Research Network. https://doi.org/10.2139/ssrn.3638373

[32] Brosius, P.J., A.L. Tsing and C. Zerner. 2005. Communities and Conservation: Histories and Politics of Community-Based Natural Resource Management. Rowman Altamira.

Index

For Product Safety Concerns and Information please contact our EU
representative GPSR@taylorandfrancis.com
Taylor & Francis Verlag GmbH, Kaufingerstraße 24, 80331 München, Germany

www.ingramcontent.com/pod-product-compliance
Lightning Source LLC
Chambersburg PA
CBHW061349210326
41598CB00035B/5934

* 9 7 8 1 0 3 2 0 5 8 3 8 2 *